Latest Advances in Atomic Cluster Collisions

Fission, Fusion, Electron, Ion and Photon Impact

Latest Advances in Atomic Cluster Collisions

Fission, Fusion, Electron, Ion and Photon Impact

edited by

Jean-Patrick Connerade

The Blackett Laboratory,
Imperial College London, London, UK

Andrey V. Solov'yov

A. F. Ioffe Physical-Technical Institute,
Russian Academy of Sciences, St. Petersburg, Russia

ICP Imperial College Press

Published by

Imperial College Press
57 Shelton Street
Covent Garden
London WC2H 9HE

Distributed by

World Scientific Publishing Co. Pte. Ltd.
5 Toh Tuck Link, Singapore 596224
USA office: 27 Warren Street, Suite 401-402, Hackensack, NJ 07601
UK office: 57 Shelton Street, Covent Garden, London WC2H 9HE

British Library Cataloguing-in-Publication Data
A catalogue record for this book is available from the British Library.

LATEST ADVANCES IN ATOMIC CLUSTER COLLISIONS
Fission, Fusion, Electron, Ion and Photon Impact

ISBN 1-86094-495-7

Printed in Singapore by World Scientific Printers (S) Pte Ltd

Preface

The International Symposium "Atomic Cluster Collisions: fission, fusion, electron, ion and photon impact" (ISACC 2003, a Europhysics Conference) was held in St. Petersburg, Russia, July 18-21, 2003 as a satellite meeting of the XXIIIrd International Conference on Photonic, Electronic, and Atomic Collisions (ICPEAC 2003, Stockholm, Sweden, July 23-29, 2003). The ISACC 2003 took place at the former Palace of Grand-Duke Vladimir (nowadays used as the House of Scientists) located in the heart of St. Petersburg, near the Hermitage Museum. This international symposium promoted the growth and exchange of scientific information on the structure and properties of atomic cluster systems studied by means of photonic, electronic and atomic collisions. Particular attention during the symposium was devoted to dynamical phenomena, many-body effects taking place in cluster systems, which include problems of fusion and fission, fragmentation, collective electron excitations, phase transitions and many more. Both experimental and theoretical aspects of atomic cluster physics, which is uniquely placed between atomic and molecular physics on the one hand and solid state physics on the other, were discussed at the symposium.

St. Petersburg was a very natural location for a symposium on cluster science: much of the development of modern many-body theory in atomic physics has taken place there, and a strong school of atomic theorists, spread over several institutions and equipped with powerful computational techniques, has already made a considerable impact on the formulation of new methods of calculation for atomic and molecular clusters.

The symposium brought together more than 100 leading scientists in the field of atomic cluster physics from around the world. The special emphasis of the Symposium was devoted to the new methods of investigation of the structure and properties of atomic clusters, the collective excitations in photoabsorption and photoionization processes of atomic clusters, fission and fusion dynamics of clusters, cluster dynamics in the laser field, resonance processes in electron-cluster collisions, the interaction of ions, including multiply charged ions, with metal clusters and fullerenes and the processes of cluster deposition on a surface as well as of cluster collisions on a surface. The aim of the symposium was to present the most recent achievements in all these fields of atomic cluster science. These proceedings, we hope, bear witness that this goal has been fulfilled.

The introduction to this book surveys the general aspects of atomic cluster science and outlines some of its important new challenges. It contains an important definition of a cluster, as a new physical system possessing its own specific properties and features. This definition is important to establish that atomic cluster science is a new field of modern physics in its own right. It is highly multidisciplinary and has numerous links with traditional branches of physics and chemistry.

The first chapter of this book is devoted to recent advances in the understanding of structure and essential properties of selected atomic cluster systems, fullerenes and confined atoms. Both theoretical and experimental aspects of the field are discussed.

The second chapter covers the recent advances in the field of photo processes involving atomic clusters and fullerenes. Collective excitations of electrons as well as specific interference effects play a very significant role in the photo processes as is shown in this chapter by a number of examples.

The third chapter focuses on the problem of fission dynamics of atomic clusters. Parallels with similar processes in nuclear physics are presented. It is demonstrated that cluster and fragmentation phenomena in atomic cluster physics and in nuclear physics have many features in common. Some of the new challenges of both fields of endeavour are presented.

The fourth chapter of this book describes the problems of electron-cluster collisions. Special emphasis in this chapter is placed on the polarization and collective excitation effects. Both theoretical and experimental aspects of electron-cluster collisions are discussed.

The fifth chapter deals with the behaviour of atomic clusters in laser fields. The ionization (including multiphoton), collective dynamics of electrons in the system in the presence of the laser field and the laser induced dynamics of molecules and clusters are thoroughly described.

The sixth chapter is devoted to the physics of ionic collisions with fullerenes and metal clusters. It covers a broad spectrum of problems in this field from both experimental and theoretical points of view. The results of the very recent measurements are reported.

The last, seventh, chapter in this book is devoted to the problem of the interaction of an atomic cluster with a surface. The problems of cluster deposition and formation at a surface as well as collision processes involving clusters deposited at a surface are considered in this chapter through a number of illustrative examples. The subjects of the chapters in this book correspond to the sessions in the symposium.

The organizers of the ISACC 2003 wish to acknowledge the generous support received from the European Physical Society, the Russian Foundation for Basic Research, Government of St. Petersburg, the A. F. Ioffe Physical-Technical Institute (St. Petersburg, Russia), St. Petersburg State Technical University (Russia), St. Petersburg State University (Russia), St. Petersburg Institute for Nuclear Physics (Russia), Imperial College London (London, UK), Institute for Theoretical Physics (Frankfurt am Main University, Germany), the Alexander von Humboldt Foundation (Bonn, Germany) and the House of Scientists (St. Petersburg, Russia), which made this symposium possible and successful. The editors of this book want to express their gratitude to Dr Andrey Lyalin and Mr Ilia Solov'yov for their great help in the preparation of the manuscript of this book for the publication. Finally, we acknowledge the fruitful collaboration with Imperial College Press and World Scientific Publishing Co.

St. Petersburg, London
November 2003

Jean-Patrick Connerade
Andrey V. Solov'yov

ISACC 2003 Conference Photo

CONTENTS

VI Ion-Cluster Collisions

VII Clusters on a Surface

Introduction

ATOMIC CLUSTER SCIENCE: INTRODUCTORY NOTES

Andrey V. Solov'yov
A. F. Ioffe Physical-Technical Institute, Russian Academy of Sciences, Polytechnicheskaya 26, St. Petersburg 194021, Russia
E-mail: solovyov@th.physik.uni-frankfurt.de

Jean-Patrick Connerade
The Blackett Laboratory, Imperial College London, London SW7 2BW, UK
E-mail: j.connerade@ic.ac.uk

Walter Greiner
Institut für Theoretische Physik der Johann-Wolfgang Goethe Universität, Robert-Mayer Str. 8-10, D-60054 Frankfurt am Main, Germany
E-mail: greiner@th.physik.uni-frankfurt.de

This article is the introduction to the volume of proceedings of the "International Symposium Atomic Cluster Collisions: fission, fusion, electron, ion and photon impact" (a Europhysics Conference) held in St. Petersburg, Russia, July 18-21, 2003 (ISACC 2003) as a satellite meeting of the XXIII International Conference on Photonic, Electronic, and Atomic Collisions (ICPEAC 2003, Stockholm, Sweden, July 23-29, 2003). A brief introduction to atomic cluster physics, the interdisciplinary field, which developed rather successfully during recent years, is presented. A review of recent achievements in the detailed *ab initio* description of structure and properties of atomic clusters and complex molecules as well as the methods of their study is given. The main trends of development in the field are discussed and some of its new focuses are outlined.

1. Introduction

The "International Symposium Atomic Cluster Collisions: fission, fusion, electron, ion and photon impact" (a Europhysics Conference) was held in St. Petersburg, Russia, July 18-21, 2003 (ISACC 2003) as a satellite meeting of the XXIIIrd International Conference on Photonic, Electronic, and Atomic Collisions (ICPEAC 2003, Stockholm, Sweden, July 23-29, 2003). This international Symposium promoted the growth and exchange of scientific information on the structure and properties of atomic cluster systems studied by means of photonic, electronic and atomic collisions. Particular attention during the symposium was devoted to dynamical phenomena, many-body effects taking place in cluster systems, which include problems of fusion and fission, fragmentation, collective electron excitations, phase transitions and many more. Both experimental and theoretical aspects of atomic cluster physics uniquely placed between atomic and molecular physics on the one hand and solid state physics on the other, were discussed at the symposium.

During the last decade it was recognized, both experimentally and theoretically, that complex molecules and atomic clusters (ACs) often possess unique properties, which make them a new object of physical research, rather different from both a single atom and from the solid state (see Refs. 1, 2 and references therein). The knowledge of the detailed electronic and ionic structure of single complex molecules and nano-clusters can be essential for various practical applications, such as the formation of new materials, nano-structures, in the design of drugs and biologically active species as well as for the understanding of fundamental issues, such as the functioning of quantum and thermodynamic laws in nano-scale systems or mechanisms for the formation of complex multi-atomic systems, self-assembly and functioning.

The demand for understanding of the principles of assembly and functioning of complex multi-atomic systems such as bio-molecules or nano-clusters is tremendous, because of the potential use of this knowledge for purposes of microelectronics, of biochemistry, the drug industry etc. The problems of self-organization, of self-assembly and of the functioning of complex multi-atomic aggregates and their interactions have been addressed both theoretically and experimentally in a large number of papers from different perspectives (for a review, see Refs. 1, 2). Often, these problems can be reduced to the problem of the interaction of a limited number of atoms within a complex molecule or even to the interaction of a single atom or ion with a certain fragment of a complex molecule (an active center responsible for a certain function) or a cluster structure. Thus, in order to achieve a real breakthrough in the field, one needs to learn how to handle both theoretically and experimentally (i.e. to be able to manipulate experimentally and predict theoretically) properties of multi-atomic systems containing about 100 atoms, or maybe a little less or a little more than this limit. With this knowledge in hand, one can then move towards a detailed *ab initio* understanding of the properties of larger multi-atomic systems, bio-molecules (proteins, DNA), which typically consist of rather small fragments (amino acids or bases) whose structure and interactions do not involve more than 100 atoms.

These structures are, nowadays, subjects of very intensive theoretical and experimental studies in many physical, chemical and biological laboratories and institutions worldwide. A variety of methods have been used to investigate these objects (see, e.g., Refs. 1–9 and this book). Due to these efforts a vast amount of physical, chemical and biological data on the properties of complex multi-atomic systems and their interaction with the environment have been accumulated. However, it can be stated that until now there is no consistent theoretical approach, based solely on the fundamental principles of quantum physics, which might allow one, not only to explain systematically the known experimental data, but also, and this is quite essential, to predict new properties of the objects and new phenomena related to them. Nearly all theoretical approaches developed so far can be termed 'phenomenological' in the sense that each one of them substitutes the full quantum-mechanical description of the dynamics of constituents of a multi-atomic structure with a model theory which uses a set of parameters deduced from the experimental data. Each of these models is able to reproduce a limited number of particular properties of a complex multi-atomic system of a particular type, since the sets of the parameters involved are not of a universal nature. Thus, the model theories have severe restrictions: in each case they can explain but few of the experimental data. From them, often, one can hardly draw general conclusions or produce predictions of the properties of other structures. A more accurate description of the electronic and ionic structures, internal dynamics and interaction with external objects and fields has to be elaborated. The development of such an approach, which is multifaceted and includes not only theoretical investigations based on the first principles of quantum many-body theory but also implies a great amount of experimental work and computing, is currently the subject of joint efforts by specialists in various fields of physics and chemistry.

Theoretical approaches for the description of complex multi-atomic systems built on *ab initio* principles, model approaches and experiments will create real breakthroughs in understanding essential properties of complex multi-atomic formations, nano-clusters, bio-molecules and the mechanisms of their assembly and functioning. This will open up new possibilities for cheap computer experiments for modelling complex multi-atomic systems possessing unique properties, for example, biologically active molecules. This knowledge is demanded in various applications ranging from micro-electronics to micro-biology and medicine.

The complete theoretical description of nano-scale systems consisting of about 100 atoms is extremely difficult.[4] So far, an *ab initio* many-body quantum mechanical description accounting for all electrons in the system can be used effectively for systems of a few tens of atoms[5,6] rather than hundreds. The computer power required for such calculations grows exponentially with increasing molecular or cluster size. Therefore, one needs to invoke various simplified model approaches in order to describe complex systems of sufficiently large size.[4,7,8] However, often, the predictability of large systems by such approaches varies dramatically, particularly in the cases when modelling of complex molecules or nano-clusters neglects quantum effects.

Therefore, a careful choice of the model and accurate accounting for many-body and quantum phenomena are very important, as is demonstrated by various examples in this book. Thus, the high predictive power of a model can be achieved on the basis of detailed comparison of the predictions of the model and *ab initio* approaches with each other and with experiment for relatively small systems, consisting of tens of atoms, and by the extrapolation of the model postulates towards larger scale systems.[4–6]

In order to illustrate these ideas and some of the topics and focuses of this book, in the next sections, we briefly discuss fission, fusion and collision processes involving ACs as well as some general aspects of AC science.

2. Atomic Cluster Science

A group of atoms bound together by interatomic forces is called an atomic cluster. There is no qualitative distinction between small clusters and molecules, except perhaps that the binding forces must be such as to permit the system to grow much larger (in principle: with no upper limit to size) by stacking more and more atoms or molecules of the same type if the system is to be called a cluster. As the number of atoms in the system increases, ACs acquire more and more specific properties making them unique physical objects different from both single molecules and from the solid state.

In nature, there are many different types of AC: van der Waals clusters, metallic clusters, fullerenes, molecular, semiconductor, mixed clusters, and

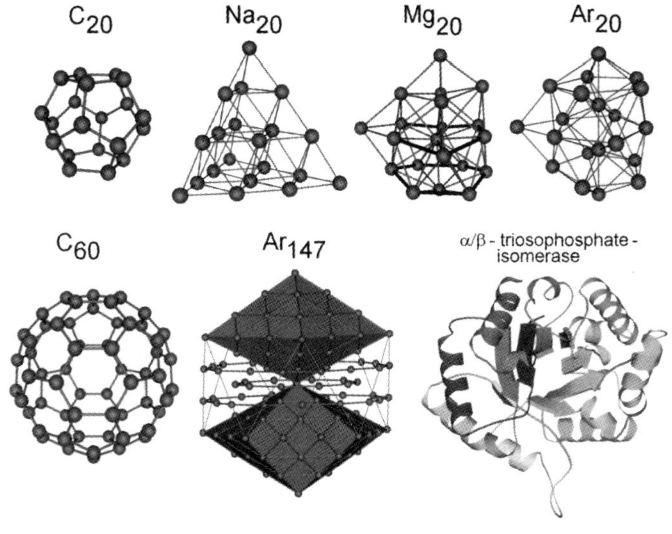

Fig. 1. The different nature of interatomic forces results in different principles for their organization within clusters and complex molecules. Geometries of the presented ACs have been calculated in Refs. 4–6 and 9 the structure of the protein globule ($\alpha/\beta-$ triosophosphate-isomerase) is taken from Ref. 3.

their shapes can depart considerably from the common spherical form: arborescent, linear, spirals, etc. Usually, one can distinguish between different types of clusters by the nature of the forces between the atoms, or by the principles of spatial organization within the clusters. Clusters can exist in all forms of matter: solid state, liquid, gases and plasmas.

In Fig. 1, we present images of a few clusters in order to show a big variety of cluster forms existing in nature. We also show the structure of the $\alpha/\beta-$ triosophosphate-isomerase globule aiming to stress that complex molecules such as proteins can be treated as clusters of subunits and that each of the subunits is a cluster on its own.

The novelty of AC physics arises mostly from the fact that cluster properties provide a better understanding of the transition from the single atom or molecule to the solid state limit. Modern experimental techniques have made it possible to study this transition. By increasing the cluster size, one can observe the emergence of the physical features in the system, such as plasmon excitations, electron conduction band formation, superconductivity and superfluidity, phase transitions, fission and many more. Most of these many-body phenomena exist in solid state but are absent for single atoms.

The science of clusters is a highly interdisciplinary field. ACs concern astrophysicists, atomic and molecular physicists, chemists, molecular biologists, solid-state physicists, nuclear physicists, plasma physicists, technologists all of whom see them as a branch of their subjects but cluster physics is a new subject in its own right.

Significant progress achieved in the field over the past two decades ushered in the understanding of ACs as new physical objects with their own distinctive properties. This became clear after such experimental successes as the discovery of the fullerene C_{60}, of the electronic shell structure in metal clusters, the observation of plasmon resonances in metal clusters and fullerenes, the observation of magic numbers for various other types of clusters, the formation of singly and doubly charged negative cluster ions and many more. A complete review of this field can be found in review papers and books, see e.g. Refs. 1, 2, 10–15 and the present book.

3. Distinctive Properties of Atomic Clusters: Cluster Magic Numbers

ACs, as new physical objects, possess some properties, which are distinctive characteristics of these systems. The cluster geometry turns out to be an important feature of clusters, influencing their stability and vice-versa. The determination of the most stable cluster forms is not a trivial task and the solution of this problem is different for various types of cluster. The stability of clusters and their transformations is a theme which does not exist at the atomic level and is not of great significance for solid state but is of crucial importance for AC systems. This problem is closely connected to the problem of cluster magic numbers.

The sequence of cluster magic numbers carries essential information about a cluster's electronic and ionic structure. Understanding the magic numbers of a cluster is pretty well equivalent to understanding its electronic and ionic structure.[4] A good example of this kind occurs for sodium clusters. In this case, the magic numbers arise from the formation of closed shells of delocalised electrons, one from each atom (see Refs. 10, 14 and references therein). Another example is the discovery of fullerenes, and in particular the C_{60} molecule,[16] by means of the mass spectroscopy of carbon clusters.

In Fig. 2, we present the mass spectra measured for Na and Ar clusters (see[10,12] and references therein), which clearly demonstrate the emergence

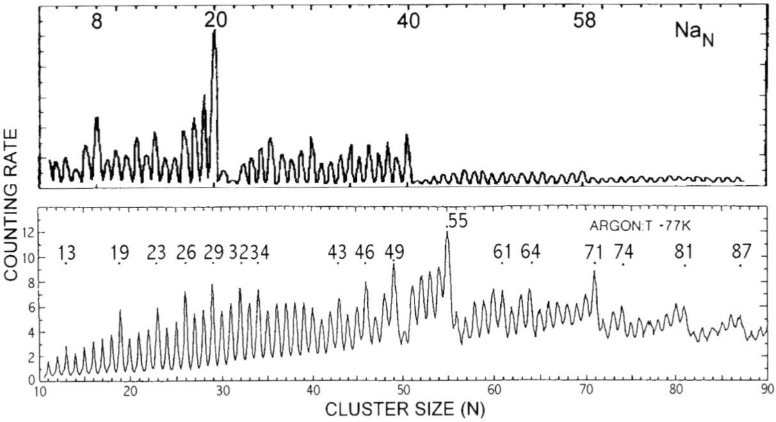

Fig. 2. Mass spectra measured for Ar and Na clusters (see Refs. 10, 12, 15 and references therein). The intense peaks indicate enhanced stability.

of magic numbers. The forces binding atoms in these two different types of clusters are different. The argon (noble gas) clusters are formed by van der Waals forces, while atoms in the sodium (alkali) clusters are bound by the delocalized valence electrons moving in the entire cluster volume. The differences in the inter-atomic potentials and pairing forces lead to significant differences in structure between Na and Ar clusters, their mass spectra and their magic numbers.

In Fig. 3, we present and compare the geometries of a few small Na and Ar clusters of the same size. It is clear from Fig. 3 that different principles of cluster organization result in different geometries of the alkali and noble gas cluster families.

Such differences can easily be explained. The van der Waals forces lead to enhanced stability of cluster geometries based on the most dense icosahedral packing. The most prominent peaks in mass spectra of argon clusters correspond to completed icosahedral shells of 13, 55, 147, 309 ets atoms. The origin of these magic numbers can be understood on the basis of the classical equations. The origin of the sodium cluster magic numbers is different and is based on the principles of quantum mechanics. In this case the cluster magic numbers 8, 20, 34, 40, 58, 92 ets correspond to the completed shells of the delocalised electrons: $1s^2 1p^6 1d^{10} 2s^2 1f^{14} 2p^6$ ets. This feature of small metal clusters make them qualitatively similar to atomic nuclei for which quantum shell effects play the crucial role in determining their properties.[17]

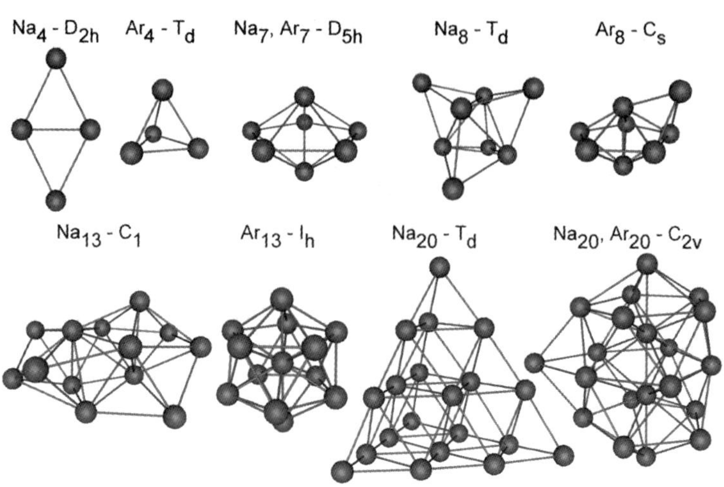

Fig. 3. Geometries and the point symmetry groups of some Na and Ar clusters calculated in Refs. 4, 5.

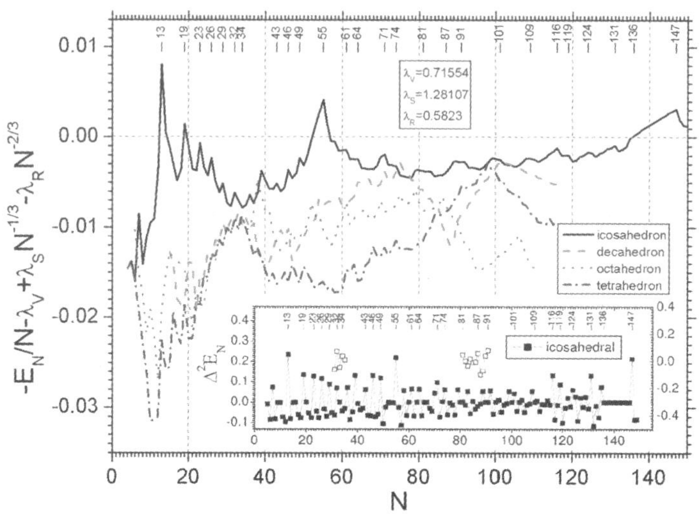

Fig. 4. Binding energies and their second differences for Ar clusters calculated in Ref. 4.

The enhanced stability of cluster systems can be characterized by computing the second differences in cluster binding energies. In Fig. 4 we present Ar clusters binding energies and their second differences calculated in Ref. 4 The correspondence of the peaks in Fig. 4 to those in the Ar clusters mass spectrum shown in Fig. 2 is readily established.

Finally, let us stress the obvious connection between AC physics and physics and chemistry of large molecules, such as proteins or DNA, which in fact can be treated as large clusters of amino acids or bases. The characteristic size of a fragment (amino acid or base) in such clusters is of the order of a few tens of atoms, i.e. the size of a small cluster. It is obvious that the knowledge gained from the AC studies is relevant for the biomolecular investigations and vice versa. A bunch of interesting phenomena can arise at the juncture of the two fields. For example, fusion of ACs with bio-molecules can create new objects which can be handled as easily as ACs or possess some specific properties and characteristics of ACs, but at the same time carry all essential features of bio-molecules and participate in bio-processes.

4. Collisions Involving Atomic Clusters

The properties of clusters can be studied by means of photon, electron and ion scattering (see Course 9 by A.V Solov'yov in Ref. 2 and this book). These methods are the traditional tools for probing properties and internal structure of various physical objects.

Interesting phenomena arise in elastic collisions of electrons with ACs. For example, the diffraction of fast electrons by the fullerene C_{60} molecule was predicted and later observed.[18] The diffraction pattern in the electron elastic scattering cross section caries important information on the electron density in the vicinity of the fullerene's surface.

Electron excitations in metal cluster systems have a profoundly collective nature (see Ref. 11 and references therein). They can be pictured as oscillations of electron density against ions, the so-called plasmon oscillations. This name is carried over from solid state physics where a similar phenomenon occurs. Collective electron excitations have also been studied for single atoms and molecules. In this case the effect is known under the name of the shape or giant resonance. The name giant resonance came to atomic physics from nuclear physics, where the collective oscillations of neutrons against protons have been investigated.[17]

The interest of plasmon excitations in small metal clusters is connected with the fact that the plasmon resonances carry a lot of useful information about cluster electronic and ionic structure. By observing plasmon excitations in clusters one can study, for example, the transition from the pure classical Mie picture of the plasmon oscillations to its quantum limit or detect cluster deformations by the value of splitting of the plasmon resonance frequencies.

The plasmon resonances can be seen in the cross sections of various collision processes: photabsorption and photoionization, electron inelastic scattering, electron attachment, bremsstrahlung (see Course 9 by A.V Solov'yov in Ref. 2). Both surface and volume plasmons can be excited. In electron collisions and in the multiphoton absorption regime, plasmons with large angular momenta play an important role in the formation of the cross sections of these processes.[19]

In Fig. 5, we present experimentally measured and theoretically calculated cross section for the photoabsoption of some Na and Mg clusters.[20] The cross sections are resonantly enhanced owing to the excitation of plasmon oscillations in the target cluster.

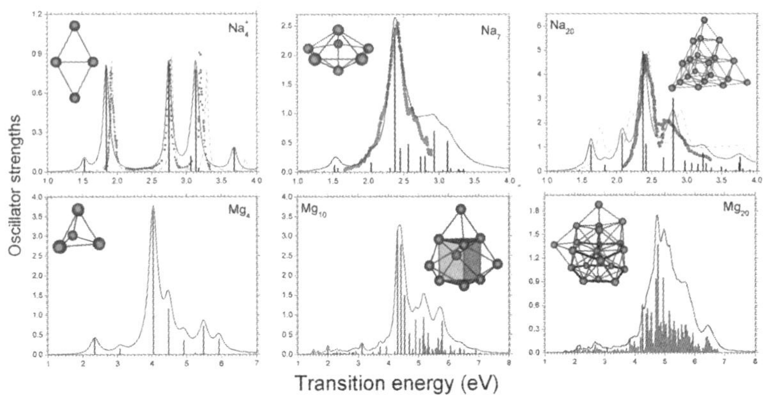

Fig. 5. Photoabsortion spectra of some Na and Mg clusters.[20]

Plasmon excitations in clusters decay via the Landau damping mechanism, while the relaxation of single electron excitations in clusters occurs via the interaction with the vibrations of ions, i.e. via the electron-phonon interaction (see Course 9 by A.V Solov'yov in Ref. 2).

Collisions involving ACs raise many more interesting physical problems.

For example, in collisions one can study phase transitions (solid-liquid or liquid-gas) in mesoscopic systems or the cluster multifragmentation process.

Another problem is linked closely to the problem of plasmon excitations in metal clusters. With increasing cluster size, the electronic energy levels of the single constituent atoms become grouped together, tending to form the conduction band, valence band etc. In this situation, the problem of localisation-delocalisation of the valence electron density in the cluster arises. This is known as the first order Mott phase transition. Plasmon excitations can be used as a probe of the Mott transition in ACs.

5. Fission Instability of Multiply Charged Clusters.

Multicharged ACs become unstable towards fission. The process of multicharged metal clusters fission is qualitatively analogous to nuclear fission. The fission instability of charged liquid droplets was first described by Lord Rayleigh in 1882 within the framework of classical electrodynamics.[21] Reviews of recent work on metallic cluster fission, can be found in Refs. 2, 7, 13, 14.

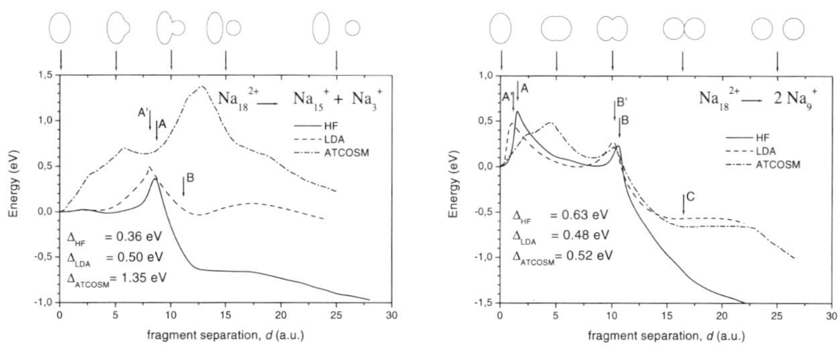

Fig. 6. Fission barriers for the asymmetric and symmetric fission channels of Na_{18}^{2+}: $Na_{18}^{2+} \to Na_{15}^{+} + Na_{3}^{+}$ and $Na_{18}^{2+} \to 2Na_{9}^{+}$ calculated in Ref. 7.

The fission process of ACs is interesting because it reveals the obvious parallel of AC studies with nuclear physics, where the fission process of nuclei has been studied for many decades.[17] The experiments on cluster fission provide a very good opportunity to test various concepts, approximations and AC models. Fission convincingly demonstrates the importance of the correct accounting for quantum and many body phenomena in the description of multi-atomic systems. Dynamical aspects of the AC fission problem are also of great interest, because, contrary to nuclear physics, in the fission of ACs all the forces in the system are known and thus one can develop the full dynamical description of the process.

To illustrate the fission of charged metal clusters we plot in Fig. 6 the fission barriers for the symmetric and asymmetric fission channels of Na_{18}^{2+} : $Na_{18}^{2+} \rightarrow Na_{15}^{+} + Na_{3}^{+}$ and $Na_{18}^{2+} \rightarrow 2Na_{9}^{+}$. The barriers plotted in Fig. 6 have been calculated in Ref. 7 within the two-center LDA and Hartree-Fock jellium model and compared with the asymmetric two-center-oscillator shell model (ATCOSM). Figure 6 demonstrates the evolution of cluster shape during the fission process, the importance of cluster deformations, many-electron correlation and shell effects.

6. Fusion Process of Atomic Clusters.

The formation of a sequence of cluster magic numbers should be closely connected to the mechanisms of cluster formation and growth. It is natural to expect that one can explain the magic numbers sequence and find the most stable cluster isomers by modelling mechanisms of cluster assembly and growth, i.e. the fusion process of ACs.[4]

The problem of magic clusters is closely connected to the problem of searching for global minima on the cluster multidimentional potential energy surface. The number of local minima on the potential energy surface increases exponentially with the growth of cluster size and is estimated[2,4] to be of the order of 10^{43} for $N = 100$. Thus, searching for global minima becomes an increasingly difficult problem for large clusters. There are different algorithms and methods of the global minimisation, which have been employed for the global minimisation of AC systems (see Refs. 2, 4 and references therein). These techniques are often based on Monte-Carlo simulations.

Alternatively, the algorithm based on dynamic searching for the most stable isomers in the cluster fusion process has been recently proposed.[4] The calculations performed with this new algorithm demonstrated that this approach is an efficient alternative to the known techniques of cluster global minimisation. The big advantage of the fusion approach consists in the fact that it allows one to study not just the optimized cluster geometries, but also their formation mechanisms.

In the recent work,[4] the fusion algorithm was formulated in a most simple, but general form. In the most simple scenario, it was assumed that

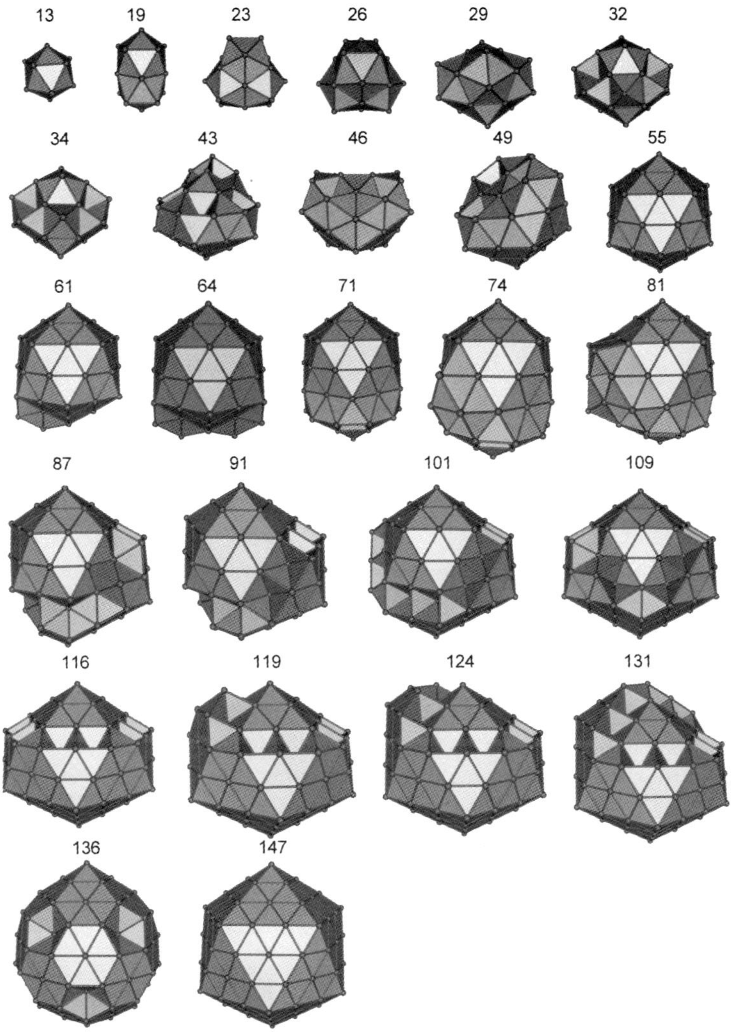

Fig. 7. Images of the Lennard-Jonnes global energy minimum cluster isomers.[4,9] The mass numbers of the pictured clusters correspond to the magic numbers of the noble gas (Ar, Kr, Xe) clusters.

atoms in a cluster are bound by Lennard-Jones potentials and the cluster fusion takes place atom by atom. In this process, new atoms are placed on the cluster surface in the middle of the cluster faces. Then, all atoms in the system are allowed to move, while the energy of the system is decreased.

The motion of the atoms is stopped when the energy minimum is reached. The geometries and energies of all cluster isomers found in this way are stored and analysed. The most stable cluster configuration (cluster isomer) is then used as a starting configuration for the next step of the cluster growing process.

Starting from the initial tetrahedral cluster configuration and using the strategy described in Ref. 4 cluster fusion paths have been analysed up to the cluster sizes of more than 150 atoms. We have found that in this way practically all known global energy minimum structures of the Lennard-Jonnes clusters can be determined. Figure 7 shows the images of the Lennard-Jonnes global energy minimum cluster isomers.[4] The mass numbers of the clusters represented correspond to the magic numbers of the noble gas (Ar, Kr, Xe) clusters.

So far, the cluster fusion algorithm has been applied to the noble gas clusters which are based on the LJ type of the inter-atomic interaction. However, the fusion process can be generated in a similar way for systems, like metal clusters, held together by quantum forces. This technique can also be used for the simulation of the fusion process of complex bio-molecules (proteins and DNA) or for the study of protein folding. It would be interesting to see to which extent the parameters of inter-atomic interactions can influence the cluster fusion process and the corresponding sequence of magic numbers or whether the clusterization in nuclear matter consisting of alpha particles and/or nucleons is possible. Studying cluster thermodynamic characteristics with the use of the technique developed is another interesting theme which is left open for future considerations.

7. Conclusions

In recent years, AC physics has made very significant progress, but a large number of problems in the field are still open. The transition of matter from the atomic to the solid state implies changes of organization which turn out to be a good deal more subtle and complex than was originally supposed. Different types of clusters, composite clusters, various size ranges, cluster geometries, complex molecules (including biological), clusters on a surface and in plasmas, all provide additional themes which make this field of science very rich and varied. Collisions involving ACs, mass spectroscopy and laser techniques provide tools for experimental studies of the AC structure and properties.

However, what are the experimental limitations? Where should the theory go next? Where does the future lie? Could clusters one day become the smallest devices or be used to make the smallest devices? Could one manipulate cluster isomers for the production of new materials and nanostructures? What is the difference between a cluster and a bio-molecule or a virus? Could molecules as complex as proteins or DNA and their functions be understood on the basis of classical mechanics or does one ultimately need to invoke the quantum theory? What are the principles of the self-organization of matter, of self-assembling and functioning on the nanoscale? We merely mention such intriguing questions here, but we hope that at least some of them will be resolved during the future development of the topics described in this book.

References

1. J.P. Connerade, A.V. Solov'yov and W. Greiner, *Europhysicsnews* 33 , 200 (2002).
2. C. Guet, P. Hobza, F. Spiegelman and F. David (eds.), *NATO Advanced Study Institute, Session LXXIII, Summer School "Atomic Clusters and Nanoparticles"*, Les Houches, France, July 2-28, 2000, EDP Sciences and Springer Verlag, Berlin, New York, London, Paris, Tokyo, (2001).
3. A.V. Finkelshtein and O.B. Ptizin, *Physics of Proteins*, University, Moscow, (2002).
4. I.A. Solov'yov, A.V. Solov'yov, W. Greiner, A. Koshelev and A. Shutovich, *Phys.Rev.Lett.* 90 (2003) 053401; *Journal of Chemical Physics* (2003) in print; physics/0306185, v.1, 26 Jun (2003).
5. I.A. Solov'yov, A.V. Solov'yov and W. Greiner, *Phys. Rev. A* 65, 053203, (2002).
6. A.G.Lyalin, I.A. Solov'yov, A.V. Solov'yov and W. Greiner, *Phys. Rev. A* 67, 063203 (2003).
7. A.G. Lyalin, A.V. Solov'yov and W. Greiner, *Phys. Rev. A* 65, 043202 (2002).
8. A. Matveentsev, A.G. Lyalin, I.A. Solov'yov, A.V. Solov'yov and W. Greiner, *Int. J. Mod. Phys. E* 12, 81 (2003).
9. A. Koshelev, A. Shutovich, I.A. Solov'yov, A.V. Solov'yov, W. Greiner, *Proceedings of International Workshop "From Atomic to Nano-scale"*, Old Dominion University, December 12th-14 th, 2002, Norfolk, Virginia, USA (2002), editors Colm T. Whelan and Jim Mc Guire, Old Dominion University, 184-194 (2003).
10. W.A. de Heer, *Rev. Mod. Phys.* 65, 611 (1993).
11. C. Bréchignac, J.P. Connerade, *J.Phys.B: At. Mol. Opt. Phys.* 27, 3795 (1994).
12. H. Haberland (ed.), *Clusters of Atoms and Molecules, Theory, Experiment and Clusters of Atoms*, Springer Series in Chemical Physics 52, Berlin, Heidelberg, New York, Springer (1994).
13. U. Näher, S. Bjørnholm, S. Frauendorf, F. Garcias and C. Guet, *Physics*

Reports 285, 245 (1997).
14. W. Ekardt (ed.), *Metal Clusters*, Wiley, New York, (1999).
15. S.Sugano and H.Koizumi, *Microcluster Physics*, Second Edition, Springer, Berlin, Heidelberg, London, (1998).
16. H.W. Kroto et al., *Nature*, 318, 163 (1985).
17. J.M. Eisenberg and W. Greiner, *Nuclear Theory*, North Holland, Amsterdam, (1987).
18. L.G. Gerchikov, P.V. Efimov, V.M. Mikoushkin and A.V. Solov'yov, *Phys.Rev.Lett.* 81, 2707 (1998).
19. J.P. Connerade and A.V. Solov'yov, *Phys. Rev. A* 66, 013207 (2002).
20. I.A. Solov'yov, A.V. Solov'yov and W. Greiner, in *Latest Advances in Atomic Clusters Collision: Fission, Fusion, Electron, Ion and Photon Impact*, Editors J.P. Connerade and A.V. Solov'yov, Imperial College Press and World Scientific, London, (2003).
21. Lord Rayleigh, *Philos. Mag* 14, 185 (1882).

Structure and Properties of Atomic Clusters

CONFINED ATOMS IN BUBBLES, CLUSTERS, FULLERENES, QUANTUM DOTS AND SOLIDS

Jean-Patrick Connerade

Quantum Optics and Laser Science Group, Physics Department, Blackett Laboratory
Imperial College London UK
E-mail: j.connerade@ic.ac.uk

Prasert Kengkan

Physics Department, Khon Kaen University, Khon Kaen 4002 Thailand
E-mail: prasertk@kku.ac.th

A general review of the subject of Confined Atoms is presented. This subject has a long history, extending back almost to the origins of Quantum Mechanics, but has remained fairly quiescent until recent times, because suitable experimental examples of confinement did not exist. Today, with the advent of cluster physics, the discovery of endohedral metallofullerenes and the fabrication of quantum dots, examples in fact abound, and the theory of confined atoms is undergoing a revival. The confined atom is, in some sense, the first rung of a ladder which leads up from the free atom to nanoscale physics. Indeed, it can be seen as the smallest device available in nanoscience. Endohedral atoms are very topical today: they have been proposed as suitable building blocks for the register of a quantum computer because a spherical cage can have the effect of isolating the spin of an atom confined at the center from the outside world. Atoms under extreme pressure also provide examples of quantum confinement, and have practical applications, for example in the diagnosis of metal fatigue in the walls of a nuclear reactor. Thus, quantum confinement is of general interest not only to atomic physicists, but also to a wide range of scientists from many different disciplines.

1. Introduction

The purpose of the present review is to provide a very brief general introduction to a subject which has been somewhat neglected in the past, but which is fundamental to the understanding of clusters, dots, atoms in bubbles, etc, all of which are of topical interest. This subject is the quantum confinement of atoms, or 'confined atoms' for short. What we mean by quantum confinement is that an atom is trapped inside a volume whose size is comparable to its own dimensions. This definition is useful to distinguish such atoms from atoms trapped in larger cavities, for example in microwave cavities, whose properties are rather different from the ones considered here. Another technical term often used (especially for atoms trapped in fullerenes) is the expression: 'endohedral' from the Greek words 'endo' meaning 'inside' and 'hedron' as in 'polyhedron'. This word distinguished a confined atom A, which is trapped inside a fullerene (for example C_{60}) from an atom substituting one of the atoms in the carbon shell or else hanging onto the outside. The endohedral atom is often denoted as $A@C_{60}$.

2. Confined atoms

In fundamental physics, there are a number of model problems which are, in one way or another, idealizations of a real situation, and have become important because they can actually be solved analytically or 'exactly'. However, if one interprets the analytic solutions too literally, they correspond to situations which are not quite real. Some simple examples make this clear: the infinitely deep square well is routinely solved in introductions to quantum mechanics, and accompanies the teaching of elementary courses, but we tend to forget that, in fact, it is unobservable, since the particle it contains is isolated from the outside world. Another example is the hydrogen atom, which can reputedly be solved exactly, except that the radiation field is left out of the equations, which again renders the atom unobservable. Black-body radiation is a similar case: to be ideal, a black-body should allow no radiation to escape, and this condition, a simple analytic formula for the radiation is obtained.

In the same way, the confined atom possesses a limiting or ideal situation which is not truly accessible, but which can be solved completely. This is the case of hydrogen confined in an impenetrable sphere, as first pointed out by

Sommerfeld and Welker[1] in a paper which was their offering for Pauli's sixtieth birthday. As they realized, the solutions for hydrogen confined at the center of a spherical cavity with impenetrable walls are obtained from the excited states of hydrogen, provided only that the radius of the cavity coincides with the radius of a node. In this way, one can obtain special solutions and, indeed, a general formula for the ground state energy as a function of cavity radius in the form of a series expansion. By studying the dependence of energy on cavity radius, Sommerfeld and Welker[1] noticed that the energy rises above the binding energy of free hydrogen once the cavity radius becomes smaller than the innermost node. At this point, the electron is no longer bound by the proton, but is confined only by the cavity. Sommerfeld and Welker[1] considered this process as a form of ionization, and gave it as the origin of the conduction band of a solid, when the atom is confined within a Wigner-Seitz cell. Today, we would rather describe the phenomenon as a form of delocalisation induced by confinement.

The paper just cited is of course a seminal contribution involving one of the great masters of the subject, but it is not, in fact, the first paper to discuss the problem. That honour probably belongs to Michels *et al*[2] whose paper, by an interesting coincidence, was also a birthday offering, this time destined for van der Waals. The confined atoms, in this case, were not those of the solid, but atoms under extreme pressure, high enough for quantum confinement effects to appear.

The names of Pauli and van der Waals are intimately connected with fundamentals of the subject, as the following discussion will bring out. Apart from a few notable contributions,[3,4,5] the subject of confined atoms remained dormant for some time, until a sudden explosive growth [see e.g. Refs. 6-16] in recent years, for reasons alluded to in the introduction. It is perhaps useful to list a few of the new developments, and comment on their implications.

3. Atoms Under Pressure

One area which has remained active over many years is the subject of atoms under very high pressure. This can be physical pressure, as occurs for example when a solid is compressed under a diamond anvil, or a chemical 'pseudo-pressure' induced when a polaronic distortion is formed inside a solid by the insertion of an impurity atom. In both cases, the pressures can be high enough to induce changes in atomic properties. Seen from this perspective, the case of hydrogen is hardly the most interesting one. The electron in hydrogen has but two options under pressure. Either it remains attached to the proton, or it delocalizes. Much more interesting are the cases of heavier atoms such as the transition metals or rare-earths, for which shell filling occurs in a very remarkable way, by the process known as 'orbital collapse'.[17] What is found in such cases is that the electronic configuration of the ground state is pressure-sensitive or, to put it in another way, that the Periodic Table of atoms under extreme pressure is not the same as that of free atoms[18]. When one reflects that the Periodic Table expresses the chemistry of the elements, then it becomes clear that the changes in behaviour induced by strong pressure are far-reaching and that atomic physics must be adapted to situations in which very high pressures occur.

The manner in which high pressures are introduced into atomic physics is also an interesting one. This is achieved[19] by considering the compressibility of atoms, and requires us to revisit the fundamentals of thermodynamics. In principle, pressure is introduced through the kinetic theory of gases, in which atoms or molecules are treated as projectiles which carry momentum and impinge on a plate, or on a boundary which confine the gas. Thus, pressure is thought of as a consequence of kinematics, but atoms themselves are regarded as incompressible. The situation was changed somewhat when van der Waals introduced corrections to the ideal gas equation which include allowing for the finite volume of the atoms in the gas. In principle, once the volume of atoms is introduced into the equations, the idea that this volume might change as a function of the pressure is not far away. However, this step is not taken in elementary thermodynamics, and the van der Waals correction still treats the atomic volume as a constant.

When one solves the Schrödinger equation for a confined atom, it emerges, however, that the volume of the atom (i.e. the volume occupied by the wavefunctions) changes as a function of confinement at the same time as the energy. Strictly, this volume is not the volume of the cavity, whose walls may anyway be more or less penetrable, but comes from the expectation value of the operator r^3, so as to yield a legitimate quantum mechanical observable, linked to the energy E obtained from the same equation. The same calculation can be performed for the free atom, yielding the energy E_0 and the volume V_0. We can then write $\Delta V = V_0 - V$ and $\Delta E = E_0 - E$, which allows us to obtain the pressure P from the standard relation $\Delta E = P\Delta V$. We see that P is then the ratio of two quantum mechanical observables, and is therefore a legitimate quantity in the sense of quantum mechanics, which can be computed by solving the Schrödinger equation. One may then ask whether the pressure obtained in this way is indeed the same as the physical pressure as determined in experiments. The following example shows that this is indeed the case. In Figure 1, we show a comparison between experiment and theory for caesium under pressure.[20]

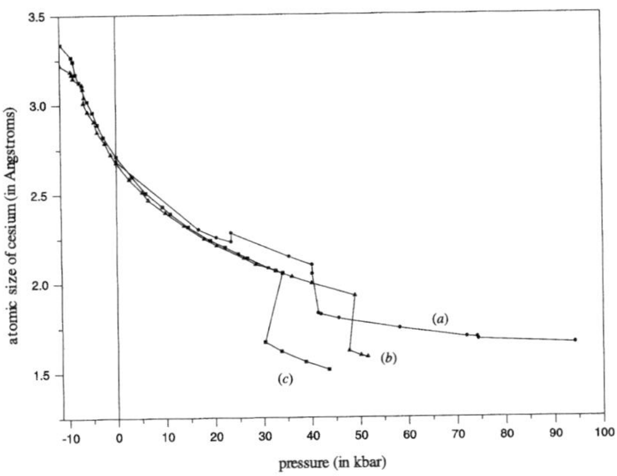

Fig. 1. The atomic size of caesium versus the pressure (a) experimental curve (b) relativistic calculation for a confined caesium atom and (c) nonrelativistic calculation with the same computer code and boundary conditions.

The experimental curve in Figure 1 exhibits two discontinuities. The first of these, a small one on the left hand side of the figure, is a solid state effect, due to a change in crystal structure under pressure, and appears only on the experimental curve. The second, much more pronounced, towards the center of the figure, is an atomic effect, and corresponds to structure of similar magnitude in the two theoretical curves displayed, which are computed for a confined caesium atom. These two calculations are performed by enclosing the atom in an impenetrable sphere, in the manner already described. For curve (b), the Dirac-Fock method is used, and for curve (c), the same method is used, allowing the velocity of light to be a very large number, which simulates the non-relativistic case.

A number of comments can immediately be made concerning the comparison. First, the cause of the discontinuity is a change in the configuration of the atom under pressure, when the *6s* electron of free caesium becomes a *5d* electron under high pressure, because of the changes in shell-filling referred to above, and the orbital collapse of the *5d* electron under pressure. Second, we note that this is a purely atomic effect, since parameter-free calculations from the Dirac equation reproduce it. Of course, it is possible to simulate the effect by band theory, but only by introducing parameters which subsume the physics of orbital collapse. Third, we note that relativistic effects are rather large: the calculation sweeps across the range of interest as the velocity of light is changed. The remaining discrepancy between the observed discontinuity and relativistic calculations can be attributed to a number of causes, of which the neglect of solid-state effects is probably the largest.

It is remarkable that calculations for the confined atom work so well for the atom in the solid, but care is needed before generalizing this conclusion. A solid can be formed in two ways. In the first, the atoms are loosely packed, with large spacings between them, in which case compression will lead to a change in the lattice spacing, with not much change in the volume of the atoms. In the second situation, the atoms are densely packed, with little separation between them, and compression of the solid results in the atoms themselves becoming compressed. The latter situation is the one illustrated by the data of Fig. 1.

Caesium, of course, is a very ‘soft’ atom, occupying a large volume, possessing a low ionization potential, and is very susceptible to quantum compression. At the opposite extreme, we have helium, which is very compact, has a high ionization potential, and is much more difficult to compress. Nevertheless, the compression of helium has been observed. Alpha particles from a nuclear reactor, when they traverse the walls of the reactor made from a special steel alloy, have a small but finite probability of being stopped on a structural defect of the solid and converting to helium. Once this process has started, it tends to repeat in the same place, and thus a helium bubble is formed, which grows inside the solid. Observations of such bubbles by electron energy loss spectroscopy[21] have revealed that the energy level strucure of the helium atoms in these bubbles is not the same as that of free helium atoms. It can be recovered, however, by computing the energy levels of helium under pressure[22] and, by matching the computed level structure to the observations, the pressure inside the bubbles can be inferred. Quite clearly, the pressure inside the bubbles grows with time and, when it exceeds the tensile strength of the material, the walls begin to crack. Thus, monitoring the bubbles spectroscopically is a diagnostic tool for the ageing of reactor walls. In all these examples, the force which confines atoms is due to the exclusion principle, and is sometimes for this reason referred to as the Pauli force.

Pressure, as already noted above, need not be physical, but can also be a chemical 'pseudo-pressure'. A situation where this approach is useful is the study of host materials for reversible 'rocking chair' lithium ion batteries. Lithium has the best electrochemical properties for the realization of rechargeable batteries, leading to the optimum power/weight ratio, but lithium metal is potentially dangerous as an electrode material, leading to a fire hazard or a risk of explosion if a battery is recharged. The solution to this problem is to insert lithium ions into a solid, which then acts as the electrode, and this is the basis of the Li-ion batteries which are now used in portable computers, etc. For the insertion to be effective, host materials must be found which allow the small, incompressible, Li+ ions to migrate throughout the electrode, and which are compressible enough to distort, so that Li+ ions can be stored inside. Furthermore, the storage process must be reversible. This requires that there should be no phase change, or recrystallisation of the lattice structure (so-called 'topotactic' insertion). In other words, the solid must preserve its structure under chemical pressure, but the atoms themselves must be compressible.[23] It is indeed remarkable that materials used in the fabrication of electrodes for Li+ ion batteries often involve transition elements and rare-earths, for which orbital collapse can be controlled by chemical pressure. This allows one to model reversible insertion by applying the principles of atomic confinement.[24]

4. Dimensionless Representation or 'Universal Curves'

Before leaving the subject of atoms under pressure, we turn to the issue of how the compressibilities of different atoms can be compared in a quantitative way, and their properties put onto a common scale. It has been shown that the compressibility is in fact made up of two parts. The first is a scaling factor which can be calculated for free atoms and varies widely from atom to atom, the softest (caesium) being some two thousand times more compressible than the hardest (helium). The second is a dimensionless quantity, which is obtained by defining a reduced volume (the ratio of the volume of the compressed atom to the original volume of the free atom) and a reduced energy (the ratio of the binding energy of the compressed atom to the binding energy of the free atom). In terms of these dimensionless quantities, the compression curves of all atoms can be plotted on the same scale, and this graph reveals that the nonlinear variation in the compressibility is in fact very similar for all atoms, hard or soft, as demonstrated by the curves of Fig. 2.[25]

If we analyse these plots, we see that the curves for caesium and the curves for helium are rather similar in form. All other atoms, in fact, lie between caesium (a large alkali) and helium (the smallest noble gas), and are therefore contained between these curves.

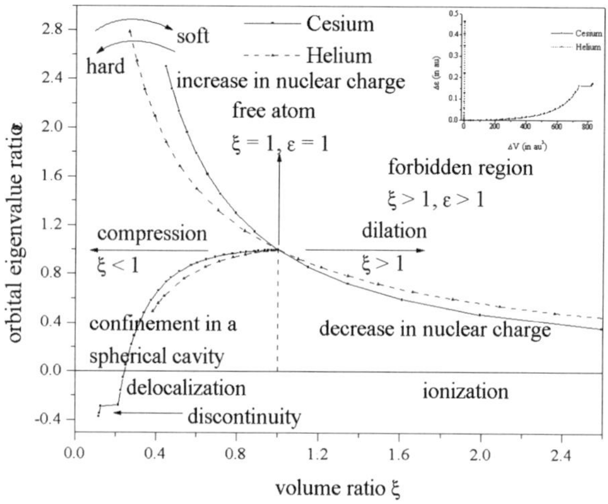

Fig. 2. Universal compressibility plot, showing different branches according to the nature of the confining potential. Note how similar the behaviour is for hard atoms (helium, dashed curves) and for soft atoms (caesium, full curves).

The different branches of these curves correspond to different forms of compression. Atoms can be compressed externally (by confinement in a cavity with repulsive walls) or internally (by inner shell excitation, or artificially in calculations by increasing the nuclear charge). They can also be dilated by confinement in a cavity with attractive walls. Each one of these paths, in principle, can generate a branch in the plots of Fig. 2, but all these branches must return through the point (1,1) which is simply the position of the free atom. These curves have many interesting properties,[25] and the scaling shows that most of the compressibility arises through coulombic interactions, which vary systematically with the size of the atoms. Recently, we have shown that pseudoatoms with one-dimensional potentials related to the coulombic form also lie on such dimensionless plots.

5. Quantum Dots

The quantum dot is closely related to the problem of the atom under pressure, or the atom confined inside a cage with repulsive walls. Essentially, the Hamiltonian is the same, with minor variations due to the nature of the material, and the degree of sophistication required in the representation of the confining potential. A good example of the similarity of approaches can be found in Ref. 14.

6. Endohedral Confinement Inside an Attractive Cage

The possibility of dilating atoms by confinement within an attractive cavity might at first sight seem far-fetched, but in fact is achievable. An attractive molecular cage exists, in the form of the fullerene, which is capable of forming negative ions, and is therefore known to attract electrons. Experiments on electron scattering have revealed[26] that the fullerene cage can be modeled as a spherical attractive shell, with well-understood quantum interference properties. To confine an atom in such a shell theoretically, it is sufficient to add to the self-consistent field potential of the free atom an attractive shell, whose properties (geometry and depth) are adjusted to match experimental values.[27] Of course, when doing this, one must not forget a fundamental difference between confinement within repulsive and attractive shells. In a repulsive shell, it is clear that the endohedral atom will tend to 'sit' at the center, whereas in an attractive shell, it is more likely that the atom will be pulled off-centre.[28] However, in all such situations in which central symmetry is broken slightly but not completely, it is appropriate to start by considering the much simpler symmetric case, and then to expand the real situation within the basis provided by the spherical case. This, after all, is the whole basis of the standard expansions of atomic physics, and the problems are tackled in the same spirit.

For certain atoms (an example is N), confinement occurs with the atom very close to the center of the C_{60} cage. Such examples turn out to be important, because the spin-orbit interaction then turns out to be very small, and the spin is isolated, or screened by the cage. Confined atoms of this type are candidates as building blocks for the register of a 'quantum computer', with the advantage that they are much easier to contain in a real device than atoms in RF traps.[11]

Confinement within an attractive shell is thus found to possess the following characteristics: (i) the resulting system possesses the quantum signature of both constituent species, ie. There are cavity states and resonances and, in addition, bound states and resonances originating from the confined atom. Where they overlap in energy, avoided crossings occur.[29] (ii) Resonances are in general of three types [30], namely resonances originating from the cavity, which would be present even if the atom were removed, resonances due to the atom, which would persist even if the cavity were not present, and finally a class of resonances which only exist because both the cavity and the atom are present. To understand how the latter (which we term molecular) resonances can occur, it is useful to consider how the system behaves if the cavity does not have complete spherical symmetry. As is well-known, a perfectly spherical shell does not rotate in Quantum Mechanics, and complete spherical symmetry also implies that angular momentum is a conserved quantity. Both of these statements cease to be true when the cavity is made up, not of a perfectly spherical charge distribution, but of sixty atoms placed symmetrically on a sphere. Under these circumstances, angular momenta are mixed, and new channels open up which only exist because both the endohedral atom and its confining cage are present at the same time.

Finally, it is useful to comment on the significance of relativity in dealing with confined atoms. As we have stressed, the most important and illuminating situations are those in which orbital collapse is possible, and this effect is highly sensitive to small changes in the atomic potential. It also occurs for rather heavy atoms, and the combination of the two means that relativity must be included when computing the behaviour of confined atoms. Thus, a natural starting point is the Dirac-Fock method. However, there is a technical point in this case concerning boundary conditions, especially in the case of atoms under pressure, when it is tempting to introduce impenetrable cavities. One must remember that, strictly, the imposition of Dirichlet boundary conditions for an impenetrable cavity would violate relativity, since both the large and the small component of the Dirac spinor cannot be made zero at the same point. Strictly, therefore, one should choose other boundary conditions for this case, and our investigations suggest that the MIT bag model provides the most appropriate ones.[31]

7. Conclusion

We have presented a very brief introduction to the subject of quantum confinement. Of necessity, we have left out a number of other interesting situations, such as confinement of impurity atoms in mixed clusters, confinement in zeolites, the influence of super-strong laser fields, etc. In connection with some of these more complex problems, we are currently developing a model for one-dimensional quantum confinement, which will also allow chains of atoms to be treated. We also omitted any discussion of the chemical pressure effects, which occur as a function of size in certain clusters, and are also capable of inducing orbital collapse.

Despite these omissions, we hope to have conveyed the current sense of excitement of this subject, which is a rapidly expanding field. Although much of the work is theoretical, there is a general feeling that, once certain problems of production and detection have been mastered, experiments on confined atoms will develop quickly.

References

1. A. Sommerfeld and H. Welker, *Ann. Phys.* **32**, 56 (1938).
2. A. Michels, J.de Boer and A. Bijl, *Physica* **4**, 991 (1937) (van der Waals-Festschrift)
3. R.E. Watson, *Phys. Rev.* **111**, 1108 (1958).
4. C. Zikovich-Wilson J.H. Planelles and W. Jaskólski, *Int. J. Quantum. Chem.* **50**, 429, (1994).
5. W. Jaskólski, *Phys. Rept.* **271**, 1, (1996).
6. Y.P. Varshni, *J. Phys. B.* **31**, 2849, (1998).
7. J.-P. Connerade V.K. Dolmatov P.A. Lakshmi and S.T. Manson, *J. Phys B* **32,** L239, (1999).
8. J.-P. Connerade V.K. Dolmatov and S.T. Manson, *J. Phys. B* **32**, L395, (1999).
9. Shi Ting-yung Qiao Hao-xue and Li Bai-wen, *J. Phys. B* **33**, L349, (2000).
10. A.S. Baltenkov V.K. Dolmatov & S.T. Manson, *Phys. Rev. A* **64**, 62707, (2001).
11. W. Harneit, *Phys Rev A* **65**, 032322, (2002).
12. C. Laughlin B.L. Burrows and M. Cohen, *J. Phys. B* **35**, 701, (2002).
13. L. Forró and L. Mihály, *Rep. Progr. Phys.* **64**, 649, (2001).
14. T. Sako and G.H.F. Diercksen, *J. Phys. B* **36**, 1681, (2003).
15. J.-P. Connerade P. Kengkan and R. Semaoune, *J. Chinese Chem Soc* **48**, 265, (2001).
16. J.-P. Connerade, *Indian J. Phys* **76B**, 359, (2002).
17. J.-P. Connerade in *Highly Excited Atoms* Cambridge University Press (1998).
18. J.-P. Connerade V.K. Dolmatov and P. Anantha Lakshmi, *J. Phys. B* **33**, 251, (2000).
19. J.-P. Connerade, *J. Phys. C* **15**, L367, (1982).

20. J.-P. Connerade and R. Semaoune, *J. Phys. B* **33**, 3467, (2000).
21. C.A. Walsh, J. Yuan and L.M. Brown, *Phil. Mag.* **80**, 1507, (2000).
22. C.T. Whelan Private Communication, (2000).
23. J.-P. Connerade, *J. of Alloys and Compounds* **255**, 79, (1997).
24. J.-P. Connerade J.-C. Jumas and J. Olivier-Fourcade, *J. of Solid State Chemistry* **152**, 533, (2000).
25. J.-P. Connerade P. Kengkan A Lakshmi and R. Semaoune, *J. Phys. B* **33**, L847, (2000).
26. Y.B. Xu M.Q. Tan and U. Becker, *Phys. Rev. Lett.* **76**, (1996).
27. J.-P. Connerade and R. Semaoune, *J. Phys. B* **33**, 869, (2000).
28. V.I. Pupyshev, *J. Phys. B* **33**, 961, (2000).
29. J.-P. Connerade V.K. Dolmatov P.A. Lakshmi and S.T. Manson, *J. Phys. B* **32**, L239, (1999).
30. J.-P. Connerade V.K. Dolmatov and S.T. Manson, *J. Phys. B* **33**, 2279, (2000).
31. V. Alonzo and S. De Vincenzo, *J. Phys. A* **30**, 8573, (1997).

SIMULATION OF MELTING AND IONIZATION POTENTIAL OF METAL CLUSTERS

M. Manninen, K. Manninen and A. Rytkönen
Department of Physics, University of Jyväskylä, Finland
E-mail: matti.manninen@phys.jyu.fi

We have used classical and *ab initio* molecular dynamics to study the melting of sodium clusters in order to see the effects of the geometric and electronic magic numbers on the melting temperature as a function of the cluster size. It seems that classical many-atom interactions can not explain the experimentally observed size-dependence of the melting temperature. For selected cluster sizes we have used *ab initio* molecular dynamics to study the effects of the electronic structure on the melting and on the ionization potential. The results reveal no correlation between the vertical ionization potential and the degree of surface disorder, melting, or the total energy of the cluster.

1. Introduction

Since the discovery of the reduced melting temperature of metal clusters[1,2] the melting of clusters have been an intensive area of theoretical research.[3–9] While the overall picture of the decrease of the melting temperature with the cluster size is reproduced in the theoretical studies, the recent experiments of Schmidt *et al*[10–12] for sodium clusters show that for small clusters the melting temperature varies strongly and nonmonotonously with the cluster size. Theoretical studies have not been able to explain the experimental findings.

Martin *et al*[13] studied the effect of the cluster temperature on the ionization properties of large sodium clusters and observed a phase transition from icosahedral to molten state. The size-dependence of this transition did not follow the $1/R$-dependence predicted by simple theoretical arguments, a possible explanation being the surface melting. Also for this observation a quantitative theoretical explanation is still missing.

In this paper we will report our recent work on the melting of sodium clusters and on its effect on the ionization potential. This work turned out to be computationally very demanding and we are not yet in the position to perform *ab initio* molecular dynamics of melting for a large number of cluster sizes. For this reason we combine classical and *ab initio* molecular dynamics.

2. Theoretical Methods

The classical molecular dynamics simulations were performed with a semi-classical many-body potential

$$V = \sum_{i=1}^{N} \left\{ \sqrt{\sum_{j=1 \neq i}^{N} \zeta_0^2 \exp\left[-2q\left(\frac{r_{ij}}{r_0} - 1\right)\right]} - \sum_{j=1 \neq i}^{N} \epsilon_0 \exp\left[-p\left(\frac{r_{ij}}{r_0} - 1\right)\right] \right\} \tag{1}$$

where $\epsilon_0 = 15.956$ meV, ζ_0=291.14 meV, r_0=6.99 a_0, q=1.30 and p=10.13. This potential is based on the tight-binding approximation[14] and the parametrization used here is taken from Ref. 15. In this work all the simulations were performed at a constant temperature using the Nosé-Hoover thermostat.[16] The time step in the simulations was 5 fs.

The caloric curve for small clusters was determined by performing constant temperature simulations at several different temperatures and fitting a continuous function to the observed total energy versus temperature. A fermi function was used to fit between the assumed linear temperature dependences of the solid and liquid phases.

The *ab initio* molecular dynamics simulations were done with the BO-LSD-MD method of Barnett and Landman.[17] Naturally, the *ab initio* simulations are limited to much shorter time scales than classical simulations and similar statistical accuracy can not be obtained.

The ionization potential is determined as a total energy difference between the positive ion and the neutral cluster.

3. Melting with Classical Tight-binding Potential

Using the model potential of Eq. (1) we calculated the caloric curve for selected cluster sizes between 40 and 350 atoms. The melting temperature and the latent heat were determined by fitting the simulation data to a continuous curve as explained above. Figure 1 shows the melting temperatures in comparison with the experimental results.[12] The simulated results show as large variation as a function of the size as the experimental ones. However, in detail the results do not agree. The overall melting temperature of

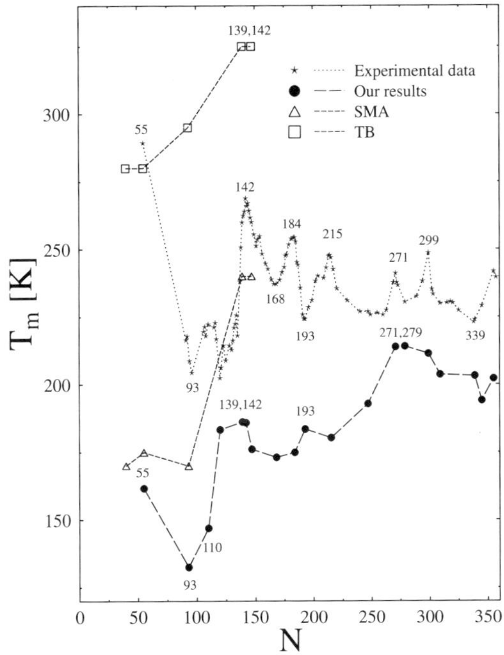

Fig. 1. Melting temperature of sodium clusters as a function of the clustrer size. The black dots show results simulated using the tight-binding potential. The experimental data are from the Freiburg group.[10–12] The SMA and TB theoretical results are from Calvo and Spiegelmann.[9]

the simulations is much smaller than the experimental one and there is a clear increase of T_m with the cluster size. Moreover, while the simulations give qualitatively correctly the size-dependence in cluster 55-93-142, it fails for sizes 184-193-215. Note that only selected sizes are calculated and the dashed lines are only guides for the eye: the actual curve could have much more minima and maxima. Figure 1 shows for comparison also the theoretical results of Calvo and Spiegelmann.[9] The results denoted by SMA are based on the same potential as ours but the determination of the melting temperature is different (our method is closely related to the experimental method). The TB result is based on quantum mechanical tight-binding calculation, pointing out the importance of the electronic structure.

Figure 2 shows the latent heat as a function of the cluster size. In average, our simulations give a fair agreement with the experiments. Again, however, the detailed variations as a function of the cluster size are different.

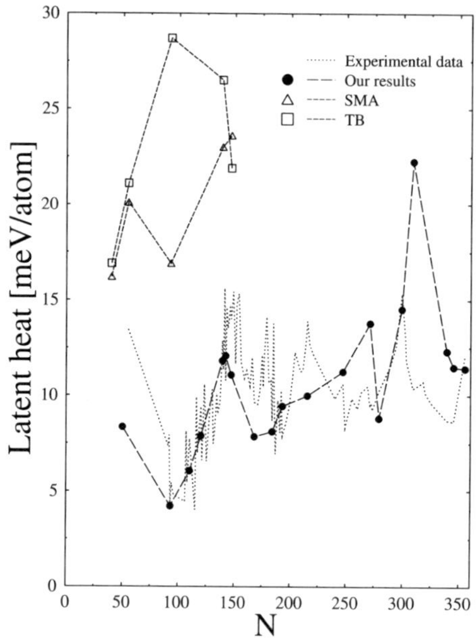

Fig. 2. Latent heat as a function of the cluster size. The different symbols are the same as in Fig. 1.

4. Surface Melting

Surface melting was studied[18] for icosahedral clusters including from 147 to 1415 atoms, using the classical model potential. The average mean square displacement (MSD) in 30 ps was determined separately for the bulk and surface atoms. The difference of these two MSDs is shown in Fig. 3 as a function of the cluster temperature. Independent of the cluster size the difference ΔMSD starts to increase at about 200 K, which we interpret as the onset of surface melting. For different sizes ΔMSD drops to zero at different temperatures, as an indication of the bulk melting temperature (bulk atoms become as mobile as the surface atoms). The icosahedron with 147 atoms does not show any surface melting since its bulk melting temperature, in our simulations, is only about 175 K.

5. Melting with BO-LSD-MD

We have earlier[19,20] studied melting of 40 and 55 atom sodium clusters using *ab initio* molecular dynamics. In these simulations it was found that

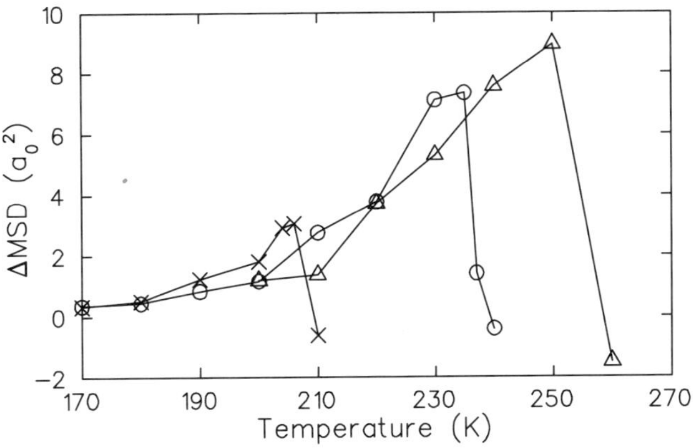

Fig. 3. Difference between the mean square displacement (in 30 ps) of the surface atoms and the bulk atoms. Crosses show the results for 309 atoms, circles for 923 atoms and triangles for 1415 atoms.

both clusters show a clear melting transition which can be detected either from the caloric curve or studying the T-dependence of the MSD. The very limited statistics of the *ab initio* molecular dynamics give only a rough estimate of the melting temperature: it was estimated to be between 300-350 for Na_{40} and 310-360 for Na_{55}.

In this work we have extended the BO-LDA-MD simulations to the Na_{93}^+ ion. Up to now, we have made simulations only at three different constant temperatures (248, 262, and 274 K). Clearly, this is not enough to estimate the caloric curve. However, the determined diffusion constants indicate that the cluster is liquid already at the lowest simulated temperature of 248 K. This is in qualitative agreement with the experiment: the electronically magic Na_{93}^+ melts at a clearly lower temperature than Na_{55}. The electronic shell structure, with a gap at the Fermi surface, is clearly seen in the liquid cluster.

6. Ionization Potential

We studied[18] the effect of the melting on the ionization potential of clusters with 147 and 142 atoms. The latter is a complete icosahedron and in the former some of the corner atoms are missing. We performed classical simulations with the tight-binding model and selected a random set of 10 atomic positions (corresponding to different times) for each temperature and cluster size. These atomic coordinates were then used for calculating

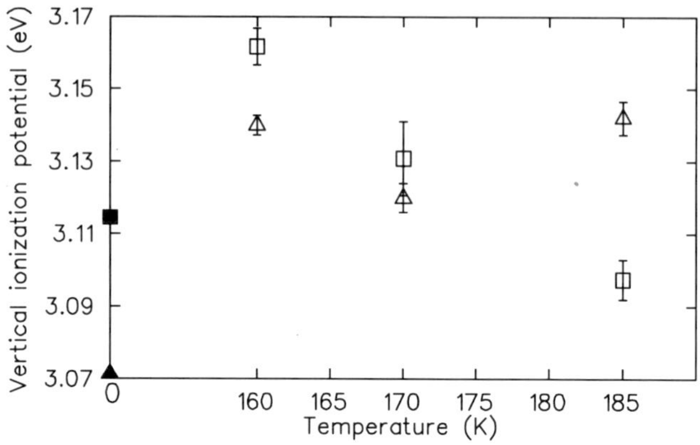

Fig. 4. Vertical ionizaiton potential as a function of the temperature. The triangles and squares represent averages over 10 random atomic comfigurations for Na_{142} and Na_{147}, respectively. The error bars show the standard deviations.

the vertical ionization potential using *ab initio* LSDA electronic structure calculations. (Note that at finite temperatures the adiabatic ionization potential does not have any meaning).

The results are shown in Fig. 4. The temperature dependence of the ionization potential is small and does not show clear systematics. For both cluster sizes the IP increases when the temperature is increased from zero to 160 K at which temperature the clusters are still solid. Increasing further the temperature, the IP of Na_{147} decreases while that of Na_{142} seems not to be monotonous. The highest temperature shown corresponds already to a liquid cluster. The differences in IP are so small that no effect associated with the melting transition can be pinpointed.

7. Conclusions

We have studied the melting and its effect on the ionization potential by combining classical tight-binding molecular dynamics and *ab initio* electronic structure calculations. The simulated melting temperature shows large nonmonotonous dependence on the cluster's size, but the detailed size-dependence does not agree with the experimental results. Clusters with more than about 300 atoms show also surface melting at a temperature which seems to be nearly independent of the cluster size.

With the *ab initio* molecular dynamics we are not yet able to get enough statistics to determine the melting temperature reliably. However, the simulations for 40, 55, and 93 atom clusters do not indicate disagreement with the experiments.

The ionization potential depends on the temperature but the melting transition does not seem to have any marked effect on it.

Acknowledgments

This work has been supported by the Academy of Finland under the Finnish Centre of Excellence Programme 2000-2005 (Project No. 44875, Nuclear and Condensed Matter Programme at JYFL).

References

1. Ph. Buffat and J.-P. Borel, *Phys. Rev.* A 13, 2287 (1976).
2. J.-P. Borel, *Surf. Sci.* 106, 1 (1981).
3. I.L. Garzón and J. Jellinek, *Z. Phys.* D 20, 235 (1991).
4. S. Valkealahti and M. Manninen, *Comp. Mater. Sci.* 1, 123, (1993).
5. N. Ju and A. Bulgac, *Phys. Rev.* B 48, 2721 (1993).
6. R. Poteau, F. Spiegelmann, and P. Labastie, *Z. Phys.* D 30, 57 (1994).
7. H. Grönbeck, D. Tománek, S.G. Kim, and A. Rosen, *Z. Phys.* D 40, 469 (1997).
8. C.L. Cleveland, W.D. Luedtke, and U. Landman, *Phys. Rev. Lett.* 81, 2036 (1998).
9. F. Calvo and F. Spiegelmann, *J. Chem. Phys.* 112, 2888 (2000).
10. M. Schmidt, R. Kusche, W. Kronmüller, B. von Issendorff, and H. Haberland, *Phys. Rev. Lett.* 79, 99 (1997).
11. M. Schmidt, R. Kusche, W. Kronmüller, B. von Issendorff, and H. Haberland, *Nature* 393, 212 (1998).
12. R. Kusche, Th. Hippler, M. Schmidt, B.V. von Issendorff, and H. Haberland, *Eur. Phys. J.* D 9, 1 (1999).
13. T.P. Martin, U. Näher, H. Schaber, and U. Zimmermann, *J. Chem. Phys.* 100, 2322 (1994).
14. D. Tomanék, S. Mukherjee, and K.H. Bennemann, *Phys. Rev.* B 28, 665 (1983).
15. Y. Li, E. Blaisten-Barojas, and D.A. Papaconstantopoulos, *Phys. Rev.* B 57, 15519 (1998).
16. S. Nosé, Mol. Phys. 52, 255 (1984); W.G. Hoover, *Phys. Rev.* A 31, 1695 (1985).
17. R.N. Barnett and U. Landman, *Phys. Rev.* B 48, 2081 (1993).
18. A. Rytkönen and M. Manninen, *Eur. Phys. J.* D 23, 351 (2003).
19. A. Rytkönen, H. Häkkinen, and M. Manninen, *Phys. Rev. Lett.* 80, 3940 (1998).
20. A. Rytkönen, H. Häkkinen, and M. Manninen, *Eur. Phys. J.* D 9, 451 (1999).

NEW APPROACH TO DENSITY FUNCTIONAL THEORY AND DESCRIPTION OF SPECTRA OF FINITE ELECTRON SYSTEMS

M.Ya. Amusia[a,b], A.Z. Msezane[c], and V.R. Shaginyan[c,d]

[a] *The Racah Institute of Physics, the Hebrew University, Jerusalem 91904, Israel*

[b] *Physical-Technical Institute, 194021 St. Petersburg, Russia*

[c] *CTSPS, Clark Atlanta University, Atlanta, Georgia 30314, USA*

[d] *Petersburg Nuclear Physics Institute, Gatchina, 188300, Russia*

E-mail: vrshag@thd.pnpi.spb.ru

The self consistent version of the density functional theory is presented, which allows to calculate the ground state and dynamic properties of finite multi-electron systems. An exact functional equation for the effective interaction, from which one can construct the action functional, density functional, the response functions, and excitation spectra of the considered systems, is outlined. In the context of the density functional theory we consider the single particle excitation spectra of electron systems and relate the single particle spectrum to the eigenvalues of the corresponding Kohn-Sham equations. We find that the single particle spectrum coincides neither with the eigenvalues of the Kohn-Sham equations nor with those of the Hartree-Fock equations.

1. Introduction

The density functional theory (DFT), that originated from the pioneering work of Hohenberg and Kohn[1], has been extremely effective in describing the ground state of finite many-electron systems. Such a success gave birth to many papers concerned with the generalization of DFT, which would permit the description of the excitation spectra also. The generalization, on theoretical grounds, originated mainly from the Runge-Gross theorem, which helped to transform DFT into the time-dependent density functional theory TDDFT.[3] Both, DFT and TDDFT, are based on the one-to-one correspondence between particle densities of the considered systems and external potentials acting upon these particles. Unfortunately, the one-to-one correspondence establishes only the existence of the functionals in principle, leaving aside a very important question on how one can construct them in reality. This is why the successes of DFT and TDDFT strongly depend upon the availability of good approximations for the functionals. This shortcoming was resolved to a large extent in[2,4,5] where exact equations connecting the action functional, effective interaction and linear response function were derived. But the linear response function, containing information of the particle-hole and collective excitations, does not directly present information about the single particle spectrum.

In this Report, the self consistent version of the density functional theory is outlined, which allows to calculate the ground state and dynamic properties of finite multi-electron systems starting with the Coulomb interaction. An exact functional equation for the effective interaction, from which one can construct the action functional, density functional, the response functions, and excitation spectra of the considered systems, is presented. The effective interaction relating the linear response function of non-interacting particles to the exact linear response function is of finite radius and density dependent. We derive equations describing single particle excitations of multi-electron systems, using as a basis the exact functional equations, and show that single particle spectra do not coincide either with the eigenvalues of the Kohn-Sham equations or with those of the Hartree-Fock equations.

2. Exact Equation for the Functional

Let us briefly outline the equations for the exchange-correlation functional $E_{xc}[\rho]$ of the ground state energy and exchange-correlation functional $A_{xc}[\rho]$ of the action $A[\rho]$ in the case when the system in question is not perturbed by an external field. In that case an equality holds [2,4]

$$E_{xc}[\rho] = A_{xc}[\rho]|_{\rho(r,\omega=0)}, \tag{1}$$

since A_{xc} is also defined in the static densities domain. The exchange-correlation functional $E_{xc}[\rho]$ is defined by the total energy functional $E[\rho]$ as

$$E[\rho] = T_k[\rho] + \frac{1}{2}\int \frac{\rho(\mathbf{r}_1)\rho(\mathbf{r}_2)}{|\mathbf{r}_1 - \mathbf{r}_2|} d\mathbf{r}_1 d\mathbf{r}_2 + E_{xc}[\rho], \tag{2}$$

where $T_k[\rho]$ is the functional of the kinetic energy of the non-interacting Kohn-Sham particles. The atomic system of units $e = m = \hbar = 1$ is used in this paper. The exchange-correlation functional may be obtained from [2]

$$E_{xc}[\rho] = -\frac{1}{4\pi}\int \left[\chi(\mathbf{r}_1, \mathbf{r}_2, iw, g') + 2\pi\rho(\mathbf{r}_1)\delta(w)\delta(\mathbf{r}_1 - \mathbf{r}_2)\right] \frac{dw dg d\mathbf{r}_1 d\mathbf{r}_2}{|\mathbf{r}_1 - \mathbf{r}_2|}. \tag{3}$$

Equation (3) represents the expression for the exchange-correlation energy of a system [2], expressed via the linear response function $\chi(\mathbf{r}_1, \mathbf{r}_2, iw, g')$, with g' being the coupling constant. For Eq. (3) to describe $A_{xc}[\rho]$ and $E_{xc}[\rho]$ the only thing we need is the ability to calculate the functional derivatives of $E_{xc}[\rho]$ with respect to the density. According to Eq. (3), it means an ability to calculate the functional derivatives of the linear response function χ with respect to the density $\rho(\mathbf{r}, \omega)$ which was developed in [2,5,6]. The linear response function is given by the integral equation

$$\chi(\mathbf{r}_1, \mathbf{r}_2, \omega) = \chi_0(\mathbf{r}_1, \mathbf{r}_2, \omega) + \int \chi_0(\mathbf{r}_1, \mathbf{r}'_1, \omega) R(\mathbf{r}'_1, \mathbf{r}'_2, \omega) \chi(\mathbf{r}'_2, \mathbf{r}_2, \omega) d\mathbf{r}'_1 d\mathbf{r}'_2, \tag{4}$$

with χ_0 being the linear response function of non-interacting particles, moving in the single particle time-independent field [2,5]. It is evident that the linear response function $\chi(g)$ tends to the linear response function of the system in question as g goes to 1. The exact functional equation for $R(\mathbf{r}_1, \mathbf{r}_2, \omega, g)$ is [2,5]

$$R(\mathbf{r}_1, \mathbf{r}_2, \omega, g) = \frac{g}{|\mathbf{r}_1 - \mathbf{r}_2|} \tag{5}$$

$$-\frac{1}{2} \frac{\delta^2}{\delta\rho(\mathbf{r}_1, \omega)\delta\rho(\mathbf{r}_2, -\omega)} \int \int_0^g \chi(\mathbf{r}'_1, \mathbf{r}'_2, iw, g') \frac{1}{|\mathbf{r}'_1 - \mathbf{r}'_2|} d\mathbf{r}_1{}' d\mathbf{r}_2{}' \frac{dw}{2\pi} dg'.$$

Here $R(\mathbf{r}_1, \mathbf{r}_2, \omega, g)$ is the effective interaction depending on the coupling constant g of the Coulomb interaction. The coupling constant g in Eq. (5) is in the range $(0-1)$. The single particle potential v_{xc}, being time-independent, is determined by the relation [2,5],

$$v_{xc}(\mathbf{r}) = \frac{\delta}{\delta\rho(\mathbf{r})} E_{xc}[\rho]. \tag{6}$$

Here the functional derivative is calculated at $\rho = \rho_0$ with ρ_0 being the equilibrium density. By substituting (3) into (6), it can be shown that the single particle potential v_{xc} has the proper asymptotic behavior [5,6],

$$v_{xc}(r \to \infty) \to v_x(r \to \infty) \to -\frac{1}{r}. \tag{7}$$

The potential v_{xc} determines the energies ε_i and the wave functions ϕ_i

$$\left(-\frac{\nabla^2}{2} + V_H(\mathbf{r}) + V_{ext}(\mathbf{r}) + v_{xc}(\mathbf{r})\right) \phi_i(\mathbf{r}) = \varepsilon_i \phi_i(\mathbf{r}). \tag{8}$$

These constitute the linear response function $\chi_0(\mathbf{r}_1, \mathbf{r}_2, \omega)$ entering Eq. (4)

$$\chi_0 = \sum_{i,k} n_i(1-n_k)\phi_i^*(\mathbf{r}_1)\phi_i(\mathbf{r}_2)\phi_k^*(\mathbf{r}_2)\phi_k(\mathbf{r}_1)\left[\frac{1}{\omega - \omega_{ik} + i\eta} - \frac{1}{\omega + \omega_{ik} - i\eta}\right] \tag{9}$$

and the real density of the system ρ,

$$\rho(\mathbf{r}) = \sum_i n_i |\phi_i(\mathbf{r})|^2. \tag{10}$$

Here n_i are the occupation numbers, V_{ext} contains all external single particle potentials of the system, viz. the Coulomb potentials of the nuclei. E_H is the Hartree energy

$$E_H = \frac{1}{2}\int \frac{\rho(\mathbf{r}_1)\rho(\mathbf{r}_2)}{|\mathbf{r}_1 - \mathbf{r}_2|} d\mathbf{r}_1 d\mathbf{r}_2, \tag{11}$$

with the Hartree potential $V_H(\mathbf{r}) = \delta E_H/\delta\rho(\mathbf{r})$, and ω_{ik} is the one-particle excitation energy $\omega_{ik} = \varepsilon_k - \varepsilon_i$, and η is an infinitesimally small positive number.

3. The Effective Interaction

The above equations (2-5) solve the problem of calculating E_{xc}, the ground state energy and the particle-hole and collective excitation spectra of a system without resorting to approximations for E_{xc}, based on additional and foreign inputs to the considered problem, such as found in calculations such as Monte Carlo simulations. We note, that using these approximations, one faces difficulties in constructing the effective interaction of finite radius and the linear response functions [1]. On the basis of the suggested approach, one can solve these problems. For instance, in the case of a homogeneous electron liquid it is possible to determine analytically an efficient approximate expression R_{RPAE} for the effective interaction R, which essentially improves the well-known Random Phase Approximation by taking into account the exchange interaction of the electrons properly, thus forming the Random Phase Approximation with Exchange [4,5]. The corresponding expression for R_{RPAE} is

$$R_{RPAE}(q, g, \rho) = \frac{4\pi g}{q^2} + \frac{\delta E_x}{\delta\rho} = \frac{4\pi g}{q^2} + R_E(q, g, \rho), \tag{12}$$

where

$$R_E(q, g, \rho) = -\frac{g\pi}{p_F^2}\left[\frac{q^2}{12p_F^2}\ln\left|1 - \frac{4p_F^2}{q^2}\right| - \frac{2p_F}{3q}\ln\left|\frac{2p_F - q}{2p_F + q}\right| + \frac{1}{3}\right]. \tag{13}$$

Here E_x is the exchange energy given by Eq. (3) when χ is replaced by χ_0. The electron density ρ is connected to the Fermi momentum by the ordinary relation $\rho = p_F^3/3\pi^2$. Having in hand the effective interaction in $R_{RPAE}(q, g, \rho)$, one can calculate the correlation energy ε^c per electron of an electron gas with the density r_s. The dimensionless parameter $r_s = r_0/a_B$ is usually introduced to characterize the density, with r_0 being the average distance between electrons, and a_B is the Bohr radius.

r_s	ε^c_M	ε^c_{RPA}	ε^c_{RPAE}
1	-1.62	-2.14	-1.62
3	-1.01	-1.44	-1.02
5	-0.77	-1.16	-0.80
10	-0.51	-0.84	-0.56
20	-0.31	-0.58	-0.38
50	-0.16	-0.35	-0.22

In the above table, Monte Carlo results [7] ε^c_M are compared with the results of the RPA calculation ε^c_{RPA}, and ε^c_{RPAE} when the effective interaction R was approximated by R_{RPAE} [2,4]. The energies per electron are given in eV. Note that the effective interaction $R_{RPAE}(q, \rho)$ permits the description of the electron gas correlation energy ε^c in an extremely broad range of the variation of the density. At $r_s = 10$ the error is no more than 10% of the Monte Carlo calculations, while the result becomes almost exact at $r_s = 1$ and is exact when $r_s \to 0$ [2,4].

4. Single-Particle Spectrum

Now let us calculate the single particle energies ϵ_i, that, generally speaking, do not coincide with the eigenvalues ε_i of Eq. (8). Note that these eigenvalues ε_i do not have a physical meaning and cannot be regarded as the single-particle energies (see e.g. [1]). To calculate the single particle energies one can use the Landau equation [8]

$$\frac{\delta E}{\delta n_i} = \epsilon_i. \tag{14}$$

In order to illustrate how to calculate the single-particle energies ϵ_i within the DFT, we choose the simplest case when the functional E_{xc} is approximated by E_x. As we shall see, the single-particle energies ϵ_i coincide neither with the eigenvalues calculated within the Hartree-Fock (HF) method nor with ε_i of Eq. (8). To proceed, we use a method developed in [5]. The linear response function χ_0 and density $\rho(\mathbf{r})$, given by Eqs. (9) and (10) respectively, depend upon the occupation numbers. Thus, one can consider the ground state energy E as a functional of the density and the occupation numbers

$$E[\rho(\mathbf{r}), n_i] = T_k[\rho(\mathbf{r}), n_i] + E_H[\rho(\mathbf{r}), n_i] + E_x[\rho(\mathbf{r}), n_i] + \int V_{ext}(\mathbf{r})\rho(\mathbf{r})d\mathbf{r}. \tag{15}$$

Here T_k is the functional of the kinetic energy of noninteracting particles. As it follows from Eq. (3), the functional E_x is given by [6]

$$E_x[\rho] = -\frac{1}{4\pi}\int [\chi_0(\mathbf{r}_1, \mathbf{r}_2, iw) + 2\pi\rho(\mathbf{r}_1)\delta(w)\delta(\mathbf{r}_1 - \mathbf{r}_2)]\frac{dw d\mathbf{r}_1 d\mathbf{r}_2}{|\mathbf{r}_1 - \mathbf{r}_2|}. \tag{16}$$

Upon using Eq. (16), the exact exchange potential $v_x(\mathbf{r}) = \delta E_x/\delta\rho(\mathbf{r})$ of DFT can be calculated explicitly [6]. Substituting Eq. (15) into Eq. (14) and remembering that the single-particle wave functions ϕ_i and eigenvalues ε_i are given by Eq. (8) with $v_{xc}(\mathbf{r}) = v_x(\mathbf{r})$, we see that the single particle spectrum ϵ_i can be represented by the expression

$$\epsilon_i = \varepsilon_i - <\phi_i|v_x|\phi_i> + \frac{\delta E_x}{\delta n_i}. \tag{17}$$

The first and second terms on the right hand side in Eq. (17) are determined by the derivative of the functional T_k with respect to the occupation numbers n_i. To calculate the derivative we consider an auxiliary system of non-interacting particles in a field $U(\mathbf{r})$. The ground state energy E_0^U of this system is given by the following equation

$$E_0^U = T_k + \int U(\mathbf{r})\rho(\mathbf{r})d\mathbf{r}. \tag{18}$$

Varying E_0^U with respect to the occupation numbers, one gets the desired result

$$\varepsilon_i = \frac{\delta E_0^U}{\delta n_i} = \frac{\delta T_k}{\delta n_i} + <\phi_i|U|\phi_i>, \tag{19}$$

provided $U = V_H + v_x + V_{ext}$. The third term on the right hand side of Eq. (17) is related to the contribution coming from E_x defined by Eq. (16). In the considered simplest case when we approximate the functional E_{xc} by E_x, the coupling constant g enters E_x as a linear factor. If we omit the inter-electron interaction, $g \to 0$, that is, we put $E_x \to 0$, we directly get from Eq. (17) $\epsilon_i = \varepsilon_i$ as it must be in the case of a noninteracting system of electrons. Note that it is not difficult to include the correlation energy in the simplest local density approximation

$$E_c[\rho, n_i] = \int \rho(\mathbf{r})\varepsilon_c(\rho(\mathbf{r}))d\mathbf{r}. \tag{20}$$

Here the density $\rho(\mathbf{r})$ is given by Eq. (10) and the correlation potential is defined as

$$V_c(\mathbf{r}) = \frac{\delta E_c[\rho]}{\delta \rho(\mathbf{r})}. \tag{21}$$

Varying $E[\rho(\mathbf{r}), n_i]$ with respect to the occupation numbers n_i and after some straightforward calculations, we obtain the rather simple expression for the single particle spectrum

$$\epsilon_i = \varepsilon_i - < \phi_i|v_x|\phi_i > - \sum_k n_k \int \left[\frac{\phi_i^*(\mathbf{r}_1)\phi_i(\mathbf{r}_2)\phi_k^*(\mathbf{r}_2)\phi_k(\mathbf{r}_1)}{|\mathbf{r}_1 - \mathbf{r}_2|}\right] d\mathbf{r}_1 d\mathbf{r}_2. \tag{22}$$

Here ε_i are the eigenvalues of Eq. (8) with $v_{xc} = v_x + V_c$. We employ Eq. (19) and choose the potential U as $U = V_H + v_x + V_c + V_{ext}$ to calculate the derivative $\delta T_k/\delta n_i$. Approximating the correlation functional $E_c[\rho, n_i]$ by Eq. (20), we simplify the calculations a lot, preserving at the same time the asymptotic condition, $(v_x + V_c)_{r\to\infty} \to -1/r$. This condition is of crucial importance when calculating the wave functions and eigenvalues of vacant states within the framework of the DFT approach [5]. Note, that these functions and eigenvalues that enter Eq. (22) determine the single particle spectrum ϵ_i. This spectrum has to be compared with the experimental results. The single particle levels ϵ_i, given by Eq. (22), resemble the eigenvalues ε_i^{HF} that are obtained within the HF approximation. If the wave functions ϕ_i would be solutions of the HF equations and the correlation potential $V_c(\mathbf{r})$ would be omitted, the energies ϵ_i would exactly coincide with the HF eigenvalues ε_i^{HF}. But this is not the case, since ϕ_i are solutions of Eq. (8), and the energies ϵ_i do not coincide with either ε_i^{HF} or with the eigenvalues ε_i of Eq. (8). We also anticipate that Eq. (22), when applied to calculations of many-electron systems such as atoms, clusters and molecules, will produce reasonable results for the energy gap separating the occupied and empty states. In the case of solids, we expect that the energy gap at various high-symmetry points in the Brillouin zone of semiconductors and dielectrics can also be reproduced.

5. Conclusions

We have presented the self consistent version of the density functional theory, which allows calculation of the ground state and dynamic properties of finite multi-electron systems. An exact functional equation for the effective interaction, from which one can construct the action functional, density functional, the response functions and excitation spectra of the considered systems, has been outlined. We have shown that it is possible to calculate the single particle excitations within the framework of DFT. The developed equations permit the calculations of the single particle excitation spectra of any multielectron system such as atoms, molecules and clusters. We also anticipate also that these equations when applied to solids will produce quite reasonable results for the single particle spectra and energy gap at various high-symmetry points in the Brillouin zone of semiconductors and dielectrics. We have related the eigenvalues of the single particle Kohn-Sham equations to the real single particle spectrum. In the most straightforward case, when the exchange functional is treated rigorously while the correlation functional is taken in the local density approximation, the coupling equations are very simple. The single particle spectra do not coincide either with the eigenvalues of the Kohn-Sham equations or with those of the Hartree-Fock equations, even when the contribution coming from the correlation functional is omitted.

Acknowledgments

The visit of VRS to Clark Atlanta University has been supported by NSF through a grant to CTSPS. MYaA is grateful to the S.A. Shonbrunn Research Fund for support of his research. AZM is supported by US DOE, Division of Chemical Sciences, Office of Basic Energy Sciences, Office of Energy Research.

References

1. P. Hohenberg and W. Kohn, Phys. Rev. B 136, 864 (1965); W. Kohn and L.J. Sham, Phys. Rev. A 140, 1133 (1965); W. Kohn, P. Washishta, in: *Theory of the Inhomogeneous Electron Gas*, eds. by S. Lundqvist, N.H. March (Plenum, New York and London, 1983) p. 79; T. Garbo, T. Kreibich, S. Kurht, E.K.U. Gross, in: *Strong Coulomb Correlations in Electronic Structure: Beyond the Local Density Approximation*, ed. by V.I. Anisimov (Gordon and Breach, Tokyo, 1998).
2. V.A. Khodel, V.R. Shaginyan, and V.V. Khodel, Phys. Rep. 249, 1 (1994).
3. E. Runge and E.K.U. Gross, Phys. Rev. Lett. 52, 997 (1984).

4. V.R. Shaginyan, Solid State Comm. 55, 9 (1985); M.Ya. Amusia and V.R. Shaginyan, J. Phys. B 25, L345 (1992); M.Ya. Amusia and V.R. Shaginyan, J. Phys. II France 3, 449 (1993).
5. M.Ya. Amusia and V.R. Shaginyan, Phys. Lett. A 269, 337 (2000); M.Ya. Amusia and V.R. Shaginyan, Physica Scripta 68, C10 (2003); M.Ya. Amusia, V.R. Shaginyan, and A.Z. Msezane, Physica Scripta TXX, 1 (2003).
6. V.R. Shaginyan, Phys. Rev. A 47, 1507 (1993).
7. D. Ceperly and B. Alder, Phys. Rev. Lett. 45, 566 (1980).
8. L.D. Landau, Sov. Phys. JETP 3, 920 (1957).

AB INITIO CALCULATIONS AND MODELLING OF ATOMIC CLUSTER STRUCTURE

Ilia A. Solov'yov

A. F. Ioffe Physical-Technical Institute, 194021 St. Petersburg, Russia
and
Institut für Theoretische Physik der Johann-Wolfgang Goethe Universität, Robert-Mayer Str. 8-10, D-60054 Frankfurt am Main, Germany
E-mail: ilia@th.physik.uni-frankfurt.de

Andrey Lyalin

Institute of Physics, St Petersburg State University, 198504 St Petersburg, Petrodvorez, Russia
and
Institut für Theoretische Physik der Johann-Wolfgang Goethe Universität, Robert-Mayer Str. 8-10, D-60054 Frankfurt am Main, Germany

Andrey V. Solov'yov

A. F. Ioffe Physical-Technical Institute, 194021 St. Petersburg, Russia
and
Institut für Theoretische Physik der Johann-Wolfgang Goethe Universität, Robert-Mayer Str. 8-10, D-60054 Frankfurt am Main, Germany
E-mail: solovyov@th.physik.uni-frankfurt.de

Walter Greiner

Institut für Theoretische Physik der Johann-Wolfgang Goethe Universität, Robert-Mayer Str. 8-10, D-60054 Frankfurt am Main, Germany

The optimized structure and electronic properties of small sodium and magnesium clusters have been investigated using *ab initio* theoretical methods based on density-functional theory and post-Hartree-Fock many-body perturbation theory accounting for all electrons in the system. A new theoretical framework for modelling the fusion process of noble gas clusters is presented. We report the striking correspondence of the peaks in the experimentally measured abundance mass spectra with the peaks in the size-dependence of the second derivative of the binding energy per atom calculated for the chain of the noble gas clusters up to 150 atoms.

1. Introduction

There are many different types of clusters, such as metallic clusters, fullerenes, molecular clusters, semiconductor clusters, organic clusters, quantum dots, positively and negatively charged clusters. All have their own features and properties. Comprehensive survey of the field can be found in review papers and books.[1–7] Usually, one can distinguish between different types of clusters by the nature of forces bonding the atoms, or by the principles of spatial organization within the clusters. In our paper we want to demonstrate this feature on a few examples. Namely, we will discuss sodium, magnesium and noble gas clusters and will show the principal differences in their structure and properties.

In this work we consider the optimized ionic structure and the electronic properties of small sodium[8] and magnesium[9] clusters within the size range $N \leq 21$ calculated using *ab initio* theoretical framework based on the density functional theory and the perturbation theory on many-electron correlation interaction. On the basis of comparison of *ab initio* theoretical results with those derived from experiment and within the jellium model[10–13] we elucidate the applicability of the jellium model for the description of alkali and alkali earth cluster properties.

We also present a new theoretical framework[14–16] for modelling the fusion process of noble gas clusters. Starting from the initial tetrahedral cluster configuration, adding new atoms to the system and absorbing its energy at each step, we find cluster growing paths up to the cluster size of 150 atoms. We report the striking correspondence of the peaks in the experimentally measured abundance mass spectra with the peaks in the size-dependence of the second derivative of the binding energy per atom calculated for the chain of the noble gas clusters.

2. Sodium and Magnesium Clusters

During the last decade, there were performed numerous experimental and theoretical investigations of the properties of small alkali metal clusters as well as the processes with their involvement. Particular attention was paid to the investigation of sodium clusters, because namely the sodium clusters were used in such important experimental work as the discovery of metal cluster electron shell structure and the observation of plasmon resonances.[17–19]

In the present work we concentrate on the exploration of the properties of sodium and magnesium clusters with the number of atoms $N \leq 21$ using the density-functional theory based on the hybrid Becke-type three-parameter exchange functional paired with the gradient-corrected Lee, Yang and Parr correlation functional ($B3LYP$),[20] as well as the gradient-corrected Perdew-Wang 91 correlation functional ($B3PW91$).[21,22] Alternatively, we use a direct *ab initio* method for the description of electronic properties of sodium clusters, which is based on the consistent post-Hartree-Fock many-body perturbation theory of the fourth order ($MP4$). Our calculations have been performed with the use of the Gaussian 98 software package.[23] We have utilized the $6-31G(d)$ and $6-311G(d)$ basis sets of primitive Gaussian functions to expand the cluster orbitals.[23,24]

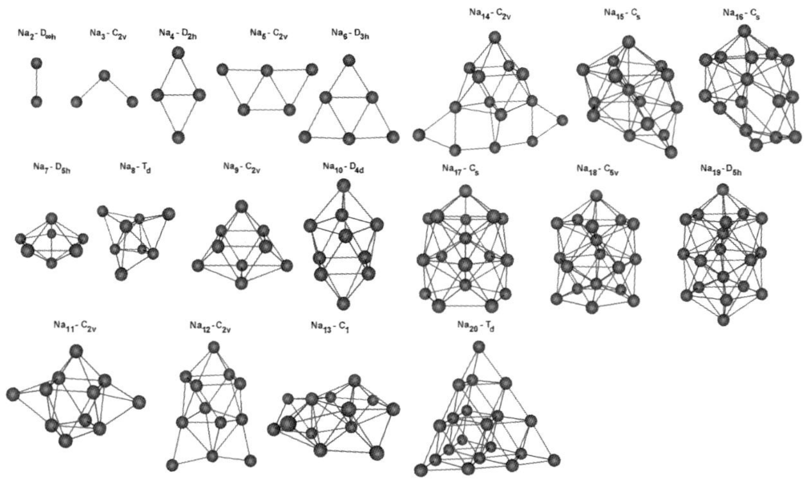

Fig. 1. Optimized geometries of neutral sodium clusters $Na_2 - Na_{20}$. The label above each cluster image indicates its point symmetry group.

Results of the cluster geometry optimization for neutral sodium clusters consisting of up to 20 atoms are shown in Fig. 1. The cluster geometries have been determined using the methodology described in Ref. 8.

Figure 1 shows that the clusters Na_8 and Na_{20} have the higher point symmetry group T_d as compared to the other clusters. This result is in a qualitative agreement with the jellium model. According to the jellium model (see Refs. 8, 10–13 and references therein), clusters with closed shells of delocalized electrons have the spherical shape, while clusters with opened electron shells are deformed. The jellium model predicts spherical shapes for the clusters with the magic numbers $N = 8, 20, 34, 40...$, having respectively the following electronic shells filled: $1s^2 1p^6, 1d^{10}2s^2, 1f^{14}, 2p^6, ...$

Let us now consider how the ionization potentials of sodium clusters evolve with increasing cluster size. Experimentally, such dependence has been measured for sodium clusters in Refs. 1, 25. The ionization potential of a cluster is defined as a difference between the energy of the singly-charged and neutral clusters. Figure 2 shows the dependence of the clusters ionization potential on N. It demonstrates that the results of the $B3LYP$ calculation[8] are in a reasonable agreement with the experimental data.[1]

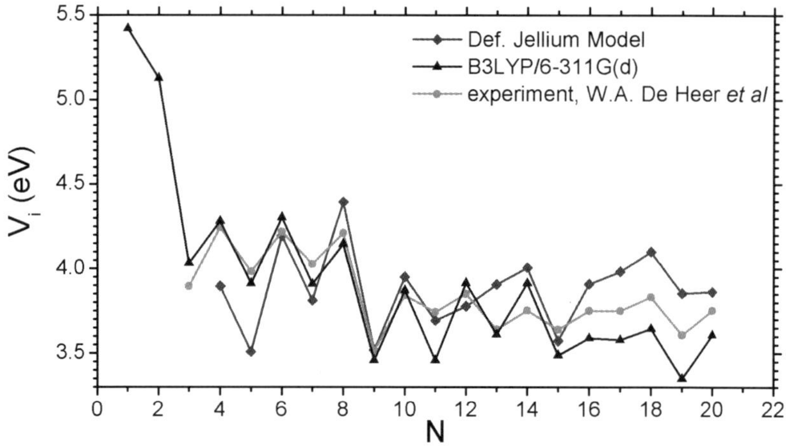

Fig. 2. Ionization potentials of neutral sodium clusters calculated in the deformed jellium model[12,13] (rhomboids) and compared with *ab initio* results[8] (triangles) and with experiment[1] (circles).

The dependencies derived by the $B3LYP$ method as well as the experimental one have a prominent odd-even oscillatory tendency. The maxima in these dependences correspond to the even-N-clusters, which means their higher stability as compared to the neighbouring odd-N-clusters. This happens because the multiplicities of the even and odd-N-clusters are different, being equal to one and two correspondingly. A significant step-like decrease in the ionization potential value happens at the transition from the dimer to the trimer cluster and also in the transition from Na_8 to Na_9. Such an irregular behaviour can be explained by the closure of the electronic 1s- and 1p-shells of the delocalized electrons in the clusters Na_2 and Na_8 respectively.

In Fig. 2, we also present the ionization potential of neutral sodium clusters calculated within the jellium model[12,13] as a function of cluster size. The comparison of the jellium model result with the *ab initio* calculation[8] demonstrates that the jellium model reproduces correctly most of the essential features of the ionization potential dependence on N. Some discrepancy, like in the region $11 \leq N \leq 14$, can be attributed to the neglection of the tri-axial deformation in the axially symmetric jellium model. In spite of the fact that *ab initio* results are closer to the experimental points, one can state quite satisfactory agreement of the jellium model results with the experimental data, which illustrates correctness of the jellium model assumptions and its applicability to the description of sodium clusters.

Let us now discuss clusters neighbouring the sodium element - the magnesium clusters. Magnesium is a divalent element. Clusters of divalent metals are expected to differ from the jellium model predictions at least at small cluster sizes. In this case, bonding between atoms is expected to have some features of the van der Waals type of bonding, because the electronic shells in the divalent atoms are filled. Thus, clusters of divalent metals are very appropriate for studying non-metal to metal transition, testing different theoretical methodologies and conceptual developments of atomic cluster physics.

The structural and electronic properties of neutral and anionic magnesium clusters Mg_N with N up to 22 have been studied[26,27] using gradient-corrected DFT and pseudopotential.

Recently calculations of various properties of neutral and cationic magnesium clusters with number of atoms N up to 21 have been performed with accounting for all-electrons in the system.[9] The optimization of the cluster geometries has been performed with the use of the $B3PW91$ and $B3LYP$ methods mentioned above and described in detail in Refs. 8, 9. The results of cluster geometry optimization for neutral magnesium clusters consisting of up to 21 atoms are shown in Fig. 3. In Fig. 3, we present only the lowest energy configurations optimized by the $B3PW91$ method.

It is worth to note that the optimized geometry structures for small neutral magnesium clusters differ significantly from those obtained for sodium clusters discussed above. Thus, the optimized sodium clusters with $N \leq 6$ have the plane structure. For Na_6, both plane and spatial isomers with very

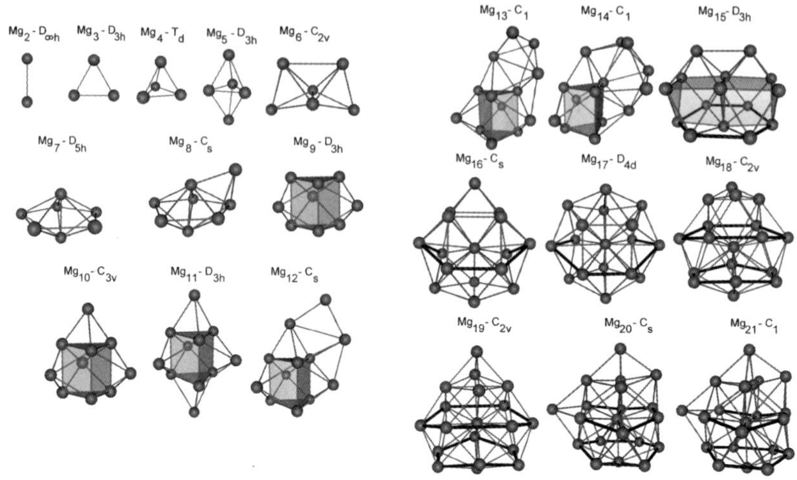

Fig. 3. Optimized geometries of the neutral magnesium clusters $Mg_2 - Mg_{21}$ calculated in the $B3PW91$ approximation. The label above each cluster image indicates the point symmetry group of the cluster.

close total energies exist. The planar behaviour of small sodium clusters has been explained as a result of the successive filling of the 1σ and 1π symmetry orbitals by delocalized valence electrons,[28] which is fully consistent with the deformed jellium model calculations.[10–13] Contrary to the small sodium clusters, the magnesium clusters are three-dimensional already at $N = 4$, forming the structures nearly the same as the van der Waals bonded clusters (see our discussion below).

Starting from Mg_9 a new element appears in the magnesium cluster structures. This is the six atom trigonal prism core, which is marked out in Fig. 3. The formation of the trigonal prism plays the important role in the magnesium cluster growth process. Adding an atom to one of the triangular faces of the trigonal prism of the Mg_9 cluster results in the Mg_{10} structure, while adding an atom to the remaining triangular face of the prism within the Mg_{10} cluster leads to the structure of Mg_{11}, as shown in Fig. 3. Further growth of the magnesium clusters for $12 \leq N \leq 14$ leads to the formation of the low symmetry ground state cluster. In spite of their low symmetry, all these clusters have the trigonal prism core. The structural rearrangement occurs for the Mg_{15} cluster, which results in the high symmetry structure of the two connected Mg_9 clusters.

Starting from Mg_{15} another motif based on the hexagonal ring structure which is marked out in Fig. 3 dominates the cluster growth. Overall, obtained structures agree with those from Refs. 26 and 27, where the Wadt-Hay pseudopotential has been used for the treatment of the magnesium ionic core. However, the most stable structures for Mg_{13}, Mg_{15}, Mg_{16}, Mg_{19} and Mg_{20} clusters obtained in Ref. 29 emerge as higher-energy isomers in our calculations. It is worth noting that the formation of hexagonal ring for $N = 15$ plays the important role in the evolution of the magnesium cluster structure towards to the bulk lattice, because the hexagonal ring is one of the basic elements of the hexagonal closest-packing (*hcp*) lattice which is the lattice of bulk magnesium. A single deformed hexagonal ring is the common element in the structures of the Mg_{16} and Mg_{17} clusters. For the Mg_{18-21} clusters, two deformed hexagonal rings appear.

3. Noble Gas Clusters

Both sodium and magnesium clusters are metal clusters. For their description it is important to take into account the quantum effects. The situation is different for noble gas clusters, for example Ar, Kr, Xe.

Noble gas clusters are formed by the long range van der Waals forces. This fact allows one to describe geometries of such systems using classical molecular dynamics approach. Relatively simple interaction between atoms in the system allows one to investigate clusters within the much larger size range, up to several hundreds atoms in a cluster, and to tackle the more sophisticated problems. Within the classical approximation, the motion of the atoms in a cluster is described by the Newton motion equations with a pairing potential. In our work we use the Lennard-Jones (LJ) potential.

With the growth of the atom number of atoms in the system the problem of searching for the global energy minimum on the cluster multidimensional potential energy surface becomes more complicated. The number of local minima on the potential energy surface increases exponentially with the growth cluster size and is estimated[5] to be of the order of 10^{43} for $N = 100$. There are different algorithms and methods of the global minimization, which have been employed for the global minimization of atomic cluster systems.[5,15,30–32]

In the present work we use the new algorithm[14–16] based on the dynamic searching for the most stable cluster isomers in the cluster fusion process. We assume that atoms in a cluster are bound by Lennard-Jones potentials and the cluster fusion takes place atom by atom. At each step of the fusion process all atoms in the system are allowed to move, while the energy of the system is decreased. The motion of the atoms is stopped when the energy minimum is reached. The geometries and energies of all cluster isomers found in this way are analysed. We have analysed cluster fusion process within the cluster size range of $N = 150$ atoms. For algorithmic details we refer to our recent papers.[14–16]

The growth of the most stable LJ cluster configurations possessing the absolute energy minimum, the so-called global minimum cluster structures, is illustrated in Fig. 4. In this figure we present the cluster geometries within the size range $4 \leq N \leq 66$ and determine the transition path from smaller clusters to larger ones. We do not show cluster structures with $N \geq 67$ in this paper and refer to our recent work.[14–16]

Figure 4 demonstrates that most of global energy minimum cluster geometries can be obtained from the preceding cluster configurations by fusing a single atom to the cluster surface. Such situations take place for about 75% of the clusters considered.

In the simplest scenario clusters of $N+1$ atoms are generated from the N-atomic cluster with the lowest energy by adding one atom to the center of mass of all the faces laying on the cluster surface. Here, the cluster surface is considered as a polyhedron, so that the vertices of the polyhedron are the atoms and two vertices are connected by an edge, if the distance between them is less than the given value.

It is interesting that all the cluster geometries calculated have the structure, in which a number of completed and open polygons round the cluster axis. The maximum possible number of atoms in polygons depends on the cluster size. In the cluster with $N \leq 150$ the pentagonal, decagonal and pentadecagonal polygons are present, which is closely related to the fact that most of the cluster configurations are based on the icosahedral type of symmetry.

Our simulations demonstrate that the fusion of a single atom to the global energy minimum cluster structure of N atoms, in some cases, does not lead to the global energy minimum of $N+1$ atomic cluster.[14–16] This happens for N = 18, 27, 30, 35, 38, 51, 65, 66, 68, 69, 73, 76, 78, 84, 86, 87, 88, 93, 96, 98, 102, 105, 111, 113, 115, 121, 123, 126, 128, 130, 133. For finding the global energy minimum of these clusters, one needs to perform their additional optimization and to rearrange one or a few atoms at the cluster surface. In Fig. 4 we mark the atoms that are rearranged and their initial positions by grey circles and rings respectively. This figure shows that the rearrangement takes place always in the vicinity of the cluster surface. This fact has a simple physical interpretation. The surface atoms

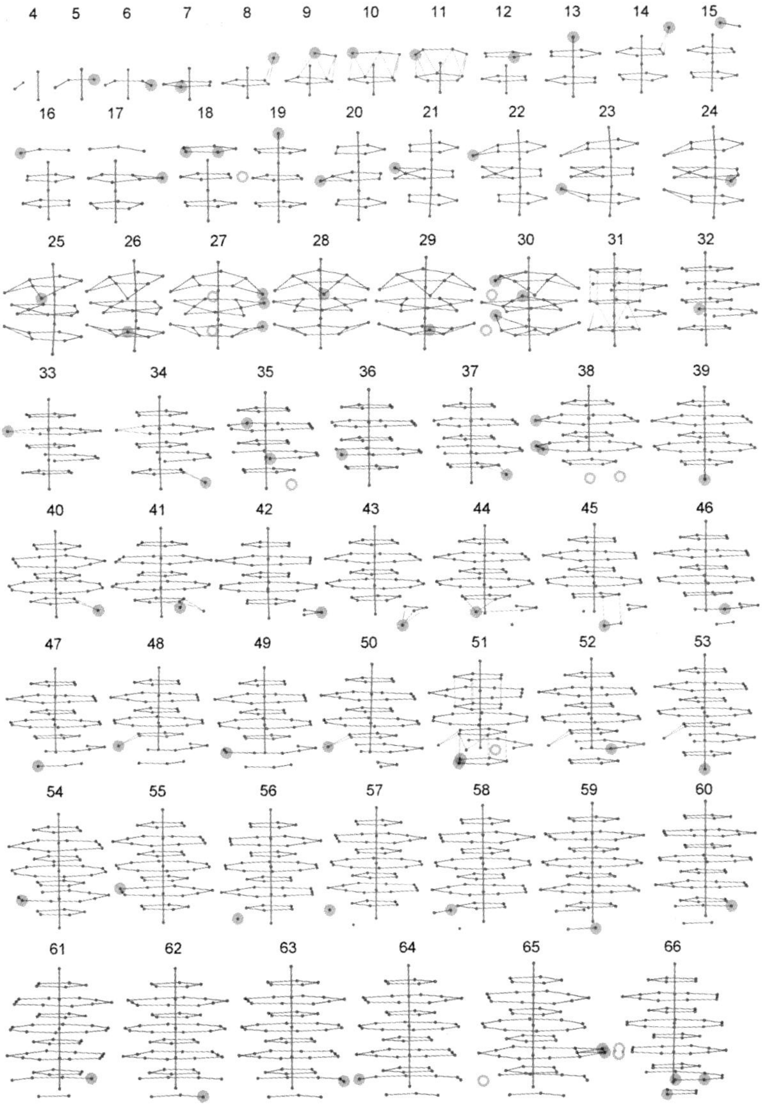

Fig. 4. Growth of noble gas global energy minimum cluster structures with $N \leq 66$. The new atoms added to the clusters are marked with a grey circle, while the grey rings demonstrate the atom removal.

are bound weaker than those inside the cluster volume, and thus they are more movable and allow the surface atomic rearrangement. The surface rearrangement of atoms should be an essential component of the cluster growth process.

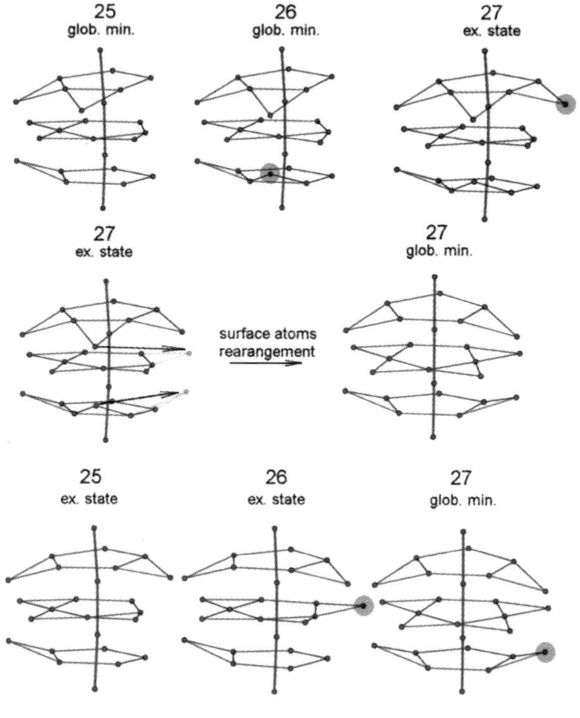

Fig. 5. Fusion of a single atom to the global energy minimum cluster structure of 26 atoms does not lead to the global energy minimum of the LJ_{27} cluster (first row). Rearrangement of surface atoms in the LJ_{27} cluster leading to the formation of the global energy minimum cluster structure is needed (second row). The result of such rearrangement can be obtained if one starts the cluster growth from the excited state of the LJ_{25} cluster (third row).

In the first row of Fig. 5 we demonstrate that the fusion of a single atom to the global energy minimum cluster structure of 26 atoms does not lead to the global energy minimum of the LJ_{27} cluster. Thus, the rearrangement of surface atoms in the LJ_{27} cluster leading to the formation of the global energy minimum cluster structure is needed. The necessary rearrangement of atoms is shown in the second row of Fig. 5. The result of such rearrangement can be obtained if one starts the cluster growth from the excited state of the LJ_{25} cluster. This is illustrated in the third row of Fig. 5.

In this particular example only two smaller cluster sizes are involved in the fusion process of the LJ_{27} cluster. However, the cluster fusion via excited states is not always that simple and evident. In some cases, it involves more then 10 intermediate steps. Such a situation occurs, for example, for the fusion of LJ_{69} cluster, which can be obtained from the excited state of LJ_{55} cluster. In this case the core structure of the cluster remains the same. However, in some cases it changes radically. Below, we call such radical rearrangements of the cluster structure as lattice rearrangements.

The first lattice rearrangement takes place in the transition from LJ_{30} to LJ_{31} cluster. It is clear from Fig. 4 that the structure of these two neighbouring clusters differs significantly and that it is impossible to obtain the structure of the LJ_{31} cluster by simple surface rearrangement of atoms. The LJ_{31} cluster structure emerges in the cluster growing process and involves a long chain of excited states of the clusters with $N \geq 13$.

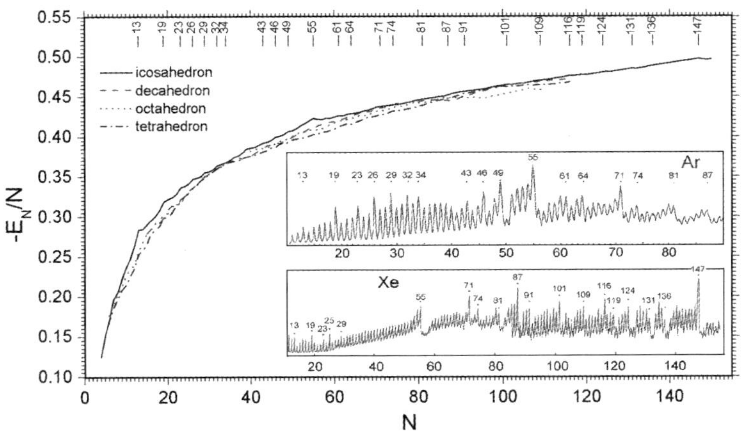

Fig. 6. Binding energy per atom for LJ-clusters as a function of cluster size calculated for the cluster chains based on the icosahedral, octahedral, tetrahedral and decahedral symmetry.[14–16] In the inset we present the experimentally measured abundance mass spectrum for the Ar and Xe clusters.[33, 34]

The binding energies per atom as a function of cluster size for the calculated cluster chains are shown in Fig. 6. In the insertion to Fig. 6 we present the experimentally measured abundance mass spectrum for the Ar and Xe clusters.[33, 34] Figure 6 shows that the most stable clusters are obtained on the basis of the icosahedral symmetry configurations with exceptions for $N = 38, 75 \leq N \leq 77$ and $N = 98$. In these cases the octahedral, decahedral and tetrahedral cluster symmetries become more favourable respectively.

The main trend of the energy curves plotted in Fig. 6 can be understood on the basis of the liquid drop model, according to which the cluster energy is the sum of the volume and the surface energy contributions:

$$E_N = -\lambda_V N + \lambda_S N^{2/3} - \lambda_R N^{1/3} \tag{1}$$

Here the first and the second terms describe the volume, and the surface cluster energy correspondingly. The third term is the cluster energy arising due to the curvature of the cluster surface. Choosing constants in (1) as $\lambda_V = 0.71554$, $\lambda_S = 1.28107$ and $\lambda_R = 0.5823$, one can fit the global energy minimum curve plotted in Fig. 6 with the accuracy less than one percent. The deviations of the energy curves calculated for various chains of cluster isomers from the liquid drop model (1) are plotted in Fig. 7. The curves for the icosahedral and the global energy minimum cluster chains go very close to each other and the peaks on these dependences indicate the increased stability of the corresponding magic clusters.

The dependence of the binding energies per atom for the most stable cluster configurations on N allows one to generate the sequence of the cluster magic numbers. In the inset to Fig. 7 we plot the second derivatives ΔE_N^2 for the chain of icosahedral isomers. We compare the obtained dependence with the experimentally measured abundance mass spectrum for the Ar and Xe clusters[33,34] (see inset to Fig. 6) and establish the striking correspondence of the peaks in the measured mass spectrum with those in the ΔE_n^2 dependence. Indeed, the magic numbers determined from $\Delta^2 E_N$ are in a very good agreement with the numbers experimentally measured for the Ar and Xe clusters: 13, 19, 23, 26, 29, 32, 34, 43, 46, 49, 55, 61, 64, 71, 74, 81, 87, 91, 101, 109, 116, 119, 124, 131, 136, 147.[33,34] The most prominent peaks in this sequence, 13, 55 and 147, correspond to the closed icosahedral shells, while other numbers correspond to the filling of various parts of the icosahedral shell.

4. Conclusion

The optimized geometries and electronic properties of sodium and magnesium clusters consisting of up to 21 atoms have been investigated using the $B3PW91$, $B3LYP$ and $MP4$ methods accounting for the all electrons in the system. We compared the results of our calculations with the results obtained within the jellium model and with the available experimental data.

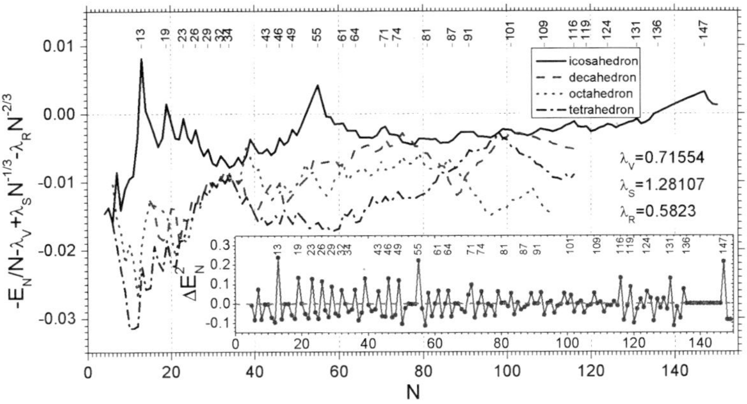

Fig. 7. Energy curves deviations from the liquid drop model (7) calculated for various cluster isomers chains. In the inset we plot the second derivative $\Delta^2 E_N$ calculated for the icosahedral cluster isomers chain.

From these comparisons, we have elucidated the level of applicability of the jellium model to the description of sodium and magnesium clusters.

We have also developed a new algorithm for modelling the cluster growth process based on the dynamic searching for the most stable cluster isomers. This algorithm can be considered an efficient alternative to the known cluster global minimization techniques. We have demonstrated that the majority of energetically favourable cluster structures can be obtained from the preceding cluster configurations by fusion of a single atom to the cluster surface. However, in some cases the surface and lattice rearrangements of the cluster occur. For the energetically favourable cluster configurations we report the striking correspondence of the peaks in the dependence of the second derivative of the binding energy per atom on cluster size with the peaks in the mass abundance spectra measured for the noble gas clusters.

The results of this work can be extended in various directions. One can use the similar methods to study structure and properties of various types of clusters. It is interesting to extend calculations towards larger cluster sizes and to perform more advanced comparison of model and *ab initio* approaches, as well as to study collisions and electron excitations in clusters with the optimized geometries. These and many more other problems of atomic cluster physics can be tackled with the use of the methods considered in our work.

Acknowledgments

The authors acknowledge support of this work by the Studienstiftung des Deutschen Volkes, Alexander von Humbolt Foundation, the INTAS (grant No 03-51-6170), Russian Foundation for Basic Research (grant No. 03-02-16415-a) and Russian Academy of Sciences (grant No. 44).

References

1. W.A. de Heer, *Rev. Mod. Phys.* 65, 611 (1993).
2. H. Haberland (ed.), *Clusters of Atoms and Molecules, Theory, Experiment and Clusters of Atoms*, Springer Series in Chemical Physics, Berlin 52 (1994).
3. U. Näher, S. Bjørnholm, S. Frauendorf, F. Garcias and C. Guet, *Physics Reports* 285, 245 (1997).
4. W. Ekardt (ed.), *Metal Clusters* Wiley, New York (1999)
5. *Atomic Clusters and Nanoparticles*, NATO Advanced Study Institute, les Houches Session LXXIII, les Houches, 2000, edited by C. Guet, P. Hobza, F. Spiegelman and F. David, EDP Sciences and Springer Verlag, Berlin (2001).
6. J. Jellinek (ed.), *Theory of Atomic and Molecular Clusters. With a Glimpse at Experiments*, Springer Series in Cluster Physics, Berlin (1999).
7. K-H. Meiwes-Broer (ed.), *Metal Clusters at Surfaces. Structure, Quantum Properties, Physical Chemistry*, Springer Series in Cluster Physics, Berlin (1999).
8. I.A. Solov'yov, A.V. Solov'yov and W. Greiner, *Phys. Rev. A* 65, 053203 (2002).
9. A. Lyalin, I.A. Solov'yov, A.V. Solov'yov and W. Greiner, *Phys. Rev. A.* 67, 063203 (2003).
10. A.G. Lyalin, S.K. Semenov, A.V. Solov'yov, N.A. Cherepkov and W. Greiner, *J. Phys. B* 33, 3653 (2000).
11. A.G. Lyalin, S.K. Semenov, A.V. Solov'yov, N.A. Cherepkov, J.-P. Connerade, and W. Greiner, *J. Chin. Chem. Soc.* (Taipei) 48, 419 (2001).
12. A. Matveentsev, A. Lyalin, I. Solov'yov, A. Solov'yov and W. Greiner, *Int. J. Mod. Phys. E* 12, 81 (2003).
13. A. Lyalin, A. Matveentsev, I.A. Solov'yov, A.V. Solov'yov and W. Greiner, *Eur. Phys. J. D* 24, 15 (2003).
14. I.A. Solov'yov, A.V. Solov'yov, W. Greiner, A. Koshelev and A. Shutovich, *Phys. Rev. Lett.* 90, 053401 (2003).
15. I.A. Solov'yov, A.V. Solov'yov, W. Greiner, Submitted to the *J. Chem. Phys.*; LANL preprint: physics/0306185, (2003).
16. A. Koshelev, A. Shutovich, I.A. Solov'yov, A.V. Solov'yov, W. Greiner, in *Proceedings of the International Meeting* "From Atomic to the Nano-Scale" pp. 184-194, Norfolk, Virginia, USA, December 12-14 (2002), editors J. Mc Guire and C.T. Whelan, Old Dominion University, USA (2003).
17. W.D. Knight, K. Clemenger, W.A. de Heer, W.A. Saunders, M.Y. Chou and M.L. Cohen, *Phys. Rev. Lett.* 52, 2141 (1984).

18. C. Bréchignac, Ph. Cahuzac, F. Carlier, J. Leygnier, *Chem. Phys. Lett.* 164, 433 (1989).
19. K. Selby, M. Vollmer, J. Masui, V. Kresin, W.A. de Heer and W.D. Knight, *Phys. Rev.* B 40, 5417 (1989).
20. A.D. Becke, *J. Chem. Phys.* 98, 5648 (1993); A.D. Becke, *Phys. Rev. A* 38, 3098 (1988); C. Lee, W. Yang and R.G. Parr, *Phys. Rev. B* 37, 785 (1988).
21. J.P. Perdew, in *Electronic Structure of Solids '91*, edited by P. Ziesche and H. Eschrig p.11, Akademie Verlag, Berlin (1991).
22. K. Burke, J.P. Perdew and Y. Wang, in *Electronic Density Functional Theory: Recent Progress and New Directions*, edited by J.F. Dobson, G. Vignale and M.P. Das, Plenum (1998).
23. M.J. Frisch *et al*, computer code GAUSSIAN 98, Rev. A. 9, Gaussian Inc., Pittsburgh, PA (1998).
24. James B. Foresman and Æleen Frisch *Exploring Chemistry with Electronic Structure Methods*, Pittsburgh, PA: Gaussian Inc. (1996).
25. H. Akeby, I. Panas, L.G.M. Petterson, P. Siegbahn, U. Wahlgreen, *J. Chem. Phys.* 94, 5471 (1990).
26. P.H. Acioli and J. Jellinek, *Phys. Rev. Lett.* 89, 213402 (2002).
27. J. Jellinek and P.H. Acioli, *J. Phys. Chem. A* 106, 10919 (2002).
28. J.L. Martins, J. Buttet and R. Car, *Phys. Rev. B* 31, 1804 (1985).
29. A. Köhn, F. Weigend and R. Ahlrichs, *Phys. Chem. Chem. Phys.* 3, 711 (2001).
30. D.J. Wales, J.P.K. Doye, M.A. Miller, P.N. Mortenson and T.R. Walsh, *Adv. Chem. Phys.* 115, 1 (2000).
31. D.J. Wales and J.P.K. Doye, *J. Phys. Chem. A* 101, 5111 (1997).
32. R.H. Leary and J.P.K. Doye, *Phys. Rev. E* 60, 6320 (1999).
33. I.A. Harris, K.A. Norman, R.V. Mulkern, J.A. Northby, *Phys. Rev. Lett.* 53, 2390 (1984).
34. W. Miehle, O. Kandler, T. Leisner, O. Echt, *J. Chem. Phys* 91, 5940 (1991).

ELECTRIC AND MAGNETIC ORBITAL MODES IN SPHERICAL AND DEFORMED METAL CLUSTERS

V.O. Nesterenko

BLTP, Joint Inst. for Nuclear Research, Dubna, Moscow reg., 141980, Russia
and
Max Planck Inst. for Physics of Complex Systems, 01187, Dresden, Germany
E-mail: nester@thsun1.jinr.ru

W. Kleinig

BLTP, Joint Inst. for Nuclear Research, Dubna, Moscow reg., 141980, Russia
and
Technische Univ. Dresden, Inst. für Analysis, D-01062, Dresden, Germany

P.-G. Reinhard

Inst. für Theoretische Physik, Univ. Erlangen, D-91058, Erlangen, Germany

Specific properties of electric (E1, E2, E3) and orbital magnetic (scissors M1 and twist M2) modes in metal clusters are reviewed. The analysis is performed within the Kohn-Sham LDA RPA method. Possible routes for an experimental observation of the modes are discussed.

1. Introduction

Collective oscillations of valence electrons in metal clusters manifest themselves in a variety of electric and (orbital) magnetic plasmons. These various modes have analogues in other finite Fermi systems (e.g. giant resonances in atomic nuclei), where they are a topic of high current interest. In atomic clusters up to now only the electric dipole (E1) plasmon was thoroughly investigated.[1] Other plasmons are not so easily accessible and our knowledge about them is poor, but they contain a lot of useful information. We will briefly review various collective modes (plasmons) in clusters, their physical content, and possible routes for experimental access. A few comments will be made on the familiar E1 plasmon as well. Namely, we will discuss the in-

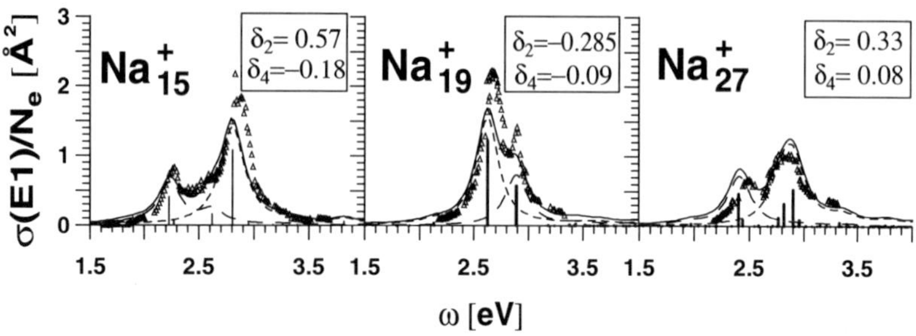

Fig. 1. Photoabsorption cross sections in Na_{15}^+, Na_{19}^+, and Na_{27}^+. The parameters of quadrupole and hexadecapole deformations are given in boxes. The experimental data[7] (triangles) are compared with our results given as vertical bars (for every RPA state) and as a strength function smoothed by the Lorentz weight (with averaging parameter 0.25 eV). Contributions of $\mu = 0$ and 1 dipole branches (the latter is twice stronger) are given by dashed curves. The bars are given in $eV\AA^2$. See Refs. 3 and 6 for more details.

terplay of deformation splitting, Landau fragmentation and shape isomers in forming the gross structure of the plasmon profile.

All modes are described in the linear regime within the random-phase-approximation (RPA) method[2,3] based on the Kohn-Sham LDA functional.[4] The ions are treated in the soft jellium approximation. The reliability of the method has been checked in diverse studies in spherical[5] and deformed[3,6] clusters.

2. Profile of E1 Plasmon

The quadrupole deformation of the cluster splits the dipole plasmon into two (axial shape) or three (triaxial shape) peaks. Thus the plasmon profile can be used to estimate magnitude and sign of the deformation. This is illustrated in Fig. 1 for prolate Na_{15}^+ and Na_{27}^+ and oblate Na_{19}^+. It is seen that the dipole plasmon is split into $\mu = 0$ and $\mu = 1$ peaks and the magnitude of the splitting is proportional to the value of the deformation parameter δ_2. The ordering of $\mu = 0$ and $\mu = 1$ peaks allows us to read off whether the cluster is prolate or oblate. Such an analysis is widely used in experiment and considered a reliable way to measure the deformation of a cluster.

It is clear that the above treatment is valid only if the deformation splitting dominates the structure of the plasmon. This is indeed the case for light-deformed clusters, where most analysis was done up to now. There

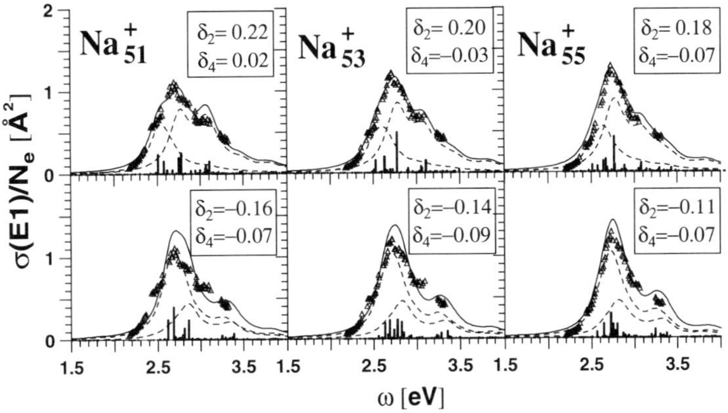

Fig. 2. As Fig. 1 but for Na^+_{51}, Na^+_{53} and Na^+_{55}. The experimental data (triangles) are from Ref. 7. RPA results are shown for the prolate ground (upper panels) and oblate isomeric (lower panels) states.

are, however, competing mechanisms spreading the plasmon spectrum. The one is Landau fragmentation where the collective strength is distributed over energetically close particle-hole states. The other is thermal activation of isomers. Both, Landau fragmentation and isomeric states, become increasingly important with growing cluster size.[3,5,6,8] To avoid misleading conclusions, one should take into account these effects as well.

This point is illustrated in Fig. 2 for deformed clusters Na^+_{51}-Na^+_{55}. They have two energetically close configurations, a prolate ground and oblate isomeric state.[3,6] The isomers have tiny energy deficits, 0.01-0.02 eV, and thus can also contribute to the dipole plasmon. The overall shape of the optical strength looks for the smaller oblate cluster Na^+_{19} like showing a bump with a shoulder at the right tail. However, it would be incorrect to treat, on these grounds, the clusters as oblate. Indeed, prolate ground state as well as oblate isomer yield optical strength of about the same form and both reproduce equally well the experimental profile. It looks like the profile is independent of cluster deformation. Besides, one sees that the right shoulder is produced not by the $\mu = 0$ mode alone, as might be expected for oblate shape, but by the $\mu = 1$ mode as well. Altogether this means that the right shoulder is not a result of the deformation splitting. Instead, our analysis has shown[6]) that it is an effect of Landau fragmentation, typical for clusters in this size region. Landau fragmentation in these clusters is very strong (the plasmons bunch many RPA states) and dominates both the profile and width of the plasmons.

This statement applies for heavier clusters as well. Indeed, the number of isomers with a small energy deficit grows with cluster size.[3] Furthermore systematic calculations show that Landau fragmentation increases with cluster size up to a maximum at $N_e \sim 10^3$ atoms, after which it decreases again.[8] Thus our conclusions for Na_{51}^+-Na_{55}^+ can be extended to a much wider size region with up to $N_e \sim 10^3$. Altogether, the optical response in heavier clusters shows up as a broad bump where both gross-structure and width are mainly determined by Landau fragmentation. Triaxiality and octupole deformation further entangle the picture. Altogether, this means that in most free deformed clusters, except small ones with $N_e < 40$, the measured profile of the dipole plasmon cannot be directly used for estimation of cluster shape. We gave the arguments for *free* clusters. Instead, heavy *supported* clusters have strongly oblate shape and show a clear deformation splitting.[9]

We note in passing that Figs. 1 and 2 display excellent agreement of our RPA results with experimental data hinting at the reliability of the approach.

3. Multipole Plasmons

3.1. *Scissors M1 plasmon*

The scissors mode (SM) in clusters is macroscopically viewed as small-amplitude rotation-like oscillations of the spheroid of valence electrons against the spheroid of the ions (hence the name SM). This is a universal mode appearing in diverse finite quantum systems. It was first found[10] in atomic nuclei and then predicted in a variety of different systems, like metal clusters,[11,12] quantum dots[13] and ultra-cold superfluid gas of fermionic atoms.[14] Besides, it was predicted[15] and observed[16] in a Bose-Einstein condensate. All these different systems have two features in common: the broken spherical symmetry and the two-component nature (neutrons and protons in nuclei, valence electrons and ions in atomic clusters, electrons and surrounding media in quantum dots, atoms and the trap in dilute Fermi gas and Bose condensate).

In axially deformed systems, the SM is generated by the orbital momentum fields L_x and L_y perpendicular to the symmetry axis z and is characterized by the quantum numbers $|\Lambda^\pi = 1^+>$ where Λ is the eigenvalue of L_z and π is the space parity. In atomic clusters, the SM energy and magnetic strength read:[11,12]

$$\omega = \frac{20.7}{r_s^2} N_e^{-1/3} \delta_2 \; eV, \quad B(M1) = 4\langle 1^+ \mid \hat{L}_x \mid 0 \rangle^2 \mu_b^2 = \simeq N_e^{4/3} \delta_2 \; \mu_b^2 \quad (1)$$

where N_e is the number of valence electrons, r_s the Wigner-Seitz radius (in $\mathring{A}$), and μ_b is the Bohr magneton. The value $B(M1)$ stands for the summed strength of the degenerated x and y-branches. The z-branch of the SM vanishes for symmetry reasons. The B(M1) strength does not depend on r_s and so is the same for different metals. It reaches impressive values in heavy clusters, e.g. 350-400 μ_b^2 for $N_e \sim 300$. The SM energy scales from 1-1.5 eV in light clusters to 0.1-0.3 eV in heavy clusters.

Both SM characteristics in (1) are proportional to the deformation parameter δ_2, which means that SM exists only in deformed systems. In atomic clusters, this holds strictly in jellium approximation. The detailed ionic structure destroys locally spherical symmetry thereby causing a finite, though very weak, M1 response even in clusters with zero global deformation.[17]

The orbital magnetic susceptibility is the sum of the Langevin diamagnetic and van Vleck paramagnetic terms:[11,19]

$$\chi_k = \chi_k^{dia} + \chi_k^{para}, \quad (2)$$

where in familiar notations

$$\chi_k^{dia} = -\mu_b^2 N_e \overline{\rho_k^2} = -\mu_b^2 \Theta_k^R, \quad (3)$$

$$\chi_k^{para} = 2\mu_b^2 \sum_{j \neq 0} \frac{\mid < j|\hat{L}_k|0 > \mid^2}{\omega_j} = \mu_b^2 \Theta_k. \quad (4)$$

Here Θ_k and Θ_k^R are respectively the cranking and rigid moments of inertia. Note that for $k = x, y$ the operator entering in the matrix element in (4) is exactly the scissors generator. Obviously, just the low-energy SM determines $\chi_{x,y}^{para}$ and the van Vleck paramagnetism.

If $\Theta_{x,y} = \Theta_{x,y}^R$, then $\chi_{x,y}^{para} = -\chi_{x,y}^{dia}$ and so a complete compensation of dia- and paramagnetic terms in $\chi_{x,y}$ takes place.[11] For symmetry reasons $\chi_z^{para} = 0$. Then the total susceptibility becomes anisotropic[11]

$$\chi_x = \chi_y = 0, \qquad \chi_z = \chi_z^{dia} \quad (5)$$

and scales from zero to diamagnetic values. Quantum shell effects (in particular for light clusters as Na_{27}^+) can destroy the fragile balance (5) and result in dia-para anisotropy when the cluster is paramagnetic in x, y-directions and diamagnetic in z-direction.[18]

The SM optical response is too weak to be measured in photoabsorption. However, the mode can probably be observed in heavy supported and strongly-oblate clusters by methods of angular-resolved electron energy-loss spectroscopy (AR-EELS). The principle point is to choose such scattering angles where E1 and other competing contributions are well suppressed in favour of the SM. The mode can also be observed in some free light clusters (Na_{11}^+ and Na_{15}^+) by means of resonant Raman scattering or resonance fluorescence. The latter is possible due to strong cross-over of the SM and quadrupole E21 low-energy mode (see discussion in Ref. 19).

3.2. *Twist M2 plasmon*

The M2 twist mode is another example of orbital-magnetic collective motion. Unlike the SM, the twist mode exists in systems of any shape, including the spherical one. The mode is viewed as small-amplitude rotation-like oscillations of different layers of a system against each other with a rotational angle proportional to z (the axis of rotation). The twist mode is also a universal collective mode. Being first proposed[20] and observed[21] in nuclei, it was then predicted in atomic clusters[22] and in trapped atomic Fermi gas.[23]

In clusters, the twist mode is characterized by low-lying (0.1-1.5 eV) 2^- states with energies and B(M2) strengths[22]

$$\omega = 17 r_s^{-2} N_e^{-1/3} \, eV \mathring{A}^2, \quad B(M2) = 0.52 r_s^2 N_e^2 \, \mu_b^2. \tag{6}$$

Twist dominates over spin M2 excitations in clusters with $N_e > 40$. In the long-wave approximation, the spin M1 response in spherical alkali metal clusters is zero. Besides, spherical clusters do not have the scissors M1 mode. For all these reasons, the twist mode becomes the strongest magnetic mode in heavy *spherical* alkali metal clusters.[22] A key feature of the twist mode is that it is built from electron-hole excitations with maximal orbital moments in the vicinity of the Fermi level.[22] AR-EELS can probably be used to search twist in clusters.

3.3. *E2 and E3 modes*

There are electric plasmons with higher multipolarity, as e.g. E2 and E3. They lie in the same energy region 2.5-4.5 eV as the dominant E1 plasmon.[2,3,24] Nevertheless, estimations[25] show that E2 and E3 plasmons may be resolved in AR-EELS at electron scattering angles 6-9°.

Interesting new results emerge for low-energy (0.2-1.5 eV) quadrupole modes (LEQM) E20, E21 and E22.[19] The modes can be measured in resonance fluorescence (RF) through the dipole plasmon. The population of

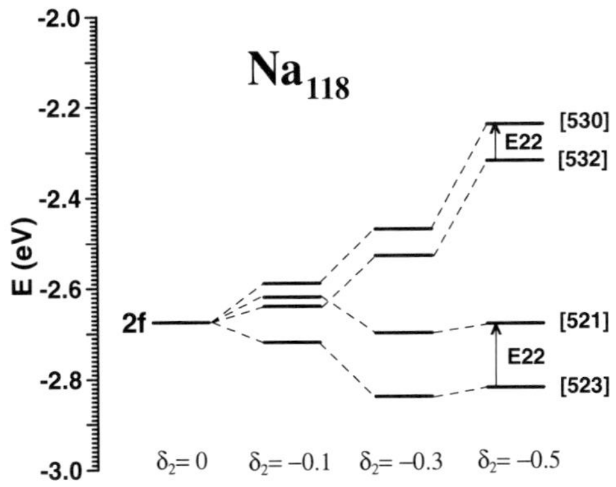

Fig. 3. Deformation splitting of $2f$-subshell in the single-particle scheme of Na_{118}. The deformation runs from δ_2=0 (spherical shape) to strongly oblate δ_2=-0.5. The levels are denoted by Nilsson-Clemenger quantum numbers $[Nn_z\Lambda]$.

LEQM can be enhanced with the stimulated Raman adiabatic passage method.[26] In strongly deformed clusters, where $\mu = 0$ and $\mu = 1$ branches of the dipole plasmon are well separated, independent RF measurements through the dipole branches are preferable. The population picture is then simpler and allows easier interpretation. For example, if RF runs only through the $\mu = 0$ dipole branch, then the population of E22 mode is forbidden and the remaining E20 and E21 modes are easier to handle.

LEQM are most interesting in two extreme cases: light deformed free clusters and heavy oblate supported clusters. In light clusters the LEQM spectrum is very dilute and can be well resolved in RF. Our calculations for Na^+_{11}, Na^+_{15}, and Na^+_{11} show that the quadrupole states can be unambiguously identified as certain electron-hole pairs.[19] The electron-hole energies from RF together with other data (Fermi energy from ionization potential measurements, etc) give a unique chance to obtain information on the single-particle spectrum in light deformed cluster. This spectrum is sensitive to many factors (deformation, temperature, ionic structure, ...) which need to be disentangled by model studies.

Heavy clusters in experiments are typically supported and are often strongly oblate. Size and shape of the supported clusters can be monitored.[9] In the case of dielectric (and porous) surfaces, the interface effect mainly reduces to shaping the cluster.[27] At first glance, the LEQM in heavy clusters looks very involved. In general, this is indeed the case. However, our calculations[19] show that LEQM have pronounced structures which can be resolved in RF experiment. The first structure is the E21 mode driven by the deformation splitting of spherical subshells. In Fig. 4 for Na_{118}, this mode corresponds to the E21 transitions [523] $\to$ [532], [521] $\to$ [532] and [521] $\to$ [530]. Obviously, the energy of this mode is proportional to the magnitude of the deformation. The E21 mode dominates in RF if it runs through the $\mu = 0$ branch of the dipole plasmon. The second structure is the E22 mode at 0.1-0.2 eV. Its strength rises with deformation while the energy does not. The E22 mode is explained in Fig. 4 as a result of bunching $[Nn_z\Lambda]$ levels with the same n_z (see E22 transitions [523] $\to$ [521] and [532] $\to$ [530]). The bunching indicates that in large systems with strong deformation the *asymptotic* Nilsson-Clemenger quantum number n_z becomes *exact*.

References

1. H. Haberland Ed., *"Clusters of atoms and molecules"*, *Springer series in chemical physics*, 52, Springer, Berlin (1994).
2. V.O. Nesterenko et al, *Phys. Rev.* A 56, 607 (1997).
3. W. Kleinig, V.O. Nesterenko, and P.-G. Reinhard, *Ann. Phys. (NY)* 297, 1 (2002).
4. O. Gunnarson and B.I. Lundqvist, *Phys. Rev.* B 13, 4274 (1976).
5. W. Kleinig, V.O. Nesterenko, P.-G. Reinhard, and Ll. Serra, *Eur. Phys. J.* D 4, 343 (1998).
6. V.O. Nesterenko, W. Kleinig, and P.-G. Reinhard, *Eur. Phys. J.* D 19, 57 (2002).
7. H. Haberland and M. Schmidt, *Eur. Phys. J.* D 6, 109 (1999).
8. J. Babst and P.-G. Reinhard, *Z. Phys.* D 42, 209 (1997).
9. T. Wenzel et al, *Appl. Phys.* B 69, 513 (1999).
10. N. Lo Iudice and F. Palumbo, *Phys. Rev. Lett.* 41, 1532 (1978).
11. E. Lipparini and S. Stringari, *Phys. Rev. Lett.* 63, 570 (1989); *Z. Phys.* D 18, 193 (1991).
12. V.O. Nesterenko, W. Kleinig, F.F. de Souza Cruz and N. Lo Iudice, *Phys. Rev. Lett.* 83, 57 (1999).
13. Ll. Serra, A.Puente, and E. Lipparini, *Phys. Rev.* B 60, R13966 (1999).
14. A. Minguzzi and M.P. Tosi, *Phys. Rev.* A 63, 023609 (2001).
15. D. Guéri and S. Stringari, *Phys. Rev. Lett.* 83, 4452 (1999).
16. O.M. Maragó et al, *Phys. Rev. Lett.* 84, 2056 (2000).
17. P.-G. Reinhard, V.O. Nesterenko, E. Suraud, S. El Gammal, and W. Kleinig, *Phys. Rev.* A 66, 013206 (2002).
18. V.O. Nesterenko et al, to be published in *Eur. Phys. J.* D; ArXiv: physics/0212084.
19. V.O. Nesterenko, W. Kleinig and P.-G. Reinhard, in preparation.
20. G. Holzward and G. Ekardt, *Z. Phys.* A 238, 1532 (1978).

21. P. von Neumann-Cosel et al, *Phys. Rev. Lett.* 82, 1105 (1999); N. Pietralla et al, *Phys. Rev.* C 58, 184 (1998).
22. V.O. Nesterenko, J.R. Marinelli, F.F. de Souza Cruz, W. Kleinig, and P.-G. Reinhard, *Phys. Rev. Lett.* 85, 3141 (2000).
23. X. Viñas et al, *Phys. Rev.* A 64, 055601 (2001).
24. P.-G. Reinhard et al, *Ann. Phys.* (Leipzig) 5, 1 (1996).
25. L. G. Gerchikov et al, *J. Phys.* B 31, 3065 (1998).
26. K. Bergmann, H. Theuer, and B.W. Shore, *Rev. Mod. Phys.* 70, 1003 (1998).
27. J.H. Parks and S.A. McPoland, *Phys. Rev. Lett.* 62, 2301 (1989).

GEOMETRIC STRUCTURE AND DYNAMICS OF MIXED CLUSTERS AND BIOMOLECULES

M. Broyer, R. Antoine, I. Compagnon, D. Rayane, P. Dugourd

Laboratoire de Spectrométrie Ionique et Moléculaire, UMR n°5579 Université Lyon I and CNRS 43 Bd du 11 Novembre 1918, 69622 Villeurbanne cedex, France

In this paper, it is demonstrated that the electric dipole of complex molecules or clusters can be measured by beam deviation in an inhomogeneous electric field. This measurement, associated to appropriate theoretical calculations and simulations, allows us to determine the geometry of these systems and their dynamical behavior as a function of temperature. Selected examples for mixed clusters (metal-fullerene and metal-benzene) and biomolecules (hydrogen bound amino acids and glycine based polypeptides) are discussed.

1. Introduction

The electric polarizability and dipole of atoms and small molecules have been studied for a long time by various methods. Only recently, complex systems started to be measured.[1] For such systems, which have very complicated excited states, the ground state properties are the most important for applications or to test theoretical methods and calculations. In this respect the measurement of the electric dipole of large molecules is a direct probe of internal charge transfers and of the geometric structure. Recently, these measurements have been applied to two families of molecules: inorganic and organic clusters and biomolecules.[2,3] The first example discussed in this paper is the electric dipole and the geometry of Bz-Metal-Bz organometallic molecules. When the benzene is replaced by a fullerene, the temperature plays a key role on the measured electric dipole. In this case the dipole allows one to probe the dynamics of the system. This interplay between dynamics and electric dipole measurement is particularly clear for non-rigid systems such as weakly bond molecular complexes or biomolecules. For biomolecules, electric dipole is one of the only experimental data that allows one to study conformations in the gas phase and auto-organization properties.

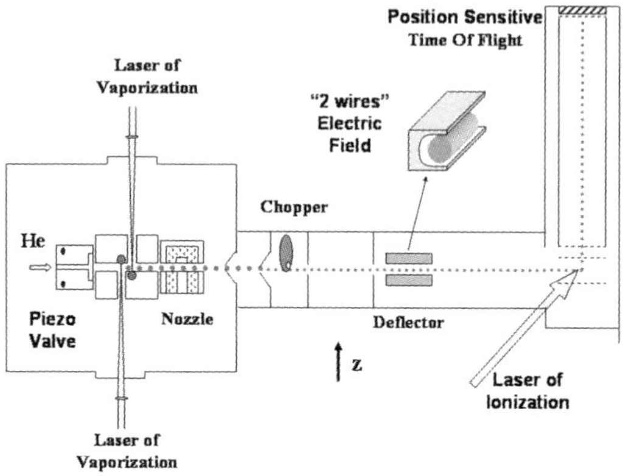

Fig. 1. Experimental setup.

2. Experimental Setup

Electric dipole measurements are performed by deflecting a well-collimated beam through a static inhomogeneous transverse electric field. The experiment consists of a laser vaporization source coupled to an electric deflector and a mass spectrometer (Fig. 1). The gas phase molecules are produced in a simple or double rod laser vaporization source (the second or third harmonics of a Nd^{3+}:YAG laser are used for the laser ablation) and are entrained by an inert gas pulse. They leave the source through a 5 cm long nozzle the temperature of which can be adjusted from 80 K to 500 K. After two skimmers, the beam is collimated by two rectangular slits (0.4 mm width) and travels through the electric deflector ("two-wires" electric field configuration). The electric field and its gradient in the deflector are $F=1.63\times10^{7}$ Vm^{-1} and $\nabla F=2.82\times10^{9}$ Vm^{-2} for a voltage of 27 kV across the two poles. Molecules are ionized one meter after the deflector in the extraction region of a position sensitive time of flight mass spectrometer. The mass of the molecule and the profile of the beam are obtained from the arrival time at the detector. The beam profile is measured as a function of the electric field in the deflector. A mechanical chopper located in front of the first slit allows one to select and measure the velocity v of the beam.

In the deflector, the instantaneous force on a molecule is:

$$\vec{f} = \vec{\mu}\nabla\vec{F} \tag{1}$$

where $\vec{\mu}$ is the electric dipole of the molecule in the field. This dipole is the sum of the permanent dipole $\vec{\mu}_0$ and of the induced dipole $\vec{\vec{\alpha}}\vec{F}$ ($\vec{\vec{\alpha}}$ is the tensor of polarizability). The average force while the molecule goes through the deflector is then given by:

$$\langle f \rangle = \langle \mu_Z \rangle \nabla F_Z = \langle (\vec{\mu}_0 + \vec{\vec{\alpha}}\vec{F})\vec{Z} \rangle \nabla F_Z \tag{2}$$

where Z is the direction of the electric field.
The deviation d for a molecule of mass m with a velocity v is given by:

$$d = K \langle f \rangle / mv^2 \tag{3}$$

where K is a geometric factor. This electric field leads to a broadening and/or a global deflection of the beam.

3. Metal Bis Benzene

In these systems the metal is in between the two benzene molecules. Two families of structures are possible: symmetric sandwich structures and C2v asymmetric complexes.[4] Figure 2 shows experimental arrival time distribution at the detector recorded for BzTiBz that belongs to the first family and for BzCoBz that belongs to the second series. For titanium, the profiles measured with and without electric field in the deflector are the same. The molecules are not deflected, which reflects the fact that the molecule is symmetrical and has no permanent dipole. For cobalt, the electric field in the deflector induces a broadening of the molecular beam. The average value of the projection of the dipole on the axis of the electric field ($\langle \mu_Z \rangle = \langle \mu \cos\theta \rangle$), depends on the rotational motion of the molecule. As in a Stern and Gerlach experiment, the broadening corresponds to the distribution of $\cos\theta$. The component of experimental dipole on the principal axis of the molecule μ_A is deduced from the arrival time (Fig. 2) by simulating the motion of the molecule in the electric field. For titanium, $\mu_A < 0.2$ D (Debye) and for cobalt 0.7±0.3 D. These results are discussed in detail in Ref. 4. This example illustrates nicely how the dipole measurement can discriminate between different calculated structures.

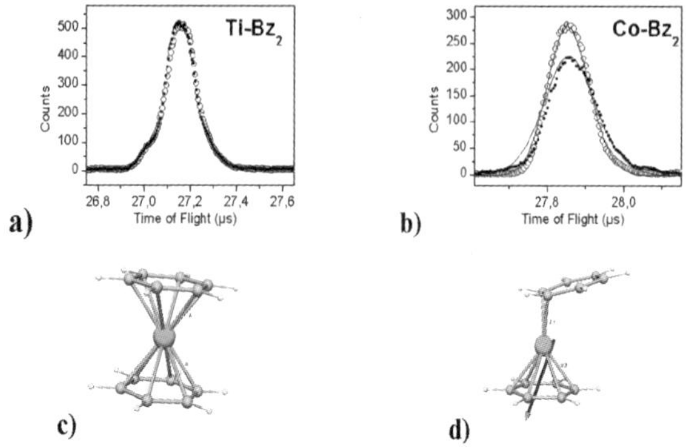

Fig. 2. Experimental arrival time distribution obtained for Ti-Bz2 (a) and Co-Bz2 (b) without electric field (empty circles) and with $F=15.1\times10^{6}$ V/m (full circles) in the deflector. The simulated profiles obtained for Co with $\mu_A=0.7$ D ($A=0.0502$ cm^{-1}, $B=C=0.0190$ cm^{-1}) are shown (full lines). Calculated structures for Ti-Bz2(a = 2.29 Å) (c) and for Co-Bz2 (d) (see Ref. 4 for details).

4. Metal Atom and Metal Cluster Fullerene

The equilibrium geometry of hexohedral metal fullerene molecules is relatively simple. For alkali metal, according to calculations,[5] the metal atom is located in front of the center of a hexagon (or a pentagon for lithium). The hexohedral molecule has a strong permanent dipole due to a charge transfer from the alkali atom to the C60. Figure 3a shows beam profiles measured with and without electric field for NaC60 at low temperature (85 K). As expected, a strong broadening of the molecular beam is observed when the electric field in the deflector is turned on. The value of the dipole deduced from the simulation of the rotational motion of the molecule in the electric field is 14.8±1.5 D.[5] This experimental result is in excellent agreement with best *ab initio* calculations (13.93 D).[5]

When the experimental temperature increases, the deflected beam profile changes completely. Figure 3b shows the beam profiles recorded at 300 K. The profile is globally deflected without any broadening. At room temperature, the alkali atom, and then the direction of the electric dipole moment, can move freely on the surface of the cage.[6] Without electric field, the average value of the dipole is zero. In a static electric field, the dipole is statistically oriented toward the direction of the electric field and its average value is given by the Debye-Langevin formula:

$$\langle \mu_Z \rangle = \left[\alpha_{e^-} + \frac{\langle \mu^2 \rangle_{T,F=0}}{3kT} \right] F_Z = \chi F_Z \tag{4}$$

where α_{e^-} is the average electronic polarizability and $<\mu^2>_{T,F=0}$ corresponds to the fluctuation of the dipole calculated at equilibrium (F =0). In this case, this average value little depends on the temperature and $< \mu^2>_{T,F=0}$ is close to the square of the electric dipole at equilibrium. The contribution of the electronic polarizability is small. A more complete description of the motion of the alkali atom on the surface of C_{60} as a function of temperature is given in Ref. 7. The dipole deduced from deflected profiles is 16.3±1.6 D, which is compatible with the value deduced from low temperature experiments.

The analysis at high temperature has been performed for LiC_{60}, NaC_{60}, KC_{60}, and CsC_{60}. The dipole moment increases with the size of the alkali atom, from 12.4 D for LiC_{60} to 21.5 D for CsC_{60}.[5] This evolution is mainly due to the decrease in the ionization potential of the alkali atom.

The same measurements were extended to clusters with several alkali atoms ($NanC_{60}$ clusters).[8] The results are given in figure 4 for n=1 to 34. These experimental results are compared to two simple models. The first model assumed that the sodium atoms form a metal shell around the fullerene. The second one corresponds to the values calculated assuming the formation of a metal droplet on the surface of the fullerene with a charge transfer of one electron between the metal cluster and the C_{60}. These two models are oversimplified. Recent semi-empirical calculations[9] show that the C_{60} may accommodate more than 1 electron and that there is a transition between wetting and non-wetting as the number of sodium atom increase. For a large number of sodium atoms, the second model is a good picture, but between 2 and 8 sodium atoms, according to the semi empirical calculation,[9] the atoms tend to be uniformly distributed around the C_{60}. Contrary to a single metal atom on the surface, $<\mu^2>_{T,F=0}$ may strongly depend on the temperature. For example for two sodium atoms, the lowest energy structure is obtained for opposite atoms on the C_{60}. The dipole for this structure is zero, but at finite temperature the two atoms move on the surface and $<\mu^2>_{T,F=0}$ is not null. This would explain the χ value

(χ=1000 Å^3) measured for Na_2C_{60}. This large χ value corresponds to an electric dipole induced by the vibronic motion of the two atoms. A similar effect is discussed in the next paragraph.

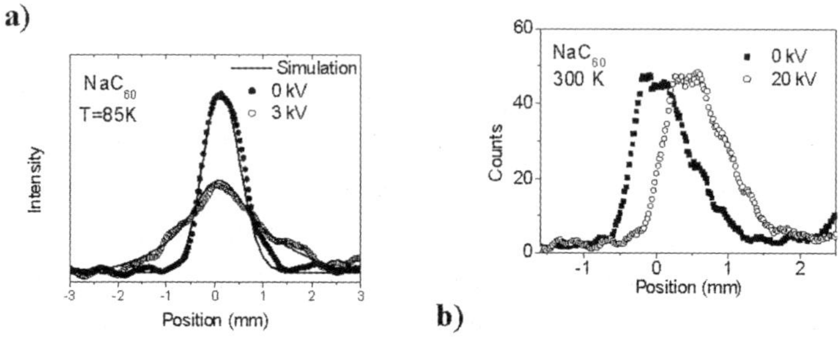

Fig. 3. Left, experimental beam profiles measured at 85 K for NaC_{60} (symbols) and calculated profiles with μ=14.8 D(full line). Right, experimental beam profiles measured at 300 K for NaC_{60}.

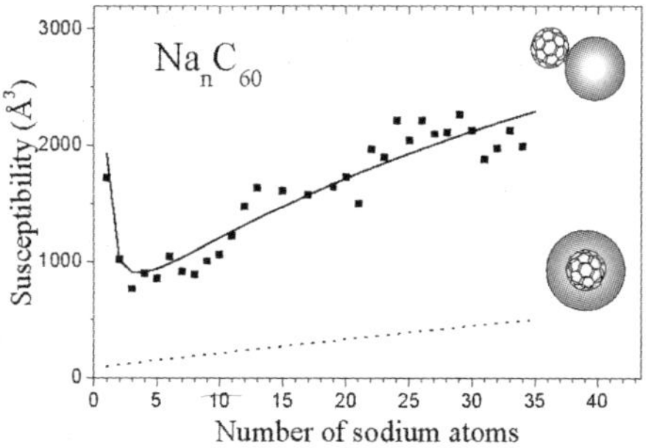

Fig. 4. Susceptibility of Na_NC_{60} clusters as a function of the number of sodium atoms. The dark squares correspond to experimental values. The full line corresponds to values calculated assuming that the sodium atoms form a metal shell around the fullerene. The dashed line corresponds to the values calculated assuming the formation of a metal droplet on the surface of the fullerene.

5. The Dimer of Para Amino Benzoic Acid $(PABA)_2$

This complex is formed of two para-aminobenzoic acid molecules (PABA) which are weakly bound by two hydrogen bonds. Each PABA molecule has a permanent dipole of 3.7 D. In the dimer (Fig. 5), the permanent dipole is zero at equilibrium. When the temperature increases, the molecule vibrates. In an altered conformation, the dipole may be non-null (see Fig. 5) and $<\mu^2>_{T,F\neq 0}$.[2] In first approximation $<\mu^2>_{T,F=0}$ is proportional to the temperature, as expected for an ensemble of uncoupled harmonic oscillators. This vibrational induced dipole is the major contribution to the susceptibility at room temperature.[2]

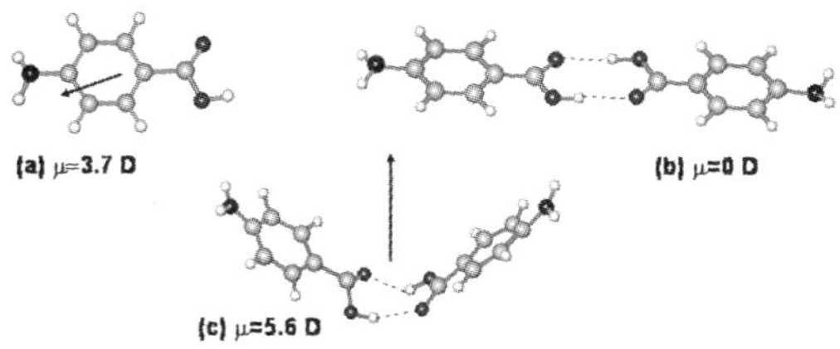

Fig. 5. (a) Para-amino-benzoic-acid molecule ($NH_2C_6H_4COOH$). (b) Equilibrium structure for the $(PABA)_2$ complex. Hydrogen bonds (between O and H atoms) are indicated by dashed lines. (c) Example of distorted structure with a dipole.

6. Glycine Based Polypeptides

In a polypeptide, the main contribution to the electric dipole is due to the peptide bond between amino acids (~3.5 D per bond). The dipole, which results of the summation of all the peptide bond dipoles, strongly depends on the conformation. For example, for organized conformations, such as helices, all the contributions add up and the molecule has a giant permanent dipole. The dipole measurement is a direct probe of the gas phase conformation of the biomolecule.[3] Figure 6 shows an example of beam profiles for WG_5 (W=tryptophane, G=glycine, molecular weight 489.5 uma). The electric susceptibility deduced from Fig. 6 is 391 Å^3, in agreement with the susceptibility calculated for a random coil. At room temperature, this molecule is very floppy and this value corresponds to an average of the square dipole on very different conformations.[3] As a comparison, WG_5 in a helix conformation would have an electric susceptibility at room temperature of 2350 Å^3. This illustrates the pertinence of the electric dipole measurement for determining the conformation of large molecules in the gas phase.

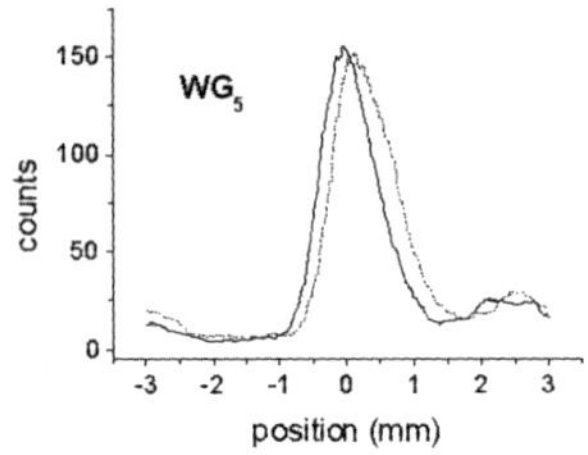

Fig. 6. Beam profiles of WG_5 peptides measured without (solid line) and with (dashed line) a voltage (25 kV) across the deflector.

Acknowledgements

The experiments on polypeptides were performed in collaboration with the group of M.F. Jarrold at Indiana University. The *ab initio* calculations on mixed clusters were performed in collaboration with A.R. Allouche and F. Rabilloud.

References

1. M. Broyer, R. Antoine, E. Benichou, I. Compagnon, P. Dugourd, D. Rayane, *C.R. Physique* **3**, 301 (2002).
2. I. Compagnon, R. Antoine, D. Rayane, M. Broyer, P. Dugourd, *Phys. Rev. Lett.* **89**, 253001 (2002).
3. R. Antoine, I. Compagnon, D. Rayane, M. Broyer, P. Dugourd, G. Breaux, F. C. Hagemeister, D. Pippen, R. R. Hudgins, M. F. Jarrold, *J. Am. Chem. Soc.* **124**, 6737 (2002).
4. D. Rayane, A.-R. Allouche, R. Antoine, M. Broyer, I. Compagnon, P. Dugourd, *Chem. Phys. Lett.* **375**, 506 (2003).
5. D. Rayane, A. R. Allouche, R. Antoine, I. Compagnon, M. Broyer, P. Dugourd, *Eur. Phys. J. D.* **24**, 9 (2003).
6. D. Rayane, R. Antoine, P. Dugourd, E. Benichou, A. R. Allouche, M. Aubert-Frécon, M. Broyer, *Phys. Rev. Lett.* **84**, 1962, (2000).
7. P. Dugourd, R. Antoine, D. Rayane, E. Benichou, M. Broyer, *Phys. Rev. A* **62**, 011201 (R) (2000).
8. P. Dugourd, R. Antoine, D. Rayane, I. Compagnon, M. Broyer, *J. Chem. Phys.* **114**, 1970 (2001).
9. J. Roques, F. Calvo, F. Spiegelman, C. Mijoule, *Phys. Rev. Lett.* **90**, 075505 (2003).

CLUSTER STUDIES IN ION TRAPS

L. Schweikhard, A. Herlert and G. Marx
Institut für Physik, Ernst-Moritz-Arndt-Universität Greifswald, Domstr. 10a, D-17487 Greifswald, Germany
E-mail: lutz.schweikhard@physik.uni-greifswald.de

K. Hansen
School of Physics and Engineering Physics, Chalmers University of Technology and Gothenburg University, SE-41296 Gothenburg, Sweden

Ion trapping is a versatile tool for cluster research. This article reviews various measurements of metal clusters stored in a Penning trap. The studies include technical aspects of ion trapping as well as properties of the clusters. Most notably the trapping of clusters allows us to investigate them over extended durations and provides the possibility for a sequence of many preparatory steps before the actual experiments. These may include the selection of cluster ensembles of a given cluster size, the adsorption of molecules and changes in charge state.

1. Introduction: Clusters in Ion Traps

Ion trapping is a valuable tool in many research areas[1,2] and has found application in a number of investigations of gas phase charged atomic and molecular clusters. These clusters bridge the gap between the field of single particles, i. e. an atom or a molecule, and the condensed phase. Ion trapping can be made use of in particular in the investigation of gas-phase clusters, i. e. clusters not embedded in matrices or supported by surfaces. The latter are technologically important but experiments are not as easily interpreted due to the interactions with the (often not easily defined) environment. In contrast, ion trapping allows us to switch on and off the interactions of interest.

In order to follow the properties of clusters on their way from the atomic to the bulk phase it is important to monitor their size. Therefore, mass spectrometry is an essential tool of cluster research. It comes in two distinct aspects: (a) Size selection, also called isolation: the clusters are investigated one cluster size at a time in order to clearly distinguish between the properties of the different species. (b) Product analysis: most reactions change the mass-over-charge ratio. In ion trapping these mass spectrometric properties are either already built in for a given experimental setup and procedure or they are easily adapted.

2. Typology of Ion Storage

Ions may be continuously kept in flight in storage rings which make use of magnetic and/or electric fields that bend an ion beam to a closed loop.[3] Alternatively, linear trapping devices which use static electric fields have also been introduced recently.[4] For storage at a given position in space with essentially no kinetic energy and confinement volumes down to a few cubic centimeters, cubic millimeters or less, ion traps have to be employed. The main two kinds are the Paul and the Penning trap. Both are in wide use in analytical chemistry where they are usually referred to as simply "ion traps"[5] and FT-ICR systems, respectively, where ICR stands for Ion Cyclotron Resonance and FT for Fourier Transform, which indicates a particular type of ion detection and mass analysis.[6]

Static electric fields are not sufficient to keep an ion at a given position, since it will always feel a force along the field lines to the electrodes where they originate or end. However, if alternating fields are used an effective potential can be achieved, where the corresponding forces keep the ions in the region of small fields. This principle of the radio-frequency (rf) or Paul trap has its analog in geometric optics where the combination of converging and diverging lenses of the same focal length (i. e. except for the sign) results in a positive net focal length. The rf trap can be a three-dimensional device with a geometry as shown in Fig. 1 composed of a ring electrode and two end caps. Alternatively, it can be constructed with four rods similar to a quadrupole mass filter, with an additional electrode at each end for axial confinement ("linear trap"). The 3-D version is unspecific with respect to the polarity of the ions. For the 2-D version the polarity is given by the choice of the sign of the voltages of the additional electrodes.

Only ions of a certain range of mass-over-charge ratio (m/z) have stable trajectories in the Paul trap. When DC fields are used in addition to the rf fields this range is further decreased. Application of defined fields thus

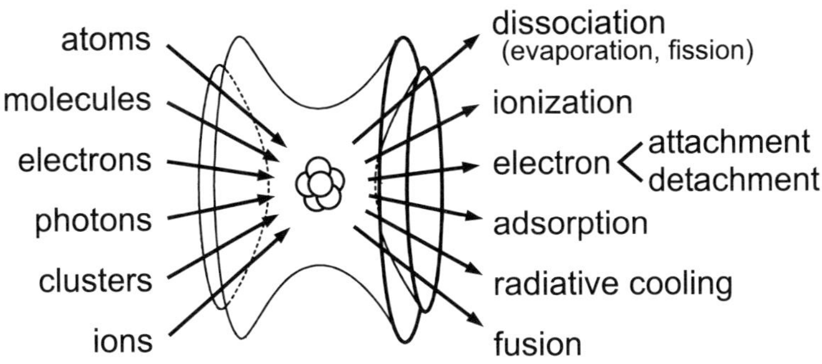

Fig. 1. Overview of interaction partners (left column) and resulting reactions (right column) of stored clusters (center, with a schematic drawing of the Penning trap electrodes).

leads to the selection of the ions of interest. On the other hand, even for broad-band storage, there are limits to the stability of the ion trajectories and, in addition, on the potential well depth. In practice this leads to ion storage which covers m/z ranges of about an order of magnitude. The position of this range can be shifted from the lightest ions to macroscopic particles (e.g. particles of 100 mm diameter or more). When quadrupolar potentials are used the ions perform harmonic motions in the trap and are an easily addressed size- (i. e. m/z-) specifically. In contrast, fields of higher multipolarity[7] have a flatter bottom, i. e. more space for "cold" ions, but there is no good size selectivity, i. e. selection and analysis has to be performed before and during cluster injection and after ejection from the trap, respectively.

In contrast, the Penning trap makes use of static fields only. Radial confinement is given by the Lorentz force of a strong magnetic field (typically of several tesla) which leads to the cyclotron motion of the ions with its characteristic m/z dependence. Along the field lines the ions are stored by a static electric potential (of about one or a few volts). The electric field leads to a second radial motional mode of the ions, centered in the trap's axis: the magnetron or drift motion. In addition, the particular combination of fields of the Penning trap results in unstable trajectories when the m/z-ratio of the ion is too large, i. e. above a "critical mass", which depends on the particular trapping parameters. However, there is no limit with respect to light ions and even electrons can be stored simultaneously with, e. g., singly charged gold clusters consisting of thirty atoms, Au_{30}^-. Thus, the particles stored at the same time can differ in mass by 7 orders of magnitude.

Since the electric fields are static, ions of only one polarity are stored in a Penning trap. There are two ways to circumvent this restriction: the hybrid form of a "combined trap" uses electric rf fields in addition to the static magnetic field. There is an alternative arrangement when both a magnetic field and the trapping of ions of both polarities is desired: the nested trap allows the storage of both cations and anions in very close vicinity, i. e. with the possibility of interactions. A step into this direction of cluster trapping, with a new method to distinguish clusters of the same size but opposite polarities, has been made recently.[8]

In the following, the Penning trap experiments of the "Cluster Trap" are considered in somewhat more detail. While these are to our knowledge the most extensive studies on various aspects of stored clusters, there have been several other reports on cluster trapping over the years. These interesting investigations will not be reviewed in this contribution due to lack of space. The reader is referred to the proceedings of the biannual "International Symposium on Small Particles and Inorganic Clusters" (ISSPIC).[9]

The experimental setup, procedure, and results of the Cluster Trap have been described on several occasions.[10–15] Thus, in the following we will in general not cite the original publications but again refer the interested reader to these reviews.

3. Experimental Setup and Procedure

The clusters are produced by laser irradiation of material from a wire in the presence of a helium gas pulse. The evaporated particles (including ions and electrons) form small aggregates. The ionic clusters are transferred by electrostatic ion-optical elements to the Penning trap. For in-flight capture the electric potential is lowered on the source-side endcap until the clusters have passed a central hole in this electrode. Similarly, at the end of any event sequence the trap's content is analyzed by ejection of the ions through the other endcap into a drift region for time-of-flight (TOF) mass spectrometry. Single-ion counting is performed by a conversion electrode detector which, since it is off-axis, allows the application of various laser beams along the trap's axis.

In general, cluster ions are injected into the trap, accumulated and size selected before a particular reaction of interest takes place. Selection, interaction and possibly a delay period may be repeatedly applied. Finally, the trap is emptied and the reaction is analyzed with respect to the resulting products. Typically, during a given sequence a few up to a few ten clusters are present in the trap. In order to increase the statistical significance of the mass spectra the sequence is repeated up to a few hundred times. For calibration or to monitor e. g. the production yield of the cluster source or the overlap of a laser beam and the cluster ensemble in the trap, the sequence with the interaction of interest can be alternated with a "reference sequence" of identical form but without the interaction under study or with a standardized interaction. In this way quasi-simultaneous measurements are performed.

4. Overview of Investigations

Figure 1 gives an overview of the various interaction partners (left column) that have been used to investigate cluster properties at the Cluster Trap and indicates the various reaction pathways observed (right column). In addition to the investigations of the clusters by use of the trap there have also been several studies that made use of the cluster ions to evaluate some Penning-trap features. The following list tries to give the flavor of what results have been achieved.

4.1. *Inflight-capture and accumulation of cluster ions from an external source and centering of the clusters of interest*

In early cluster studies with ion traps the clusters were often produced in or in close vicinity to the trap. A separation of functions, however, leads to an increased versatility. As has been shown at the Cluster Trap several cluster bunches can be accumulated and the cluster ions can be centered in the middle of the trap.

4.2. *Investigation of the stability region of the ion mass range*

As mentioned above there is a critical mass, above which storage in Penning traps is not possible. This trap characteristic has been probed and, in addition, an instability of the ion trajectory for particular trapping parameters has been found.

4.3. *Monitoring of a non-neutral electron plasma by cluster measurements*

Recently, the cluster ions have been found to be an interesting tool for the investigation and characterization of simultaneously stored electrons.[16]

Shifts of the ion cyclotron resonance indicate the electrons' space-charge density and the temporal behavior of the attachment of further electrons to anionic clusters can be related to the transfer of energy between the different motional modes of the electrons.

4.4. *Collision induced dissociation*

After capture and size selection the cluster ions' cyclotron motion is excited and a collision gas is added to the trap volume. The clusters fall apart upon collisions. Thus, the decay pathways can be followed. In addition, the dissociation energies can be estimated. Since the clusters are relatively heavy species their center-of-mass energies are rather small, even for large cyclotron radii. Thus, multicollisional excitation has to be considered. The conversion yield from kinetic energy to internal excitation energy is calibrated by comparison with photoexcitation measurements (see below).

4.5. *Photoinduced dissociation*

Similarly, a strong laser pulse can be used to break the clusters apart. The pulse energy can be increased to follow the sequential evaporation of neutral atoms and dimers of, e. g., Au_{55}^+ down to the atomic ion Au^+. In the resulting pattern of product cluster ions the signature of magic numbers (i.e. high dissociation energy and/or large drop of dissociation energy for the next bigger cluster) is observed.

4.6. *Electron induced dissociation*

In an effort to create clusters of higher charge state (see below) the clusters have been treated with electrons. Again, multiple dissociation leads to the characteristic fragment pattern as, e. g., the well-known odd-even oscillations of noble metal clusters.

4.7. *Time-resolved photoinduced dissociation*

If the collisional excitation is replaced by pulsed photoexcitation several advantages are gained: the excitation takes place at a defined time which allows the time-resolved monitoring of the delayed decay processes. The excitation energy is well-defined by the wavelength of the laser beam (modulo the number of photons absorbed). By tuning the wavelength the decay constant can be controlled.

4.8. *Electron evaporation*

Upon laser heating, the clusters may instead of fragmenting react by thermal emission of electrons. This process has been investigated (again time-resolved) for the case of anionic tungsten clusters.

4.9. *Reactions with molecules*

The reactions of small metal clusters (Au_n^+ and V_n^+) with simple molecules (N_2O^+ and H_2) have been studied. To compensate for the low resolving power of the TOF detection scheme (which is used for its very high sensitivity!) the reaction analysis made use of a selective ejection of ion species from the trap.

4.10. *IR spectroscopy of molecules attached to clusters*

Methanol molecules have been attached to gold clusters. After size selection (i. e. for defined species with a given number of gold atoms and a given number of molecules) IR laser pulses have been applied. Thus, the resonance frequency of the intramolecular CO stretching mode was probed. These experiments not only yield information on the influence of the attachment on the methanol molecule but at the same time the methanole acts as a sensor molecule: when compared with theoretical predictions the observed frequency shift between Au_8^+ and Au_9^+ can be interpreted as the transition between planar and three dimensional gold clusters.

4.11. *Radiative cooling*

In an extension of the time-resolved photodissociation experiments a two-photon excitation is performed with a delay between the two absorption processes. Since the energy of the first (pump) photon is chosen below the dissociation threshold the cluster will not break apart immediately. The second (probe) pulse leads to the decay, but only if the cluster does not cool in the meantime. Thus, by monitoring the dissociation yield as a function of the delay between the pump and probe laser pulses, the radiative cooling in the time regime of tens of microseconds to tens of milliseconds has be examined.[17]

4.12. *Model-free determination of dissociation energies*

The measurement of dissociation energies of complex particles in not as trivial as it may appear on first sight. In general, any excitation energy is rapidly distributed among the various internal degrees of freedom, and the actual decay is a statistical process – the cause of the delayed dissociation. In the case of metal clusters there is usually not enough information available on their internal structure to simulate the process and fit the dissociation energy. However, comparison of the time resolved observation of a sequential delayed photodissociation ($A \rightarrow B \rightarrow C$) with the delayed decay of the intermediate product ($B \rightarrow C$) leads directly to a rather precise value of the dissociation energy (of A). In a nutshell, in the second experiment the remaining internal energy of the first step in the sequential decay is measured. The method has been introduced for the evaporation of monomers from gold clusters[18] and has been further developed in various ways.[19]

4.13. *Electron-impact ionization*

By application of an electron beam the stored clusters can be further ionized. The collisions also lead to dissociation (see above). Thus, in addition to the abundance pattern of singly charged products, similar patterns of clusters with higher charges are observed, too. A comparison leads to direct information as to the nature of the stability of the clusters, i. e. when a given cluster size shows an extended abundance irrespective of the charge state then this is due to a favorable geometrical arrangement, whereas when (for a monovalent element) the cluster size minus charge state is the significant number, the abundance anomaly is a result of an electronic effect.

4.14. *Attachment of further electrons to anionic clusters*

Only singly charged clusters are delivered from the source. Actually, until recently no multiply charged anionic metal clusters had been reported at all. By bathing the clusters in an ensemble of electrons (see above), the clusters can be brought to higher charge states. Again, the conversion yield includes some information on the internal cluster structures.[20]

4.15. *Decay pathways of multiply charged clusters*

Once multiply charged clusters are created and stored inside the trap they can be further prepared and investigated. In general, as a first step they are isolated with respect to cluster size and charge state. They are then subjected to collisions or photoexcitation, similar to the case of the singly charged clusters. However, new decay pathways are now available: For

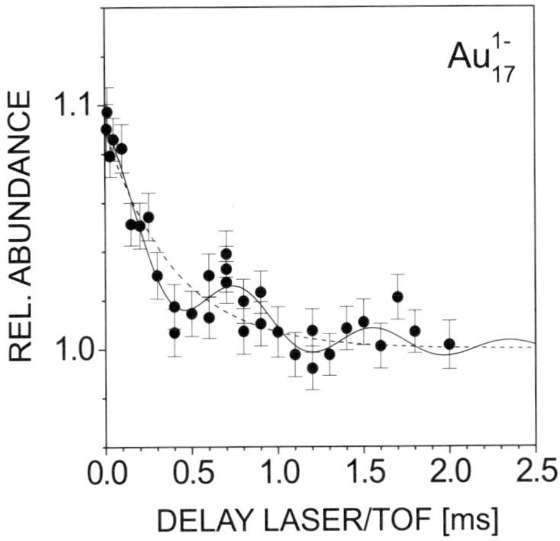

Fig. 2. Relative abundance of Au$^{1-}_{17}$ as a function of the delay between photoexcitation and TOF analysis (laser pulse 1 mJ at $\lambda = 355\,\text{nm}$). The dashed line shows the exponential decay of Au$^{1-}_{17}$ into Au$^{1-}_{16}$ and the solid line is a fit of a superposed oscillation. The data points are normalized to a reference measurement at $\Delta t = 100\,\text{ms}$.

cations a competition between the evaporation of neutral atoms and fission into charged particles has been observed. In the case of anions, again, neutral atoms or electrons are emitted. In some cases, recent measurements indicate a correlated emission of two electrons from dianionic metal clusters. In any case, all decay-pathway branching rations are functions of the cluster size.

5. Current and Future Investigations

As mentioned at the end of the last section, polyanionic metal clusters are one of the lines of research that are currently persued. Further investigations are under way. In particular, delayed processes will be studied. However, for reasons not yet known, the time-resolved investigation of anions seems to be more difficult than for cations. Figure 2 shows the relative abundancce of Au$^{1-}_{17}$ as a function of time after photoexcitation with a Nd:YAG-laser pulse of 1 mJ at $\lambda = 355\,\text{nm}$ (the signal of the product Au$^{1-}_{16}$ is complementary). In order to lower the influence of fluctuations in the ion production, all data points have been normalized to a corresponding reference measurement with a fixed delay time of 100 ms. In addition to the exponential decay of the Au_{17}^{1-}-signal (dashed line) there is an oscillatory structure at a frequency close to the magnetron frequency of the ions (solid line). For higher laser-pulse energies the amplitude of the signal oscillation is increased and the exponential decay can not be determined. So far, this phenomenon has been encountered only for negative ions.

Note, that multiply charged cationic metal clusters have also not yet been addressed by time-resolved measurements. It is planned to perform such investigations, and also to extend them to polyanionic clusters. Other ideas under consideration concern the interaction of positively and negatively charged clusters with each other. When species of opposite charge state react, which differ in the number of surplus and missing electrons, the product should retain a net charge and could thus be further stored and studied inside the trap.

As indicated in the introduction this contribution has been concerned with gas-phase clusters only. It should be noted however, that an extension to the study of supported clusters seems feasible: similar to the attachment of molecules onto metal clusters, the clusters themselves could be sitting on a chunk of substrate while the whole system (including substrate, cluster and additional molecules) is levitated in an ion trap. As an example, consider a gold cluster Au_{10}. If it is attached to a piece of carbon consisting of 150 atoms it is still not as heavy as Au_{20}. The total mass of the system is less than 4000 atomic mass units and by use of FT-ICR MS it could probably be analyzed with better than one mass unit resolution. Thus, size-selective measurements of such systems are in reach where e. g. the attachment or detachment of single atoms and molecules could be monitored by mass spectrometry. The systems could be investigated as a function of the numbers of all species involved (i. e. m gold atoms on n carbon atoms reacting with particular molecules). Thus, the cluster properties could be probed as a function of substrate size, possibly allowing for an extrapolation to the macroscopic surface.

Acknowledgments

The Cluster Trap has been supported over many years by various programs of the Deutsche Forschungsgemeinschaft and the European Union (currently within the network on "CLUSTER COOLING").

References

1. P. K. Gosh, *Ion Traps*, Oxford University Press, New York (1995).
2. G. Werth, F. G. Major, V. N. Gheorghe, *Charged Particle Traps*, Springer, Heidelberg, to be published.
3. S. P. Møller, *Nucl. Instr. and Meth. A* 394, 281 (1997).
4. A. Naaman, K. G. Bhushan, H. B. Pedersen, N. Altstein, O. Heber, M. L. Rappaport, R. Moalem, D. Zajfman, *J. Chem. Phys.* 113, 4662 (2000).
5. R. E. March, J. F. J. Todd, *Practical Aspects of Ion Trap Mass Spectrometry: Fundamentals (Vol. 1)*, CRC Press, Boca Raton (1995).
6. A. G. Marshall, *Int. J. Mass Spectrom.* 200, 331 (2000).
7. D. Gerlich, *Adv. Chem. Phys.* 82, 1 (1992).
8. L. Schweikhard, J. J. Drader, S. D.-H. Shi, C. L. Hendrickson, A. G. Marshall, *AIP Conf. Proc.* 606, 647 (2002).
9. ISSPIC 7, *Surf. Rev. Lett.* 3, 1-1226 (1996); ISSPIC 8, *Z. Phys. D* 40, 1-578 (1997); ISSPIC 9, *Eur. Phys. J. D* 9, 1-651 (1999); ISSPIC 10, *Eur. Phys. J. D* 16, 1-412 (2001); ISSPIC 11, *Eur. Phys. J. D, in press.*
10. L. Schweikhard, St. Becker, K. Dasgupta, G. Dietrich, H.-J. Kluge, D. Kreisle, S. Krückeberg, S. Kuznetsov, M. Lindinger, K. Lützenkirchen, B. Obst, C. Walther, H. Weidele, J. Ziegler, *Physica Scripta* T 59, 236 (1995).
11. St. Becker, K. Dasgupta, G. Dietrich, H.-J. Kluge, S. Kuznetsov, M. Lindinger, K. Lützenkirchen, L. Schweikhard, J. Ziegler, *Rev. Sci. Instrum.* 66, 4902 (1995).
12. L. Schweikhard, S. Krückeberg, K. Lützenkirchen, C. Walther, *Eur. Phys. J. D* 9, 15 (1999).
13. L. Schweikhard, A. Herlert, M. Vogel, *Metal Clusters as Investigated in a Penning Trap*, in *The Physics and Chemistry of Clusters (Proceedings of the Nobel Symposium 117)*, E. E. B. Campbell and M. Larsson (Eds.), World Scientific, Singapore (2001), pp. 267-277.
14. L. Schweikhard, K. Hansen, A. Herlert, M. D. Herráiz Lablanca, G. Marx, M. Vogel, *Int. J. Mass Spectrom.* 219, 363 (2002).
15. L. Schweikhard, K. Hansen, A. Herlert, G. Marx and M. Vogel, *Eur. Phys. J. D*, in press.
16. L. Schweikhard, A. Herlert, *AIP Conf. Proc.*, to be published.
17. C. Walther, G. Dietrich, W. Dostal, K. Hansen, S. Krückeberg, K. Lützenkirchen, L. Schweikhard, *Phys. Rev. Lett.* 83, 3816 (1999).
18. M. Vogel, K. Hansen, A. Herlert, L. Schweikhard, *Phys. Rev. Lett.* 87, 013401 (2001).
19. M. Vogel, K. Hansen, A. Herlert, L. Schweikhard, *J. Phys. B: At. Mol. Phys.* 36, 1073 (2003).
20. C. Yannouleas, U. Landman, A. Herlert, L. Schweikhard, *Phys. Rev. Lett.* 86, 2996 (2001).

Photoabsorption and Photoionization of Clusters

STUDY OF DELOCALIZED ELECTRON CLOUDS BY PHOTOIONIZATION OF FULLERENES IN FOURIER RECIPROCAL SPACE

S. Korica, A. Reinköster and U. Becker

Fritz-Haber-Institut der Max-Planck-Gesellschaft, Faradayweg 4-6, D-14195 Berlin, Germany. E-mail: becker_u@fhi-berlin.mpg.de

The characteristic oscillations in the partial photoionization cross sections of C_{60} are analyzed in terms of the geometrical properties of both, the cage structure and the distribution of the delocalized electron cloud of the highest occupied molecular orbitals. The analysis is based on the Fourier transform of the cross section oscillations, the results are corroborated by different theoretical models. In contrast to this good overall agreement between theory and experiment there is striking disagreement with respect to discrete resonance structure in the partial cross sections. Possible reasons for this behavior are discussed.

1. Introduction

Photoelectron spectroscopy is a versatile tool for structural studies exploiting the diffraction properties of core electron emission.[1] However, the information on the properties of the electron distributions from where they are emitted, that valence electrons carry, has been exploited only since very recently.[2] The reason for this is the fact that valence electrons are not sensitive to scattering centers, but to rapid changes of the potential energy causing their binding. In a sense, both localized centers and delocalized electron clouds may be imaged by valence electron emission if the system is spherically symmetrical. This condition is fulfilled by a whole class of systems, clusters and, more specifically, fullerenes. Particularly clusters which are well described within the jellium model are perfectly suited for such size dependent studies. Moreover, because the production of mass selected clusters is still a very difficult task, fullerenes offer an attractive alternative for the exploitation of the potential of valence photoemission measurements to extract structural information from cross section behavior as shown in Fig. 1 for a variety of photoelectron spectra.

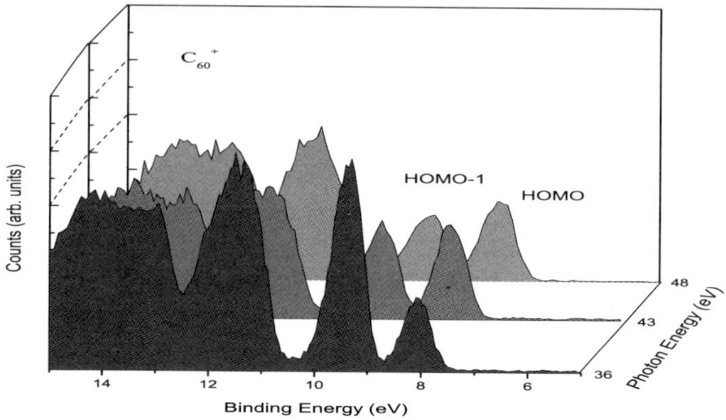

Fig. 1. The photoelectron yield (at 54.7°) as a function of the binding energy for three different photon energies.

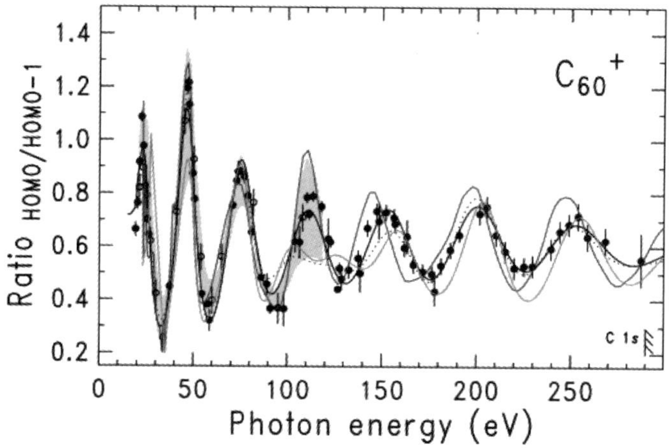

theoretical curves	—, —:	Jellium model, LDA calculations
semiempirical curves	···, —:	Shell model, Shell + Cage model
experimental data	•, ◇:	Rüdel *et al.*, Phys. Rev. Lett. **89**, 125503 (2002)

Fig. 2. Branching ratio HOMO/HOMO-1 from near threshold up to the carbon K-edge. The figure contains different experimental data sets and theoretical calculations.[2,3]

2. Results and Models

Such studies have been successfully performed during the last years on C_{60} and in part on C_{70}. The characteristic cross section behavior which exhibits the structural information on the fullerenes is the intensity modulation of the various valence photoelectron lines, in particular the HOMO and HOMO-1 lines across excitation energy. These energy dependent modulations reflect directly the carbon cage and conducting shell structure of C_{60} and C_{70}. The oscillations are alternating in phase with the angular momentum and symmetry of the final state of the outgoing electron giving rise to distinct oscillation in the branching ratio between valence lines of different symmetry. Figure 2 shows these oscillations in the branching ratio between the two outermost molecular orbitals of C_{60} together with different theoretical curves.

Recent refinements of the partial cross section data over a large energy range made it possible to analyze the observed oscillations in terms of the desired structural information. Fourier transformed cross section data directly display the radius of the fullerene and the thickness of the delocalized electron cloud.[2]

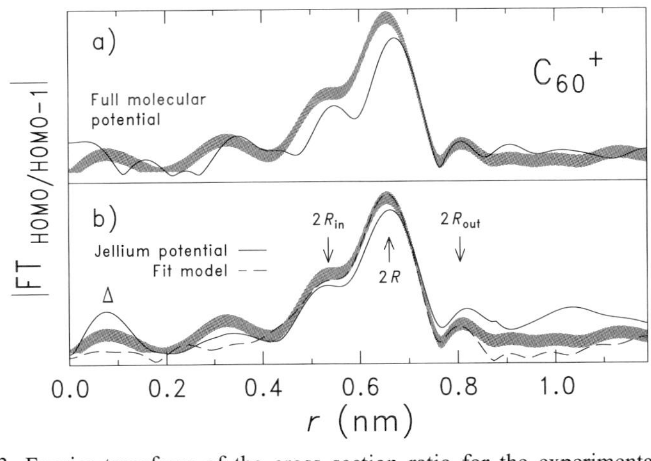

Fig. 3. Fourier transform of the cross section ratio for the experimental data (shaded area), the LDA calculation of Decleva et al. (solid line in a)), the LDA calculation using a spherical jellium potential (solid line in b)) and the result of the semi-empirical fit (dotted line).

New measurements for both gas phase and solid state show almost identical behavior of the partial cross section on large scale. However, due to the unique symmetry but also complexity of C_{60}, a deeper understanding of the excitation and ionization of its valence electrons is still a challenge for both theoreticians and experimentalists. Therefore little is known about resonant photoemission and photoelectron angular distributions. Advanced *ab initio* calculations[3] based on the local density approximation (LDA) predict pronounced resonances in the partial cross sections for the photoionization of the two outermost orbitals of C_{60}. Similar structures were predicted by using HF molecular orbitals and solving the coupled scattering equations for the ejected electron in the field of the molecular ion within a polyatomic Schwinger variational method.[4] It was therefore of great interest to experimentally prove whether these structures are present in the partial cross section and, in case they are, how well they are described by theory.

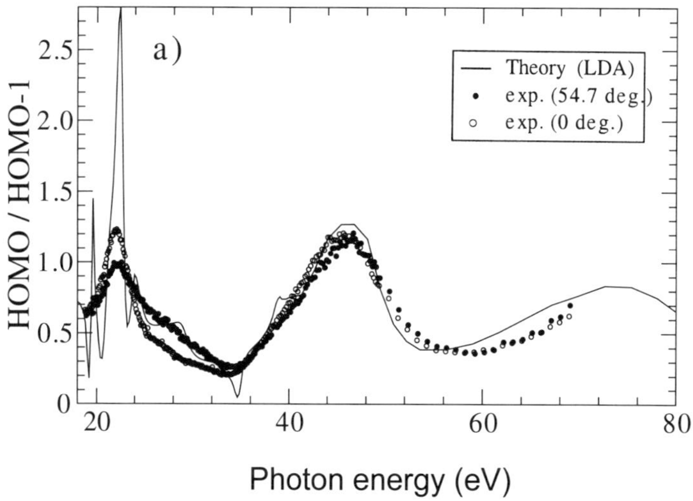

Fig. 4. Partial cross sections ratio HOMO/HOMO-1 as function of the photon energy. The figure shows experimental results (symbols) and theoretical data (solid lines).[2]

For this purpose we have performed high resolution measurements of photoelectrons emitted from C_{60} in small steps of 0.1 eV in the range of 19 up to 50eV, and in larger steps of 1 eV in the range of 50 to 70 eV. The measurements at beamline BW3 of HASYLAB at DESY were performed at two different angles with respect to the electric vector of the ionizing radiation in order to derive both the partial cross sections σ and the angular distribution asymmetry parameters β . A first analysis of the data was done deriving the branching ratios between the HOMO and HOMO- 1 photolines rather than the independent partial cross sections, because this ratio of two photolines, which lie closely together, is independent of transmission corrections.

3. Discussion

Figure 4 shows the branching ratio of the two partial cross sections HOMO/HOMO-1 derived from the small step measurements performed in the resonance region. Immediately, the comparison between the experimental and theoretical data shows two major results: (i) the broad oscillatory behavior of the σ ratio is well described by theory, but (ii) the predicted pronounced resonance structure does not exist. This is surprising, because such structures are also predicted by the calculations of Gianturco and Lucchese[4]; however their resonances are found at different positions. Because of the present measurements it seems reasonable to assume that the resonances in the partial cross sections are in a sense quenched by the vibrations of the molecule. The complex resonant wave functions of the resonantly excited states are in delicate balance between stabilization and decay, forced by the different vibrational modes, such as the breathing mode, of C_{60}. Further analysis of the independent partial cross sections rather than branching ratios, as well as angular distribution asymmetry parameters will give further insight into this challenging problem.

References

1. D. P. Woodruff and A. M. Bradshaw, *Rep. Prog. Phys.* **57**, 1029 (1994).
2. A. Rüdel, R. Hentges, U. Becker, H. S. Chakraborty, M. E. Madjet, and J. M. Rost, *Phys. Rev. Lett.* **89**, 125503 (2002) and references therein.
3. P. Decleva, S. Furlan, G. Fronzoni and M. Stener, *Chem. Phys. Lett.* **348**, 363 (2001) and private communication.
4. F. A. Gianturco and R. R. Lucchese, *Phys. Rev.* A **64**, 032706 (2001).

JELLIUM MODEL FOR PHOTOIONIZATION OF FULLERENES

V.K. Ivanov, G.Yu. Kashenock and R.G. Polozkov
St. Petersburg State Politechnical University, Politecnicheskaya 29, St. Petersburg 195251, Russia
E-mail: polozkov@tuexph.stu.neva.ru

A.V. Solov'yov
A.F.Ioffe Physical-Technical Institute, Politecnicheskaya 26, St. Petersburg 194021, Russia
E-mail: solovyov@rpro.ioffe.rssi.ru

A simple approach has been developed for the description of photoionization processes involving fullerenes. We consider the fullerenes as approximately spherical shells and use the jellium model. The local density approximation has been used for calculating initial state wave functions. The cross sections for photodetachment processes have been calculated within the self-consistent many-body theory approach including local-density (LDA) and random phase (RPA) approximations.

1. Introduction

In this paper we present results on the total and partial photoionization cross sections for C_{60} within the photon energy range up to 90 eV. Within the RPA-LDA *ab initio* approach we calculate the partial cross sections for photoionization from the highest occupied molecular orbital (HOMO) and from the next lower-lying HOMO–1 orbital of the fullerene C_{60} in the high energy region (for details see Refs. 1 and 2). The spectrum demonstrates strong oscillatory behaviour arising due to the possibility that the photoelectron forms spherical standing waves inside the hollow structure of the fullerene. The influence of the potential well shape on the behaviour of the cross section is investigated. For this purpose we introduce some parametric modification of the model. Presenting these results we hope that our findings could lend impetus to further experimental investigations. Our approach allows us to reveal the physical nature of the cross section peculiarities at different domains of photon energy: near-threshold resonance, plasmon peak, oscillatory structures. All are interpreted in the context of influence of the collective effects and the hollow molecular structure specific to fullerene.

2. Results and Discussion

2.1. *Near-threshold and plasmon resonances in photoionization of fullerenes*

The main feature of photoionization cross section of fullerenes is a giant resonance. The photoionization cross sections of fullerenes C_{60} calculated within jellium model in the RPA-LDA approach reproduced the well-known giant plasmon resonance.[3] The position (20 eV) and peak value (1700 Mb) (see Figure 1) have been found in good agreement with experimental data[4] and with the results of the other theoretical works.[5,6] This result also agrees with the simple estimate following from the Mie theory (see Ref. 7):

$$\omega_l = \sqrt{\frac{l(l+1)N_{\rm del}}{(2l+1)R_F^3}}\,. \tag{1}$$

Here, N_{del} is the number of delocalized electrons, R_F is the radius of the fullerene and l is the angular momentum. For the dipole plasmon resonance ($l = 1$) the Mie formula gives 20 eV for the position of the resonance. Thus our simple model reproduces the prominent feature of the C_{60} photoabsorption spectrum, and we may apply it with confidence to another nearly spherical fullerene C_{20}.

We have predicted for the first time the two resonances in the C_{20} photoionization spectrum[1] (see Fig. 1). The near-threshold resonance can be interpreted as a cavity resonance. It is reasonable to correlate the near-threshold resonance structure from our calculations with the existence of a resonance arising in the low-energy electron scattering on C_{20} due to the formation of the metastable negative ion C_{20}^- as it follows from quantum chemistry calculations.[8]

The second resonance is localised in the vicinity of 27 eV. Its origin is connected with the oscillations of the delocalized electron density (for details see Ref. 2). The position of the plasmon resonance is in agreement with the Mie formula which gives 27 eV. Note that the giant resonance in

C_{20} lies at higher electron energy than that for C_{60}.

Note that many-electron effects within the RPA play a crucial role in the description of the shapes and positions of the resonances arising in the photoionization cross sections of C_{20} and C_{60} (for details see Ref. 1).

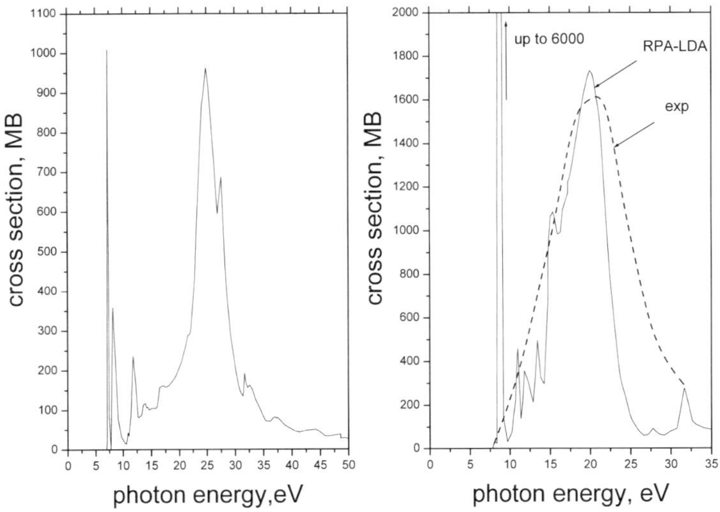

Fig. 1. Solid lines show the photoionization cross sections of the fullerene C_{20} (left figure) and of the fullerene C_{60} (right figure) calculated in the RPA-LDA and shifted on $\Delta = I_p^{\mathrm{exp}} - I_p^{\mathrm{our}}$. The dashed line on the right figure shows the experimentally measured cross section.[4]

2.2. *Oscillations of the C_{60} partial photoionization cross sections*

We present the results of the partial cross sections calculation for the photoionization from the highest occupied molecular (HOMO) and HOMO–1 orbitals of C_{60} performed within the LDA and RPA.

The partial cross sections of photoionization from the HOMO and from HOMO–1 orbitals of the fullerene C_{60} demonstrate strong oscillatory behaviour at high photon energies. The photoelectron spectroscopy data from the gas phase[9,10] C_{60} exhibit oscillations in the partial cross section. These are similar to the oscillations observed in the solid phase[11] C_{60}. In Ref. 12 this behaviour was interpreted in terms of intramolecular interference of incoming and outgoing waves. The authors[12] used the simplest square well model potential with a number of parameters for calculating the frequency

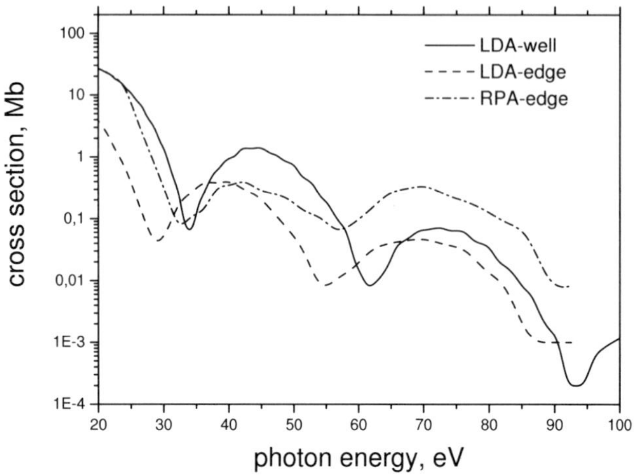

Fig. 2. Results for electron transitions from HOMO with changing orbital momentum $l \to l-1$ calculated within different methods: "LDA-well" case – one-particle approximation with self-consistent C_{60} potential evaluated with averaged core approximation; "LDA-edge" case – one-particle approximation with self-consistent C_{60} potential evaluated within self-consistent jellium model; "RPA-edge" case – with account for many-electron correlations within RPA approximation with self-consistent C_{60} potential evaluated within self-consistent jellium model.

of oscillation and the positions of minima and maxima. In the subsequent work a similar parametric potential with thickness δR was used. The finite thickness of the potential led to four possible frequencies determined by $R - \delta R/2$, $R + \delta R/2$, $2R$, and δR. It was claimed that the oscillations are beats due to interference.

In our work we use the final-state photoelectron wave function calculated with parametric self-consistent jellium-model LDA potential.[1] We also perform analogous calculations with the potential similar to the square well model one used by previous authors[12] and compare the results. We show that the frequency of oscillation is determined by only one parameter, the radius of the fullerene R_F, but positions of the minima and the maxima are governed by the shape of the fullerene potential.

We compare the partial cross sections for photoionization from HOMO and HOMO–1 orbitals of C_{60} obtained within the single-particle approximation (LDA) and with the account for many-electron correlations (RPA) (see Fig. 2). The good agreement of the oscillation frequency, the minima and the maxima positions derived in LDA and RPA demonstrates that the origin of oscillations is the single-particle effect. This proves the fact that for their analysis it is sufficient to use the simplest single-particle approximation. In the present paper we use the LDA approximation for this purpose.

Consider the dependence of the oscillatory behaviour on the shape of the fullerene potential. In the spherical jellium model the self-consistent fullerene potential has a sharp minimum at the fullerene radius. This self-consistent model has a single parameter – the fullerene radius well known from the experiment. We name this potential as the "edge case".

The spherical jellium model with only one parameter has at least two essential limitations. First, in this model, one assumes that the positive charge of the ionic core is distributed over the sphere of radius R_F. However, from the physical view point it is reasonable to introduce the width of this distribution and to consider a spherical shell for the positive background. In fact, the atoms in the fullerene C_{60} are located not exactly on the surface of a sphere even at zero temperature. Moreover, the fullerene potential is smeared due to the thermal vibrations of the atoms. Therefore, the shape of the fullerene potential with softer edges seems to be very appropriate. The calculation of photoionization cross sections accounting for both the actual symmetry of the fullerene and the thermal vibrations of atoms is a rather complicated task. In order to simplify the problem, we construct the model potential, similar to the one in Ref. 5, with the ionic core distribution averaged over a spherical shell. We call this model potential the "well-case". The width of the shell δR and the fullerene radius R_F are the parameters of this model.

The second limitation of the spherical jellium model arises from the fact that it neglects the exchange-correlation interaction between the $1s$ core electrons and the valence electrons filling σ orbitals. As a result, the single-particle energy levels of valence σ electrons broaden the valence zone. In order to take into account the interaction of the $1s$ and σ-electrons, we introduce a well-like pseudopotential with the depth V_0 and the radius equal to the thickness of the fullerene shell δR. These modifications of the potential improve the agreement of the calculated electronic structure with experiment. Thus, the width of the valence zone reduces to 32 eV. This value agrees well with the more accurate result obtained with accounting for the icosahedral splitting.[14] The ionization potential increases up to the value of 7.5 eV. Note that the LDA approximation does not exclude the self-interaction of electrons. Calculations show that if one takes the electron self interaction into account, then the ionization potential increases approximately on 2.5 eV. In the "well case", in LDA, we have found that

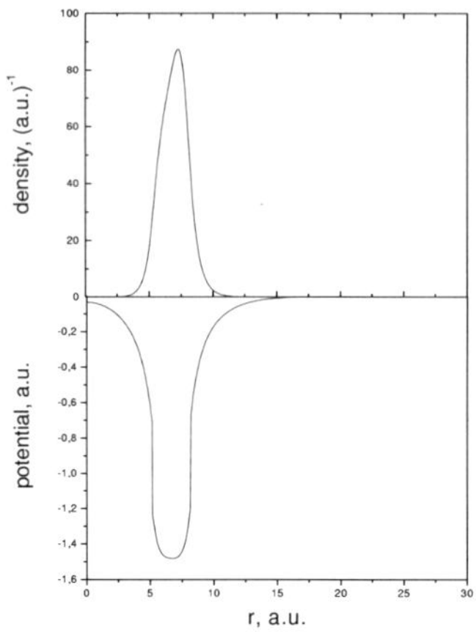

Fig. 3. Self-consistent potential and valence electron density of the fullerene C_{60} calculated with the core distribution averaged over spherical shell (well-case). Width of the shell is $\delta R = 3$ a.u. The potential depth V_0 is 0.5 a.u.

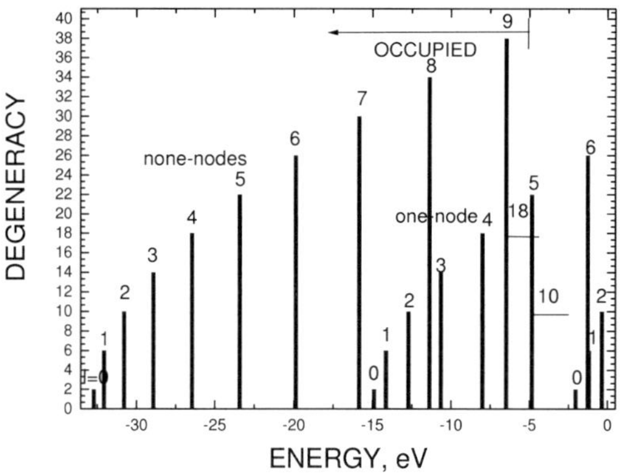

Fig. 4. Energy levels spectrum of the fullerene C_{60} calculated with the core distribution averaged over spherical shell (well-case). No-node orbitals are occupied up to $l = 9$ and one-node orbitals are occupied up to $l = 5$. We also plot the spectrum of the discrete excited states: $3s$, $8i$, $4p$, $5d$.

the ionization potential is equal to 5 eV, which with accounting for the electron self-interaction corrections provides a good agreement with the experimental value. The fullerene potential and the electronic density are shown in Fig. 3. The corresponding electronic level structure is presented in Fig. 4.

We used the ground- and the final-state photoelectron wave functions calculated with the "well case" parametric potential for the evaluation of

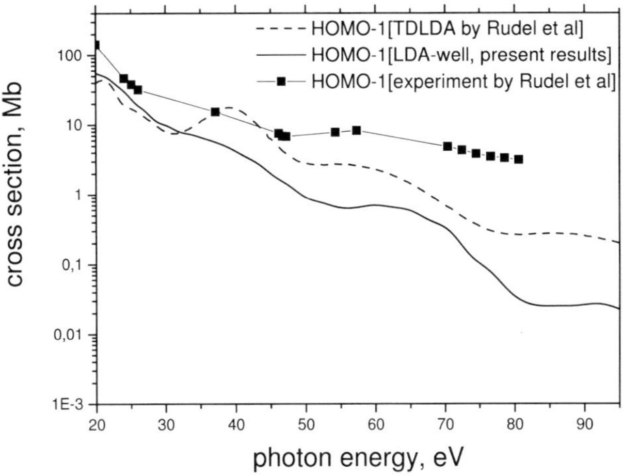

Fig. 5. Solid line shows the partial photoionization cross section of the HOMO of the fullerene C_{60} calculated within the LDA and the averaged core density approximation (well-case). The spectrum is shifted on $\Delta = I_p^{\rm exp} - I_p^{\rm pr}$. Dashed line is the result obtained within TDLDA and the averaged core density approximation.[13] Squares present the experimental results.[13]

the partial photoionization cross sections from the HOMO and HOMO–1 orbitals of the fullerene C_{60} (see figure 5). Our results are in qualitative agreement with the available experimental data and with the results of calculations performed within the TDLDA approximation.[13]

We have also studied the dependence of the oscillatory behaviour of the partial photoionization cross sections on the potential shape. In figure 2, we compare the partial cross sections for the photoionization from the HOMO-1 orbitals obtained with the two fullerene potentials: the beak-like "edge-case" and the "well-case". This figure shows that the choice of the potential influences weakly the oscillatory structure of the cross section. The period of oscillations is determined by the geometry of the fullerene, i.e its radius. The positions of the minima and the maxima in the partial cross section are shifted on 5 eV towards higher energies in the well case with respect to the edge case. This property can be interpreted within the quasi-classical approximation which is well justified due to the large value of orbital momenta $l = 6, 4$ for the partial waves considered. The phaseshift $\delta_l(k)$ arising for a short-range potential $U(r)$ is approximately equal to

$$\delta_l = -\int_{r_0}^{\infty} \frac{mU\left(r\right)dr}{\hbar^2\sqrt{k^2 - (l+1/2)^2/r^2}}$$

$$\delta_l \sim \frac{mU(r_0)r_0}{k\hbar^2}.$$

Finally, we emphasis that the results for partial photoionization cross section obtained within the single-particle LDA and many-body RPA are close. So, while at the intermediate excitation energy range the plasmon resonance arises due to the collective electron dynamics,[2] at higher excitation energies the oscillatory behaviour of the partial photoionization cross sections can be considered as a single-electron effect.

Acknowledgements

This work was supported by the Russian Foundation for Basic Research (grant No. 03-02-16415-a), Russian Academy of Sciences (grant No. 44) and INTAS.

References

1. V.K. Ivanov, G.Yu. Kashenock, R.G. Polozkov and A.V. Solovyov, *J. Phys. B: At. Mol. Opt. Phys.*, 34, L669, (2001).
2. V.K. Ivanov, G.Yu. Kashenock, R.G. Polozkov and A.V. Solovyov, *J. of Exp. And Theor. Phys.*, 96, 658, (2003).
3. C. Brchignac and J.-P. Connerade, *J. Phys. B: At. Mol. Opt. Phys.*, 27, 3795, (1994).
4. I.V. Hertel, H. Steger, J. de Vries, B. Wesser, C. Menzel, B. Kamke and W. Kamke, *Phys. Rev. Lett.*, 68, 784, (1992).
5. M.J. Puska and R.M. Nieminen, *Phys. Rev.* A, 47, 1181, (1993).
6. M.S. Hansen, J.M. Pacheco and G. Onida, *Z. Phys.* D, 35, 141, (1995).
7. J.-P. Connerade and A.V. Solov'yov, *Phys. Rev.* A 66, 013207, (2002).
8. F.A. Gianturco, G.Yu. Kashenock, R.R. Lucchese and N. Sanna, *J. Chem. Phys.*, 116, 2811, (2002).
9. T. Liebsch, O. Plotzke, F. Heiser, U. Hergenhahn, O. Hemmers, R. Wehlitz, J. Viefhaus, B. Langer, S.B. Whitfield and U. Becker, *Phys. Rev.* A, 52, 457, (1995).
10. T. Liebsch, O. Plotzke, R. Hentges, *et al*, *J. El. Spt. Rev. Ph.*, 79, 419, (1996).
11. P.J. Benning, D. Poirer, N. Troubllier, J. Martins, R. Weaver, R. Haufler, L. Chibante and L. Lamb, *Phys. Rev. B*, 44, 1962, (1991).
12. Y.B. Xu, M.Q. Tan and U. Becker, *Phys. Rev. Lett.*, 76, 3538, (1996).
13. A. Rudel, R. Hentges, U. Becker, H.S. Chakraaborty, M.E. Majiet and J.M. Rost, *Phys. Rev. Lett.*, 89, 125503, (2002).
14. K. Yabana and G.F. Bertsch, *Physica Scripta*, 48, 633, (1993).

PHOTOABSORPTION OF SMALL SODIUM AND MAGNESIUM CLUSTERS

Ilia A. Solov'yov

A. F. Ioffe Physical-Technical Institute, 194021 St. Petersburg, Russia
and
Institut für Theoretische Physik der Johann-Wolfgang Goethe Universität, Robert-Mayer Str. 8-10, D-60054 Frankfurt am Main, Germany
E-mail: ilia@th.physik.uni-frankfurt.de

Andrey V. Solov'yov

A. F. Ioffe Physical-Technical Institute, 194021 St. Petersburg, Russia
and
Institut für Theoretische Physik der Johann-Wolfgang Goethe Universität, Robert-Mayer Str. 8-10, D-60054 Frankfurt am Main, Germany
E-mail: solovyov@th.physik.uni-frankfurt.de

Walter Greiner

Institut für Theoretische Physik der Johann-Wolfgang Goethe Universität, Robert-Mayer Str. 8-10, D-60054 Frankfurt am Main, Germany

We predict a strong enhancement in the photoabsorption of small Mg clusters in the region of 4-5 eV due to the resonant excitation of the plasmon oscillations of cluster electrons. The photoabsorption spectra for neutral Mg clusters consisting of up to $N = 11$ atoms have been calculated using *ab initio* framework based on the time dependent density functional theory (TDDFT). The nature of predicted resonances has been elucidated by comparison of the results of the *ab initio* calculations with the results of the classical Mie theory. The splitting of the plasmon resonances caused by the cluster deformation is analysed. The reliability of the used calculation scheme has been proved by performing the test calculation for a number of sodium clusters and the comparison of the results obtained with the results of other methods and experiment.

1. Introduction

Optical spectroscopy is a powerful instrument for investigation of the electronic and ionic structure of clusters as well as their thermal and dynamical properties. During the last decades these issues have been intensively investigated both experimentally, by means of photodepletion and photodetachment spectroscopy, and theoretically by employing the time-dependent density functional theory (TDDFT), configuration interaction (CI) and random-phase approximation (RPA) (for review see Refs.1,2 and references therein). These methods have been used in conjunction with either jelium model[2] defined by a Hamiltonian, which treats the electrons in a cluster in the usual quantum mechanical way, but approximates the field of the ionic core, treating it as a uniform positively charged background, or with *ab initio* calculations of the electronic and ionic cluster structure, where all or at least valence electrons in the system are treated accurately.

During the last years, numerous theoretical and experimental investigations have been devoted to the study of optical response properties of alkali metal clusters. The plasmon resonances formation in Na, K and Li clusters has been studied both theoretically and experimentally.[1–5] Some attention was also devoted to the splitting and broadening of the plasmon resonances (see citations above). The mentioned metal elements belong to the first group of the periodic table, i.e. possess one s-valence electron.

The situation differs for clusters of the alkali-earth metals of the second group of the periodic table, such as Be, Mg, Ca. Study of these clusters is of particular interest, because they exhibit a transition from the weak van der Waals bonding, being the characteristic of the diatomic molecule to the metallic bonding present in the bulk. Thus, significant attention was paid to the magnesium clusters. Various properties of Mg clusters, such as their structure, the binding energy, ionization potentials, HOMO-LUMO gap, average distances, and their evolution with the cluster size have been investigated theoretically.[6–8] Recently, the mass spectrum of Mg clusters was recorded[9] and the sequence of magic numbers was determined. The investigation of optical response of small Mg clusters has not been performed so far in spite of the fact that it should carry a lot of useful information about the dynamic properties of magnesium clusters.

In this paper we predict the strong enhancement in the photoabsorption of small Mg clusters in the region of 4-5 eV due to the resonant excitation of the plasmon oscillations of the cluster electrons. Using all electron *ab initio* TDDFT we calculate the spectra for cluster structures with up to 11 atoms possessing the lowest energy. The geometries of these clusters were calculated using all electron DFT methods and described in our recent work.[6] In this work we focus on the formation of the plasmon resonances in magnesium clusters. We elucidate their nature by comparing our results with the results of the classical Mie theory and analyse the splitting of the plasmon resonances caused by the cluster deformation.

2. Theoretical Method

Theoretical methods used in our calculations are based on the density functional theory and many-body-perturbation theory. In the present work we use the gradient-corrected Becke-type three-parameter exchange functional[10] paired with the gradient-corrected Lee, Yang and Parr correlation functional[11] (B3LYP), as well as with the gradient-corrected Perdew-Wang 91 correlation functional (B3PW91).[12] We do not present the explicit forms of these functionals, because they are somewhat lengthy, and refer to the original papers.[10–14] Our calculations have been performed with the use of the Gaussian 98 software package.[15] We have utilized the 6-311+G(d) basis set of primitive Gaussian functions to expand the cluster orbitals.[13,15]

The absorption of light by small metal spheres has been investigated theoretically by Mie long ago.[16]) For particles with diameter considerably smaller than the wavelength, the absorption cross section based on the Drude dielectric function reads as:

$$\sigma(\omega) = \frac{4\pi N_e e^2}{m_e c} \frac{\omega^2 \Gamma}{\left(\omega_0^2 - \omega^2\right)^2 + \omega^2 \Gamma^2} \tag{1}$$

where ω_0 is the surface-plasma frequency of a sphere with N_e free electrons, ω is the photon frequency, Γ represents the width of the resonance, m_e is the electron mass, e is its charge and c is the light velocity. Equation (1) assumes that the dipole oscillator strengths are exhausted by the surface plasma resonance at ω_0. In metal clusters this resonance corresponds to the collective oscillation of the spherical valence-electron cloud against the positive background.

Using the sum rule one can easily show[16] that $\omega_0 = \sqrt{N_e e^2 / m_e \alpha}$, where α is the static polarizability of the cluster. For a classical metal sphere, $\alpha = N_e r_s^3$, where r_s is the Wigner-Seitz radius. With $r_s = 4.0$ a.u. for Na and $r_s = 2.66$ for Mg,[17] one derives the classical surface-plasma-resonance energies $\omega_0^{Na} = 3.40$ eV and $\omega_0^{Mg} = 6.27$ eV for Na and Mg respectively.

For small metal clusters the photoabsorption pattern differs significantly from the Mie prediction. In these systems the plasmon resonance energy is smaller as compared to the metal sphere case. The lowering of the plasmon energies in small metal clusters occurs because of the spill out effect according to which the electron density is spilled out of the cluster, increasing its volume and polarizability. For example, for spherical Na_8 and Na_{20} clusters the average static polarizability is 796.840 (a.u) and 1964.484 (a.u.) respectively.[14] Thus, the plasmon resonance energies, ω_0, read as 2.73 and 2.75 (eV) for Na_8 and Na_{20} respectively. Beside the lowering of the plasmon resonance energy in small metal clusters the photoabsorption pattern is splitted. This fragmentation arises mainly due to the cluster deformation. With the use of the sum rule, equation (1) can be generalized and written in the following form.:[16]

$$\sigma(\omega) = \frac{4\pi e^2}{m_e c} \sum_{i=1}^{n} \frac{f_i \omega^2 \Gamma_i}{\left(\omega_i^2 - \omega^2\right)^2 + \omega^2 \Gamma_i^2} \tag{2}$$

where ω_i are the transition energies, f_i and Γ_i are the corresponding oscillator strengths and widths, n is the total number of the resonant transitions.

In the case of the triaxial cluster deformation the photoabsorption cross section possesses the three peak structure. The splitting of the plasmon resonance into three peaks can easily be understood assuming the ellipsoidal form of the cluster surface. Within the framework of the deformed jelium model the ionic density is considered to be uniform within the volume confined by the ellipsoid surface defined by $\frac{x^2}{a^2} + \frac{y^2}{b^2} + \frac{z^2}{c^2} = 1$. If one assumes that the electron density fills in entirely in the interior of the ionic ellipsoid, one finds the following dipole plasmon energies corresponding to the electron density oscillations in three directions x, y, z,:[18]

$$\begin{aligned}
\omega_x &= \omega_0 \left[1 + \frac{\delta cos\gamma}{5}(1 - \sqrt{3} tan\gamma)\right] \\
\omega_y &= \omega_0 \left[1 + \frac{\delta cos\gamma}{5}(1 + \sqrt{3} tan\gamma)\right] \\
\omega_z &= \omega_0 \left[1 - \frac{2\delta cos\gamma}{5}\right]
\end{aligned} \tag{3}$$

where ω_0 is the classical Mie frequency being the average of ω_x, ω_y and ω_z, δ and γ are the deformation parameters defined by equations: $\delta cos\gamma = \frac{3}{4}\frac{2c^2-a^2-b^2}{a^2+b^2+c^2}$, $tan\gamma = \sqrt{3}\frac{a^2-b^2}{2c^2-a^2-b^2}$. Note that in the axially symmetric case one derives $\gamma = 0$ and $\omega_x = \omega_y$.

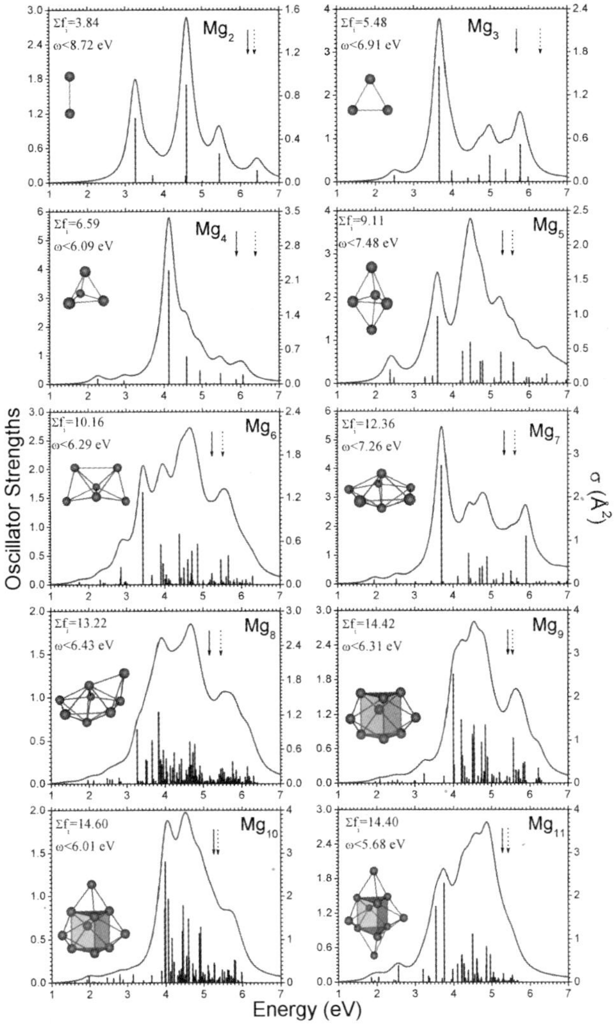

Fig. 1. Photoabsorption cross section calculated for Mg clusters with $N \leq 11$ using the $B3PW91/6-311+G(d)$ method. Vertical solid lines show the oscillator strengths for the optically allowed transitions. Their values are shown in the left hand side of the plots. The right hand side of each plot shows the scale for the corresponding photoabsorption cross section. Cluster geometries calculated in Ref.6 are shown in the insets. The label near each cluster image shows the sum of the oscillator strengths and the excitation energy range considered. By solid and dotted arrows we show the adiabatic and vertical ionization potentials respectively, calculated in Ref.6

3. Results and Discussion

In Fig. 1, we present the oscillator strengths for the dipole transitions calculated for the most stable cluster isomers of Mg_2-Mg_{11}. Cluster geometries are shown in the insets to the figure. They were calculated and discussed in Ref.6.

For sodium,[2] the plasmon resonance arises for the clusters with less than 10 atoms. Thus, it is natural to expect that for the magnesium clusters with $N \leq 10$ the formation of the plasmon resonance should be clearly seen.

Calculating the oscillator strengths f_i and substituting the found values in equation 2, we obtain the photoabsorption cross sections for magnesium clusters plotted in Fig. 1. In this calculation we have used the width $\Gamma_0 = 0.4$ eV, which is the average width for Na clusters at room temperature.[2] In this paper we do not calculate the excitation line widths for Mg clusters and do not investigate the line widths temperature dependence. These interesting problems are beyond the scope of the present paper and deserve a separate careful consideration.

In the photoabsorption spectra for Mg_2 and Mg_3 one can identify the strong resonances in the vicinity of 4 eV, which can be interpreted as the plasmon resonances splitted due to the cluster deformation. Below, we discuss this splitting in more detail. For larger clusters, the plasmon resonance energy increases slowly and evolves towards the bulk value, 6.26 eV (see dots in Fig. 2). The lowering of the plasmon resonance energy in small Mg clusters as compared to its bulk value occurs because of the spill out effect.

There are two main factors, which determine the resonance pattern of the photoabsorption spectra for magnesium clusters: collective plasmon excitations of the delocalized electrons and the resonant transitions of the electrons bound in a single magnesium atom. In the excitation energy range considered, the photoabsorption spectrum of a single Mg atom exhibits the two strong resonant excitations: $3s(^1S_0) \rightarrow 3p(^1P_1^0)$ and $3s(^1S_0) \rightarrow 4p(^1P_1^0)$ with the energies (oscillation strengths) 4.346 (1.8) and 6.118 (0.2) eV respectively.[19] The TD/B3PW91/6-311+G(d) method gives the following energies and the oscillator strengths for these lines: 4.225 (1.63) and 5.765 (0.29) eV, which are in the reasonable agreement with the data given in Ref. 19. The $3p(^1P_1^0)$ line can be easily identified in the photoabsorption spectrum for Mg_2. In terms of the plasmon resonance excitations, this line corresponds to the oscillations of the electronic density perpendicular to the cluster axis, while the strong line in the vicinity of 3 eV corresponds to the collective electron oscillations along the cluster axis. For larger clusters,

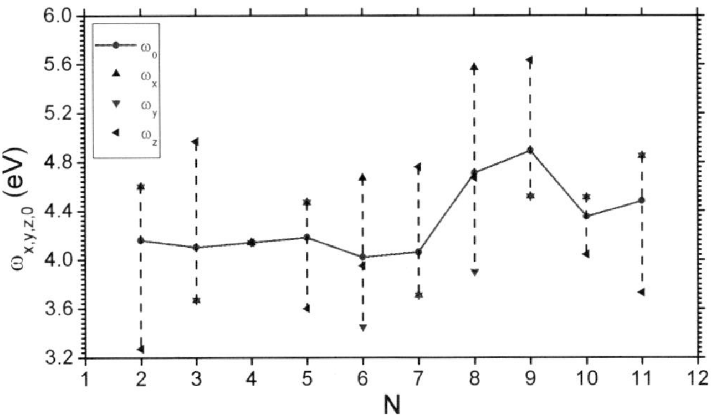

Fig. 2. Size dependence of the plasmon resonance energies ω_x, ω_y, ω_z: x (upper triangles), y (lower triangles) and z (left triangle). Circles are the Mie-frequencies ω_0 being the average of ω_x, ω_y and ω_z.

the $3p(^1P_1^0)$ line is strongly coupled with the plasmon resonance excitation occurring at the close energy. The situation is different for the $4p(^1P_1^0)$ line. Due to its higher energy, this excitation line does not couple that strongly with the plasmon resonance and can be identified in the photoabsorption spectra for the Mg_2, Mg_3, Mg_6 and Mg_7 clusters in addition to the plasmon resonances. For larger clusters (e.g. Mg_8, Mg_9, Mg_{10}), due to the growth of their plasmon resonance energies, the $4p(^1P_1^0)$ line becomes more and more of the plasmon resonance type.

For many clusters the plasmon resonance is splitted. This splitting arises mainly due to the cluster deformation. In order to illustrate this effect we plot in Fig. 2 the energies ω_x, ω_y, ω_z of the strongest resonances versus the cluster size. Using equation (3), we determine the deformation parameters δ and γ and present them in Fig. 3. One can distinguish four different cases: i) $\delta = \gamma = 0$ the cluster is spherical (see $N = 4$); ii) $\delta < 0$, $\gamma = 0$ the cluster is oblate (see $N = 3, 7, 9$); iii) $\delta > 0$, $\gamma = 0$ the cluster is prolate (see $N = 2, 5, 10, 11$); iv) $\delta \neq 0$, $\gamma \neq 0$ the cluster is triaxially deformed (see $N = 6, 8$). This analysis shows that most of the clusters considered are close to the axially symmetric form, although some clusters (Mg_6 and Mg_8) are triaxially deformed. Note that many additional satellite resonances appear in the photoabsorption spectra. The additional satellite lines are often the result of higher order cluster deformations. Thus, they are beyond the ellipsoidal model.

To show the connection between the plasmon resonance splitting and the cluster deformation we have determined the plasmon resonance energies for Mg_2 and Mg_3 from the Mie theory via the static dipole polarizabilities of the clusters and compared them with the TDDFT result. The principle values of cluster polarizability tensor α_{xx}, α_{yy},α_{zz} are 130.386, 130.386, 246.769 (a.u) for Mg_2 and 282.412, 282.412, 159.757 (a.u.) for Mg_3 respectively. Thus, the plasmon resonance energies ω_x, ω_y and ω_z read as 4.82, 4.82, 3.39 (eV) for Mg_2 and 3.8, 3.8, 5.46 (eV) for Mg_3 respectively. These values are very close to those obtained directly from the photoabsorption spectra analysis and presented in Fig. 2. This fact independently proves that the plasmon resonance is already formed in such small systems.

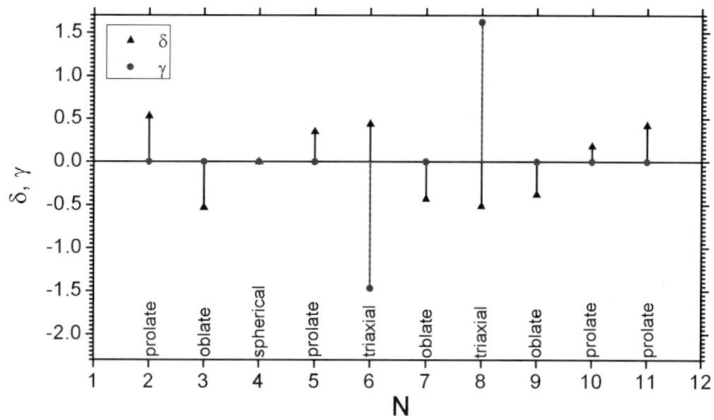

Fig. 3. Cluster deformation parameters versus the cluster size. The labels indicate the cluster deformation type.

In insets to Fig. 1, we present the sum of the oscillator strengths and the excitation energy range considered for each cluster. The sum of the oscillator strengths characterises the valence electrons delocalization rate. Note, that for many clusters it is close to the total number of valence electrons in the system. For some clusters the total sum of the oscillator strengths is significantly smaller than the number of the valence electrons (see, for example, Mg_{10}, Mg_{11}). To increase the sum of the oscillator strengths one has to calculate the photoabsorption spectra up to the higher excitation energies. The calculation of cluster excited states becomes an increasingly difficult problem with the growth of the cluster size, because of the rapid growth of the number of possible excited states in the system. In this paper we focus on the investigation of the plasmon resonances in small Mg clusters, manifesting themselves in the energy range about 4-5 eV as it is clear from our discussion. Therefore, for clusters with $N > 8$, we have calculated the photoabsorption spectra only up to the excitation energies of about 6 eV, which is significant for the elucidation of the plasmon resonance structure and at the same time it does not acquire substantial computer power.

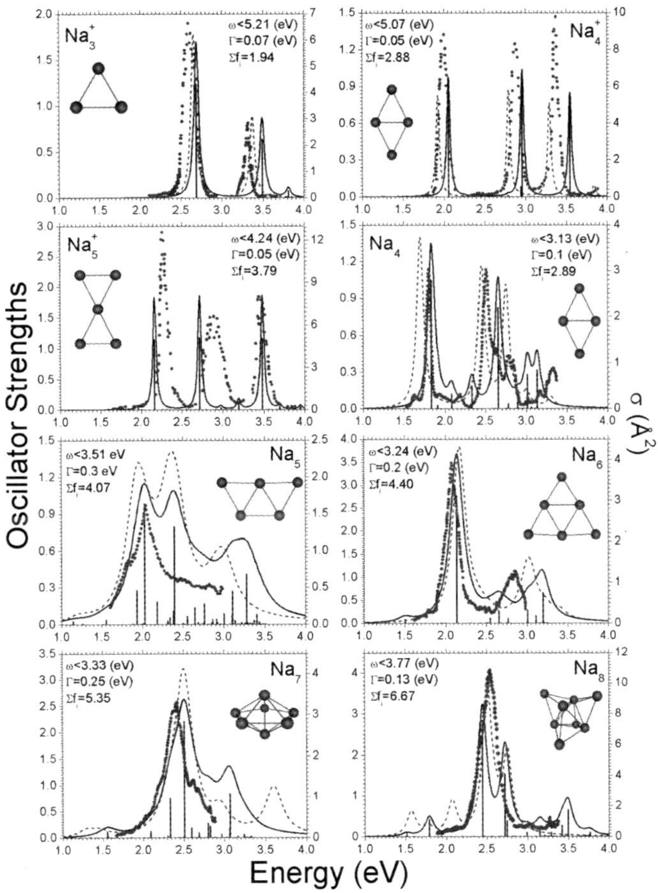

Fig. 4. Photoabsorption cross section calculated for Na^+_{3-5}, Na_{4-8} using the B3LYP functional (solid lines). Vertical lines show the oscillator strengths for the optically allowed transitions. Cluster geometries calculated in Ref.[14] are shown in the insets. The label near each cluster image shows the sum of the oscillator strengths, the excitation energy range considered and the line width. We compare our results with experimentally measured photoabsorption spectra[1,2] (dots) and with the results of previous *ab initio* CI calculation[1,2] (dashed lines).

Photoabsorption spectra for sodium clusters have been earlier investigated in a large number of papers. There were performed experimental measurements, as well as theoretical calculations[1,2] involving *ab initio* and model approaches. In order to check the level of accuracy of our calculation method, in Fig. 4, we compare the photoabsorption spectra for a few selected neutral and singly charged sodium clusters, calculated with the use of the methods described above, with the results of experimental measurements and other calculations. In Fig. 4, the experimentally measured photoabsorption spectra for Na^+_{3-5}, Na_{4-8} are plotted by dots. The results of our TDDFT calculation performed with the use of the B3LYP functional are shown by solid lines. The CI results of Bonačić-Koutecký *et al*[1,2] are shown by dashed lines.

In Ref.[14] we demonstrated that the B3LYP functional is well applicable for the description of sodium clusters. Thus, we used it for the photoabsorption spectra computations. The comparison shown in Fig. 4 demonstrates that our calculation method is a good alternative to the CI method, and our results are in a good agreement with the experimental data.

The photoabsorption spectrum of Na_5 has a prominent peak at the energy about 2.3 eV, which can be identified as a Mie plasmon resonance. This peak is also seen in the photoabsorption spectra of Na_6, Na_7 and Na_8. The plasmon resonance energy for these clusters is smaller than the bulk value, 3.4 eV, because of the spill out effect. As it is seen from Fig. 4, the resonance energy evolves slowly towards the bulk limit with increasing cluster size.

Note, that often the plasmon peaks for sodium clusters are split due to the cluster axial quadrupole deformation. Using equations (3), we have calculated the deformation parameters for axially symmetric Na_6 and Na_7. The result reads as $\delta = -0.55$ and -0.34 respectively. The deformation parameter γ vanishes for both clusters. The axially symmetric jelium model leads to the following values of δ: $\delta_{JM} = -0.48$ and -0.24 for Na_6 and Na_7 respectively.[20] Comparison shows that the splitting of the plasmon resonances can be explained by cluster deformation.

4. Conclusion

In this paper we predict the enhancement of the photoabsorption spectra for small Mg clusters in the vicinity of plasmon resonance. The photoabsorption spectra for neutral Mg clusters consisting of up to $N = 11$ atoms have been calculated using *ab initio* framework based on the time dependent density functional theory. The nature of predicted resonances have been elucidated by comparison of the results of the *ab initio* calculations with the results of the classical Mie theory. The splitting of the plasmon resonances caused by the cluster deformation is analysed. The reliability of the used calculation scheme has been proved by performing the test calculation for a number of sodium clusters and the comparison of the results obtained with the results of other methods and experiment. The calculation of the photoabsorption spectra for larger clusters requires much more computer power and is left open for further investigations.

Acknowledgements

The authors acknowledge support from the Russian Foundation for Basic Research (grant No 03-02-16415-a), Russian Academy of Sciences (grant No 44), the INTAS (grant No 03-51-6170) and the Studienstiftung des deutschen Volkes.

References

1. H. Haberland (ed.), *Clusters of Atoms and Molecules, Theory, Experiment and Clusters of Atoms* Springer Series in Chemical Physics, Berlin 52, (1994).
2. W. Ekardt (ed.), *Metal Clusters* Wiley, New York ,(1999).
3. J.-P. Connerade and A.V. Solov'yov, *Phys. Rev.* A 66, 013207, (2002).
4. W. Klenig, V.O. Nesterenko, P.G. Reinhard and L. Serra, *Eur. Phys. J.* D 4, 343, (1998).
5. M. Moseler, H. Häkkinen and U. Landman, *Phys. Rev. Lett.* 87, 053401, (2001).
6. A. Lyalin, I.A. Solov'yov, A.V. Solov'yov and W. Greiner, *Phys. Rev.* A. 67, 063203, (2003).
7. P.H. Acioli and J. Jellinek, *Phys. Rev. Lett.* 89, 213402, (2002).
8. J. Jellinek and P.H. Acioli, *J. Phys. Chem.* A 106, 10919, (2002).
9. Th. Diederich T. Döppner, J. Braune, J. Tiggesbäumker and K.-H. Meiwes-Broer, *Phys. Rev. Lett.* 86, 4807, (2001).
10. A.D. Becke, *Phys. Rev.* A 38, 30098, (1988).
11. C. Lee, W. Yang and R.G. Parr, *Phys. Rev.* B 37, 785, (1988).
12. K. Burke, J.P. Perdew and Y. Wang, in Electronic Density Functional Theory: Recent Progress and New Directions, Ed. J.F. Dobson, G. Vignale and M.P. Das Plenum, (1998).
13. James B. Foresman and Æleen Frisch *Exploring Chemistry with Electronic Structure Methods* Pittsburgh, PA: Gaussian Inc, (1996).
14. I.A. Solov'yov, A.V. Solov'yov and W. Greiner, *Phys. Rev.* A 65, 053203, (2002).
15. M.J. Frisch and *et al Gaussian 98* (Revision A.9) Gaussian Inc. Pittsburgh PA (1998).

16. U. Kreibig and M. Vollmer, *Optical Properties of Metal Clusters* Springer Series in Materials Science, Berlin 25, (1995).
17. C. Kittel, *Introduction to Solid State Physics*, 7th edn., John Wiley and Sons, New York, (1996).
18. E. Lipparini, S. Stringari, *Z.Phys.D* 18, 193, (1991).
19. A.A. Radzig and B.M. Smirnov, *Parameters of atoms and itomic ions* Energoatomizdat, Moscow, (1986).
20. A. Matveentsev, A. Lyalin, I. Solov'yov, A. Solov'yov and W. Greiner, *Int. J. of Mod. Phys.* E 12, 81, (2003).

MULTIPHOTON EXCITATION OF PLASMONS IN CLUSTERS

A.V. Solov'yov

A. F. Ioffe Physical-Technical Institute, Russian Academy of Sciences, Polytechnicheskaya 26, St. Petersburg 194021, Russia
E-mail: solovyov@th.physik.uni-frankfurt.de

J.-P. Connerade

The Blackett Laboratory, Imperial College London, London SW7 2BW, UK
E-mail: j.connerade@ic.ac.uk

We present a theoretical framework for the multiphoton excitation of plasmons. We show that, in addition to dipole plasmon excitations, multipole plasmons (quadrupole, octupole, etc.) are excited in a metallic cluster by multiphoton absorption processes, resulting in a significant difference between plasmon resonance profiles in multiphoton and single-photon absorption. The method is quite general, and applies to any system with delocalised electrons, of which the simplest are spherical metallic clusters.

1. Introduction

Plasmons are characteristic of systems containing many delocalised electrons. They occur from the quantum to the classical limit. At the quantum end, atoms do not exhibit conspicuous plasmon behaviour, because of the absence of a clear 'surface'. Metallic clusters provide excellent examples of plasmons in quantum systems, appearing for as few as eight atoms. They persist right through to very large cluster sizes, which can be considered as the solid state limit. Metallic clusters allow one to study the evolution of plasmons from quantum to classical regimes.

A feature of plasmons is their presence both in the bulk and on the surface. They possess many oscillatory modes. Dipole excitation from the ground state using a single photon has been the traditional way to explore their spectroscopy, but provides limited information on plasmon dynamics. Our purpose is to demonstrate that much more detail is accessible by multiphoton spectroscopy, and that the full dynamics of the plasmon, by coupling with more than one photon, induces a richer spectrum from which much more information can be gained.

We have developed two simple models, leaving out inessential detail to concentrate on the mechanisms by which multiphoton excitation of metallic clusters occurs. These two models are (i) a quantum and (ii) a classical picture. The first is based on the jellium approximation, in which delocalized electrons are confined within a spherical cluster, and the second treats forced oscillations in the Mie picture. We omit molecular vibrations or phonons, and consider merely collective motions of conduction electrons. This approach brings out essential features common to many systems to which jellium picture can be applied. The main conclusions about multiphoton excitation are similar in the quantum and in the semi-classical limits, so that a smooth transition from one to the other occurs. This theoretical formalism is not confined to photons. It can be used to describe any kind of higher order plasmon excitation processes, for example multiple scattering of electrons within a cluster.

Recently, a number of papers have discussed metallic clusters[1,2] and fullerenes[3] in strong laser fields. Our prime interest is in lower laser powers, for which the integrity of clusters is preserved, and multiphoton excitation just begins to intrude. In our semiclassical model, the collective flow of charge is driven by a periodic field. The results can be related to the multiphoton absorption cross section of the cluster, which takes account of quantum mechanics. In principle we could include the turn-on and turn-off of laser pulses for various power levels and initial charge distributions. However, we concentrate on a novel feature, which arises even for an infinite wavetrain interacting with a cluster (the simplest and most fundamental problem): multiple plasmon excitations driven by multi-photon excitations.

Surface plasmons are well-known in atomic clusters. Dipole surface plasmons are responsible for the formation of giant resonances in photoabsorption spectra of metal clusters.[4–7])

They determine inelastic collisions of charged particles with metal clusters,[7] where it was demonstrated that collective excitations contribute to the electron energy loss spectrum near the surface plasmon resonance. In the energy range above the ionization threshold, volume plasmons dominate the differential cross section, resulting in resonance behaviour.[7] The role of the polarization interaction and of plasmon excitations in electron attachment to metal clusters has been examined both theoretically[7] and experimentally.[8] Plasmon excitations induce resonance enhancement of the electron attachment cross section.

2. Plasmon Resonance Approximation

Our quantum analysis is based on the plasmon resonance approximation. We consider the simplest example, namely the cross section for single-photon absorption, viz:

$$\sigma_1 = \frac{4\pi^2 e^2}{c}\omega \sum_n |z_{on}|^2 \delta(\omega_{no} - \hbar\omega). \tag{1}$$

Here, e is the charge of electron, c the velocity of light, $\hbar$ is Planck's constant, $\omega_{no} = \varepsilon_n - \varepsilon_0$ the electronic excitation energy, ω the photon frequency and ez_{on} the matrix element of the z- component of the cluster dipole moment. The summation includes all final states of the excited electron, belonging to both discrete and continuous spectra.

In the jellium picture, which holds for metal clusters and to some extent for fullerenes, the main contribution to the cross section (1) arises from a small group of excited states or sometimes even a single transition of frequency close to the classical Mie resonance – also known as the plasmon resonance. For a spherical metal cluster, it is given by $\omega_l^2 = \frac{4\pi N e^2}{mV} \cdot \frac{l}{(2l+1)}$ (see e.g. Refs. 4, 5, 7). Here $V = 4\pi R^3/3$ is the cluster volume, $R = r_o N^{1/3}$ the cluster radius, r_o the Wigner-Seitz radius, N the number of delocalized electrons, l the angular momentum of the plasmon mode, m the electron's mass. Note that a single $1 - 4eV$ photon can, in practice, excite only $l = 1$ dipole plasmon oscillations.

The plasmon resonance approximation assumes that excitations near a plasmon resonance exhaust the sum rule almost completely,[4–7]) which means that $\sum_n \omega_{no}|z_{on}|^2 = \frac{N\hbar^2}{2m}$ (see e.g. Ref.[9]).

Assuming a Lorentzian distribution of width Γ_1 for the plasmon resonance states and replacing the delta function in (1) by the profile,[9] one recovers the well-known expression for the single photon absorption cross section[4,7]

$$\sigma_1 = \frac{\pi N e^2}{mc} \frac{\Gamma_1}{(\omega_1 - \omega)^2 + \frac{\Gamma_1^2}{4}} \approx \frac{4\pi N e^2}{mc} \frac{\omega^2 \Gamma_1}{(\omega_1^2 - \omega^2)^2 + \omega^2 \Gamma_1^2}. \tag{2}$$

The width Γ_1 is due to Landau damping. Its calculation is performed in Ref. 7. The cross section (2) reproduces plasmon resonances in single-photon absorption spectra of metal clusters. In the dipole approximation, the two-photon cross section is

$$\sigma_2 = \frac{32\pi^3 e^4 \hbar}{c^2}\omega^2 \sum_n \left| \sum_m \frac{z_{nm} z_{mo}}{\hbar\omega - \omega_{mo} + i\delta} \right|^2 \delta(\omega_{no} - 2\hbar\omega). \qquad (3)$$

We evaluate it in the same way as for the single-photon case. The main contribution to the sum over intermediate states m arises from virtual dipole plasmon excitations. With the use of the sum rule, it reduces to $|r_{10}|^2 \approx N\hbar/2m\omega_1$. The remaining matrix elements z_{n1} in (3) describe dipole transitions from the dipole plasmon resonance state to other excited states. Thus, for the final state, $l = 0$ or $l = 2$ only. However, there is no surface plasmon excitation with $l = 0$. Thus, only transitions to states with $l = 2$ are of interest. These arguments show that, by using two photons simultaneously, one can excite the quadrupole plasmon resonance in a metal cluster or in a fullerene. When calculating the cross section (3) near the quadrupole plasmon resonance excitation, i.e. at $2\omega \sim \omega_2$, we need consider only transitions to the resonance final state, i.e. put $\sum_n |z_{n1}|^2 \approx |z_{21}|^2$ (we use indices 1 and 2 for dipole and quadrupole plasmon resonance states) and replace the delta function, $\delta(\omega_{no} - 2\hbar\omega)$ by a Lorentzian distribution of width Γ_2, as for the single photon case.

The transition matrix element z_{21} describes the electronic transition between dipole and quadrupole plasmon resonance states. This is a collective transition which does not saturate the sum rule. Therefore, calculation of z_{21} on the basis of the sum rule overestimates this matrix element. Instead, we have calculated it by assuming that transition electron densities of both dipole and quadrupole plasmon modes are localised in a narrow layer of width ΔR near the cluster surface. This assumption is consistent with *ab initio* transition density analysis in Ref. 7 for the Na_{40} and Na_{92} clusters. For details, we refer to our paper[10] and merely quote the final result $z_{21} = -\frac{8}{3}\left(\frac{6}{5}\right)^{1/4} \frac{\hbar}{m\omega_1 \Delta R}$.

Finally, we obtain the two-photon absorption cross section:

$$\sigma_2 = \frac{8\pi^2 A^2 N e^4 \hbar}{m^3 c^2 \Delta R^2 \omega_1^3} \frac{\omega^2}{(\omega - \omega_1)^2 + \frac{\Gamma_1^2}{4}} \cdot \frac{\Gamma_2}{(\omega_2 - 2\omega)^2 + \frac{\Gamma_2^2}{4}} \qquad (4)$$

where $A = \frac{8}{3}\left(\frac{6}{5}\right)^{1/4} \approx 2.79$.

In Fig. 1 we plot cross section profiles per unit atom for single-photon (dashed line) and two-photon (solid line) absorption calculated according to (2) and (4). These profiles do not depend on the number of atoms in the cluster.

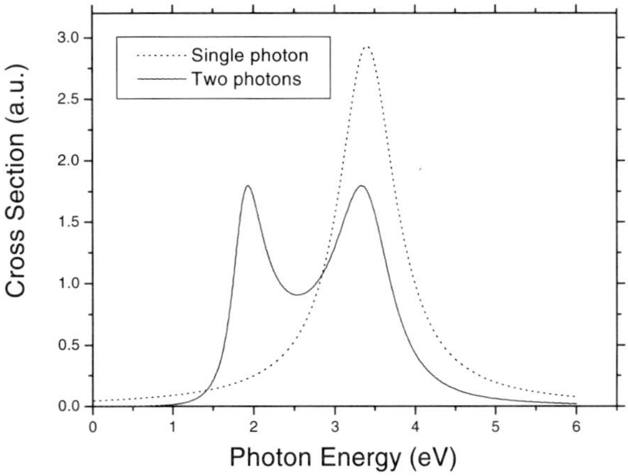

Fig. 1. The profiles of single-photon (dotted line) and two-photon (solid line) absorption calculated according to (2) and (4) and normalised per unit atom. The two-photon absorption profile is scaled by a factor 1/100. The scales are not identical for the two curves for reasons of definition of the cross sections in the single- and two-photon cases, but both are given in atomic units.

The peak in the single-photon plot gives the location of the dipole resonance. The other peak in the two-photon plot is the quadrupole resonance. This figure illustrates the contribution due to quadrupole plasmon excitation in the two-photon spectrum. In this calculation we used $r_0 = 4.0$ and $\Gamma_1 = \omega_1/4$, $\Gamma_2 = \omega_2/4$, $\Delta R = r_0$. The choice can be different for different clusters, but should always lead to qualitatively similar single- and two-photon absorption profiles. Accurate parameters are obtained only on the basis of *ab initio* calculations.

Our formalism can be used for a larger number of photons. In this case the cross section can be analysed in a similar way. This leads to the conclusion that plasmon resonances with larger angular momenta (octupole, etc) can be excited.

3. Hydrodynamic Description

We now turn to the semiclassical or hydrodynamic picture, i.e. the opposite extreme to the quantum model. A classical description of the electron density variation $\rho(\mathbf{r}, t)$ is appropriate when plasmon excitations dominate over the single-particle spectrum, since plasmons are essentially classical. We describe collective motion of $\rho(\mathbf{r},t)$ using Euler's equation and the equation of continuity.

Euler's equation couples the acceleration of the density $d\mathbf{v}/dt$ with the total local electric field $\mathbf{E}$ at the point $(\mathbf{r},t)$. It has the form:

$$\frac{\partial \mathbf{v}(\mathbf{r},t)}{\partial t} + \{\mathbf{v}(\mathbf{r},t)\cdot\nabla\}\mathbf{v}(\mathbf{r},t) = -\frac{e}{m}\nabla\varphi(\mathbf{r},t) - \frac{e}{m}\nabla\int d\mathbf{r}'\frac{\delta\rho(\mathbf{r}',t)}{|\mathbf{r}-\mathbf{r}'|}. \quad (5)$$

Here $\varphi(\mathbf{r},t)$ is the potential of the external field. The density $\rho(\mathbf{r},t) = \rho_o(\mathbf{r}) + \delta\rho(\mathbf{r},t)$, where $\rho_o(\mathbf{r})$ is the density in a free cluster without external fields and $\delta\rho(\mathbf{r},t)$ is the variation of density caused by the external field and polarization force acting together.

The motion of $\rho(\mathbf{r},t)$ obeys the equation of continuity:

$$\frac{\partial \rho(\mathbf{r},t)}{\partial t} + \nabla\cdot\{\rho(\mathbf{r},t)\mathbf{v}(\mathbf{r},t)\} = 0. \quad (6)$$

The simultaneous solution of (5), and (6) with appropriate initial conditions, and the initial distribution $\rho_o(\mathbf{r})$ determine $\delta\rho(\mathbf{r},t)$ as well as its velocity $\mathbf{v}(\mathbf{r},t)$. The full solution of this problem is given in our paper.[10] By solving a set of equations with $\varphi(\mathbf{r})$ describing the dipole electron-photon interaction up to n-th order, one can calculate $\rho(\mathbf{r},t)$ due to the field of n photons. The resulting equations describe volume and surface eigen-oscillations of $\rho(\mathbf{r},t)$ characterised by l. The surface plasmon resonance frequency ω_l is the same as above. The volume plasmon resonance frequency is $\omega_p = \sqrt{\frac{4\pi e^2 N}{mV}}$. Equations for the volume and surface plasmon oscillations can be separated from each other and solved iteratively. The solutions $\delta\rho_{lm}^{v(n)}(r)$ (for volume plasmons) and $\delta\rho_{lm}^{s(n)}$ (for surface plasmons) can be found for arbitrary large order of perturbation theory n, although the formulae become more and more tedious with increasing n. They demonstrate that, in higher orders, plasmon resonances with angular momenta larger than the angular momentum of the external field can be excited. The equations reveal a significant shift of the plasmon resonance profiles in the highest orders towards lower frequencies.

We have also applied our theory to the description of multiphoton absorption. We focus on the analysis of plasmon excitations. If surface or volume plasmon resonances are excited by photons, i.e. $\omega \sim \omega_p$, the following condition is fulfilled $\omega R/c \sim \omega_p R/c \ll 1$, which implies the dipole approximation. In this limit, volume plasmon excitations play a negligible role, resulting in a considerable simplification. In this case, one derives

$$\left((\omega n)^2 - \omega_l^2\right)\delta\rho_{l,m}^{s(n)} = -\sqrt{\frac{4\pi}{3}}\frac{Ne^2E}{mV}\delta_{n,1}\delta_{l,1}\delta_{m,0} +$$

$$+\sqrt{\frac{4\pi}{3}}\frac{eE}{mR}\sum_{l_1,m_1} I_2(l,m|l_1,m_1,|1,0)\delta\rho^{s(n-1)}_{l_1,m_1}-$$

$$-\frac{4\pi eE}{mR}\sum_{j_1=1}^{n-1}\frac{1}{n-j}\sum_{\substack{l_1,m_1\\ l_2,m_2}} I_2(l,m|l_1,m_1,|l_2,m_2)\frac{\delta\rho^{s(j)}_{l_1,m_1}\delta\rho^{s(n-j)}_{l_2,m_2}}{2l_2+1} \quad (7)$$

where $E = \sqrt{2\pi\hbar\omega/V_o}$ is the strength of the linearly polarized electric field of the photon and V_o is the normalization volume of the photon mode. The angular integral is $I_2(l,m|l_1,m_1|1,0) = \sqrt{2l_1(l_1+1))}\int d\Omega_{\mathbf{n}_r}Y^*_{l,m}(\mathbf{n}_r)\mathbf{Y}^{(1)}_{l_1,m_1}(\mathbf{n}_r)\mathbf{Y}^{(1)}_{1,0}(\mathbf{n}_r)$.

For details, we refer to Ref. 10. Equation (7) should be solved iteratively starting from $n = 1$. For $n = 1$, 2, the non-trivial solutions read as $\delta\rho^{s(1)}_{1,0} = -\sqrt{\frac{4\pi}{3}}\frac{Ne^2E}{mV(\omega^2-\omega_1^2)}$, $\delta\rho^{s(2)}_{0,0} = -\frac{\pi^{1/2}}{3m^2RV}\cdot\frac{Ne^3E^2}{(\omega^2-\omega_1^2)^2}$, $\delta\rho^{s(2)}_{2,0} = -\frac{4\pi^{1/2}}{3\sqrt{5}m^2RV}\cdot\frac{Ne^3E^2\omega^2}{(\omega^2-\omega_1^2)^2((2\omega)^2-\omega_2^2)}$

We now calculate multipole moments of the cluster induced by an external radiation field: $Q^{(n)}_{l,m} = \sqrt{\frac{4\pi}{2l+1}}R^{l+2}\delta\rho^{s(n)}_{l,m}$. Substituting here $\delta\rho^{s(n)}_{1,0}$, one obtains the dipole moment of the cluster, $D^{(1)}(\omega) = Q^{(1)}_{1,0}$ induced in the single-photon absorption process

$$D^{(1)}(\omega) = -\frac{Ne^2E}{m(\omega^2-\omega_1^2+i\omega\Gamma_1)}. \quad (8)$$

Subsituting $\delta\rho^{s(2)}_{2,0}$, into $Q^{(n)}_{l,m}$, one finds the quadrupole moment of the cluster, $Q^{(2)}(\omega) = Q^{(2)}_{2,0}$ induced in the two-photon regime:

$$Q^{(2)}(\omega) = Q^{(2)}_{2,0} = -\frac{2}{5}\frac{Ne^3E^2\omega^2}{m^2((\omega^2-\omega_1^2)^2+\omega^2\Gamma_1^2)((2\omega)^2-\omega_2^2+i2\omega\Gamma_2)}. \quad (9)$$

Here, we have introduced the plasmon resonance widths Γ_1 and Γ_2 which take into account Landau damping of the dipole and quadrupole surface plasmon resonances. They must be determined separately, e.g. by an *ab initio* approach.[7]

Three photons can induce dipole and octupole moments in the cluster, and we have derived explicit expressions for them.[10] They demonstrate that multipole moments induced in the cluster by multiphoton absorption processes possess a prominent plasmon resonance structure. The connection between $D^{(1)}(\omega)$ from (8) and the cross section σ_1 found in (2) is straightforward: $\sigma_1 = \frac{4\pi\omega}{cE}ImD^{(1)}(\omega)$.

Acknowledgments

The authors acknowledge support from the Royal Society of London, Russian Foundation for Basic Research (grant No 03-02-16415-a), Russian Academy of Sciences (grant No 44) and INTAS (grant No 03-51-6170).

References

1. L. Köller, M. Schumacher, J. Köhn, Tiggesbäumker, and K.H.Meiwes-Broer, *Phys. Rev. Lett* 82, 3783 (1999).
2. C.A. Ullrich, P.-G. Reinhard and E. Suraud, *J. Phys. B* 30, 5043 (1997).
3. S. Hunsche, T. Starczewski, A. l'Huillier, A. Persson, A. Wahlström, C-G. van Linden, B. van der Heuvell, and S. Svanberg, *Phys. Rev. Lett.* 77, 1966 (1996).
4. W. A. de Heer, *Rev.Mod.Phys.* 65, 611 (1993).
5. C. Brechignac and J.P.Connerade, *J.Phys.B: At. Mol. Opt. Phys.* 27, 3795 (1994).
6. V.K.Ivanov, G.Yu.Kashenock, R.G.Polozkov and A.V.Solov'yov, *J.Phys.B: At.Mol.Opt.Phys.* 34, 669 (2001).
7. A.V. Solov'yov, in *NATO Advanced Study Institute*, Les Houches, Session LXXIII, Summer School "Atomic Clusters and Nanoparticles", Edited by C.Guet, P.Hobza, F.Spiegelman and F.David, EDP Sciences and Springer Verlag, Berlin, Heidelberg, New York (2001).
8. S.Sentürk, J.P.Connerade, D.D.Burgess and N.J.Mason, *J.Phys.B: At.Mol.Opt.Phys.* 33, 2763 (2000).
9. L.D. Landau and E.M. Lifshitz, *Quantim Mechanics*, Pergamon, London (1965).
10. J.P. Connerade, A.V. Solov'yov, *Phys.Rev. A* 66, 013207 (2002).

Fission and Fusion Dynamics of Clusters

EXOTIC FISSION PROCESSES IN NUCLEAR PHYSICS

Walter Greiner
Institut für Theoretische Physik der Johann-Wolfgang Goethe Universität, Robert-Mayer Str. 8-10, D-60054 Frankfurt am Main, Germany
E-mail: greiner@th.physik.uni-frankfurt.de

Thomas J. Bürvenich
Theoretical Division, Los Alamos National Laboratory, Los Alamos, New Mexico 87544, USA
E-mail: tbuerven@lanl.gov

The extension of the periodic system into various new areas is investigated. Experiments for the synthesis of superheavy elements and the predictions of magic numbers with modern meson field theories are reviewed. Further on, different channels of nuclear decay are discussed including cluster radioactivity, cold fission and cold multifragmentation. A perspective for future research is given.

1. Introduction

The elements existing in nature are ordered according to their atomic (chemical) properties in the periodic system which was developed by Mendeleev and Lothar Meyer. The heaviest element of natural origin is Uranium. Its nucleus is composed of $Z = 92$ protons and a certain number of neutrons ($N = 128 - 150$). They are called the different Uranium isotopes. The transuranium elements reach from Neptunium ($Z = 93$) via Californium ($Z = 98$) and Fermium ($Z = 100$) up to Lawrencium ($Z = 103$). The heavier the elements are, the larger are their radii and their number of protons. Thus, the Coulomb repulsion in their interior increases, and they undergo fission. In other words: the transuranium elements become more instable as they get bigger.

In the late sixties the dream of the superheavy elements arose. Theoret-

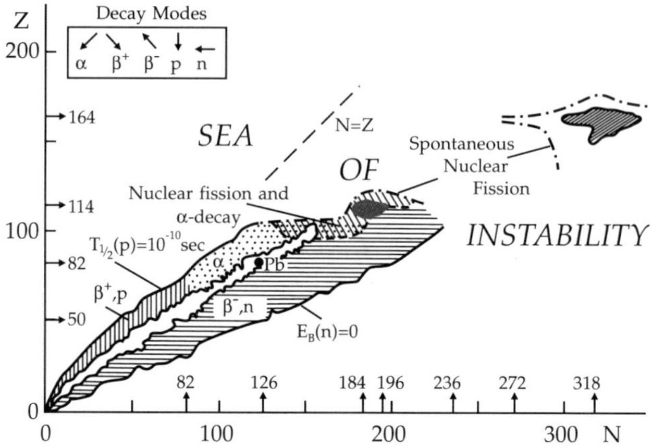

Fig. 1. The periodic system of elements as conceived by the Frankfurt school in the late sixties. The islands of superheavy elements ($Z = 114$, $N = 184$, 196 and $Z = 164$, $N = 318$) are shown as dark hatched areas.

ical nuclear physicists around S.G. Nilsson (Lund)[1] and from the Frankfurt school[2–4] predicted that so-called closed proton and neutron shells should counteract the repelling Coulomb forces. Atomic nuclei with these special "magic" proton and neutron numbers and their neighbours could again be rather stable. These magic proton (Z) and neutron (N) numbers were thought to be $Z = 114$ and $N = 184$ or 196. Typical predictions of their life times varied between seconds and many thousand years. Figure 1 summarizes the expectations at the time. One can see the islands of superheavy elements around $Z = 114$, $N = 184$ and 196, respectively, and the one around $Z = 164$, $N = 318$. The important question was how to produce these superheavy nuclei. There were many attempts, but only little progress was made. It was not until the middle of the seventies that the Frankfurt school of theoretical physics together with foreign guests (R.K. Gupta (India), A. Sandulescu (Romania))[6] theoretically understood and substantiated the concept of bombarding of double magic lead nuclei with suitable projectiles, which had been proposed intuitively by the russian nuclear physicist Y. Oganessian.[7] The two-center shell model, which is essential for the description of fission, fusion and nuclear molecules, was developed in 1969-1972 together with U. Mosel and J. Maruhn.[8] It showed that the shell structure of the two final fragments was visible far beyond the barrier into the fusioning nucleus. The collective potential energy surfaces of heavy nuclei, as they were calculated in the framework of the two-center shell model, exhibit pronounced valleys, such that these valleys provide promising doorways to the fusion of superheavy nuclei for certain projectile-target combinations (Fig. 11). If projectile and target approach each other through those "cold" valleys, they get only minimally excited and the barrier which has to be overcome (fusion barrier) is lowest (as compared to neighbouring projectile-target combinations).

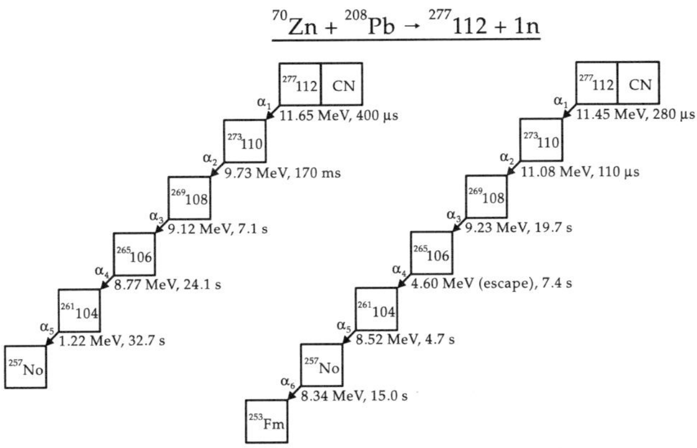

Fig. 2. The fusion of element 112 with ^{70}Zn as projectile and ^{208}Pb as target nucleus has been accomplished for the first time in 1995/96 by S. Hofmann, G. Münzenberg and their collaborators. The colliding nuclei determine an entrance to a "cold valley" as predicted as early as 1976 by Gupta, Sandulescu and Greiner. The fused nucleus 112 decays successively via α emission until finally the quasi-stable nucleus ^{253}Fm is reached. The α particles as well as the final nucleus have been observed. Combined, this renders the definite proof of the existence of a $Z = 112$ nucleus.

2. Cold Valleys in the Potential

In this way the correct projectile- and target-combinations for fusion were predicted. Indeed, Gottfried Münzenberg and Sigurd Hofmann and their group at GSI[9] have followed this approach. With the help of the SHIP mass-separator and the position sensitive detectors, which were especially developped by them, they produced the pre-superheavy elements $Z = 106$, $107, \ldots 112$, each of them with the theoretically predicted projectile-target combinations, and only with these. Everything else failed. This is an impressive success, which crowned the laborious construction work of many years. The prior example of this success, the discovery of element 112 and its long

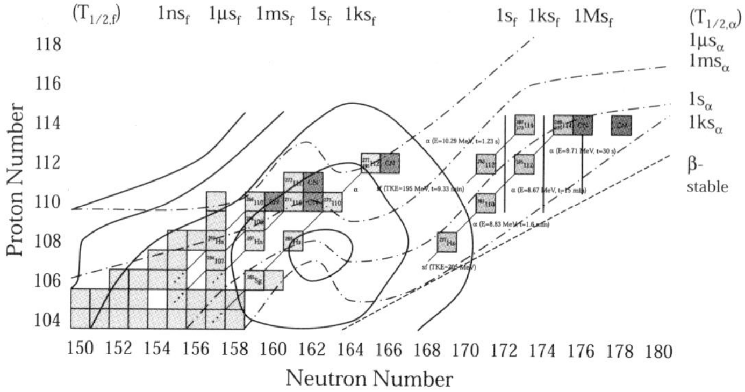

Fig. 3. The $Z = 106 - 112$ isotopes were fused by the Hofmann–Münzenberg (GSI)–group. The two $Z = 114$ isotopes and the $Z = 116$ isotope were produced by the Dubna–Livermore group. It is claimed that three neutrons are evaporated. Obviously the lifetimes of the various decay products are rather long (because they are closer to the stable valley), in crude agreement with early predictions[3,4] and in excellent agreement with the recent calculations of the Sobicevsky–group.[12]

α-decay chain, is shown in Fig. 2. Very recently the Dubna–Livermore–group produced two isotopes of $Z = 114$ element by bombarding ^{244}Pu with ^{48}Ca and also $Z = 116$ by ^{48}Ca + ^{248}C m. (Fig. 3). Furthermore these are cold–valley reactions (in this case due to the combination of a spherical and a deformed nucleus), as predicted by Gupta, Sandulescu and Greiner[10] in 1977. There exist also cold valleys for which both fragments are deformed,[11] but these have yet not been verified experimentally.

3. Shell Structure in the Superheavy Region

Studies of the shell structure of superheavy elements in the framework of the meson field theory and the Skyrme-Hartree-Fock approach have recently shown that the magic shells in the superheavy region are very isotope dependent[5,14] (see Fig. 4). According to these investigations $Z = 120$ being a magic proton number seems to be as probable as $Z = 114$. Additionally, recent investigations in a chirally symmetric mean–field theory result also in the prediction of these two magic numbers.[22,23] The corresponding magic neutron numbers are predicted to be $N = 172$ and as it seems to a lesser extend $N = 184$. Thus, this region provides an open field of research. R.A. Gherghescu et al. have calculated the potential energy surface of the $Z = 120$ nucleus. It utilizes interesting isomeric and valley structures (Fig. 5).

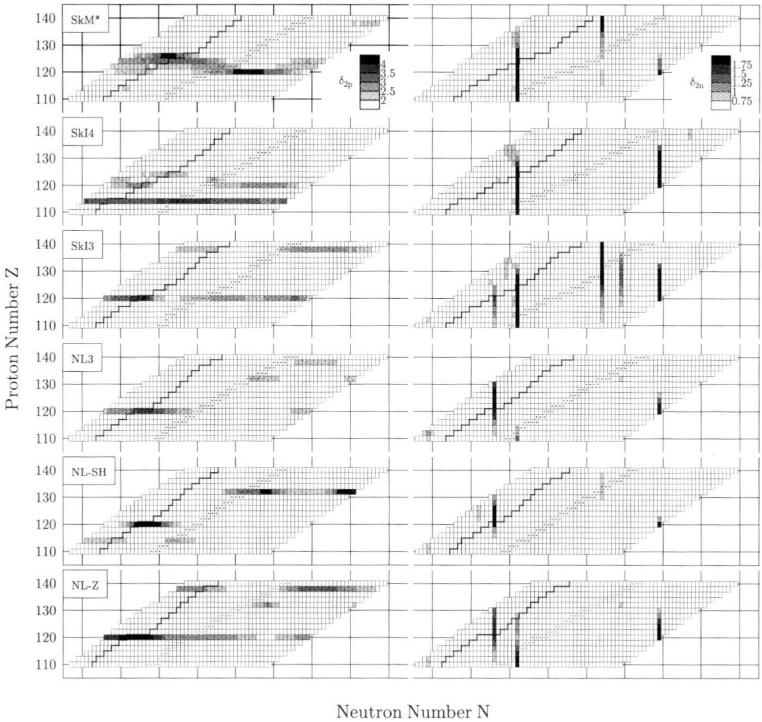

Fig. 4. Grey scale plots of proton gaps (left column) and neutron gaps (right column) in the N-Z plane for spherical calculations with the forces as indicated. The assignment of scales differs for protons and neutrons, see the uppermost boxes where the scales are indicated in units of MeV. Nuclei that are stable with respect to β decay and the two-proton dripline are emphasized. The forces with parameter sets SkI4 and NL-Z reproduce the binding energy of ${}^{264}_{156}108$ (Hassium) best, i.e. $|\delta E/E| < 0.0024$. Thus one might assume that these parameter sets could give the best predictions for the superheavies. Nevertheless, it is noticed that NL-Z predicts only $Z = 120$ as a magic number while SkI4 predicts both $Z = 114$ and $Z = 120$ as magic numbers. The magicity depends — sometimes quite strongly — on the neutron number. These studies are those of Bender, Rutz, Bürvenich, Maruhn, P.G. Reinhard *et al.*.[14]

The charge distribution of the $Z = 120, N = 172$ nucleus, calulated with mean-field models, indicates a hollow inside. This leads us to suggest that a system with 120 protons and 180 neutrons might essentially be a fullerene consisting of 60 α-particles and one additional binding neutron per alpha. This is illustrated in Fig. 6. The protons and neutrons of such a superheavy nucleus are distributed over 60 α particles and 60 neutrons.

Such an object could be expected to have interesting decay modes such as multifragmentation, spitting out many α particles. The possible condensation of α particles in light nuclei (in ground-states and exctited states) is a modern topic. It would be fascinating if such condensation could occur also in these superheavy systems. Figure 7 depicts this scenario of a nuclear fullerene structure built of α particles.

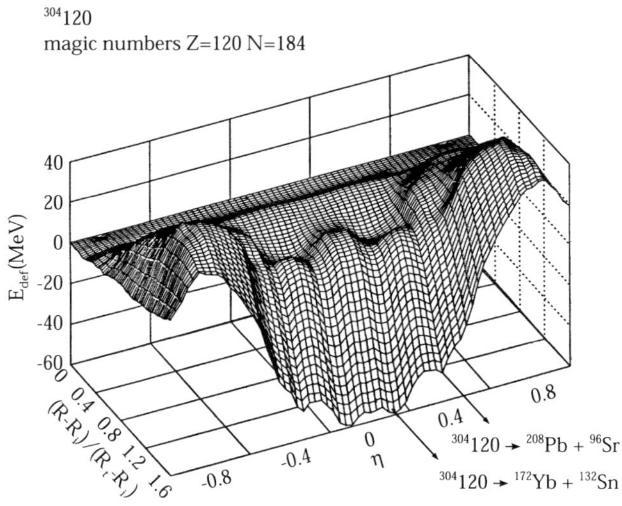

Fig. 5. Potential energy surface as a function of reduced elongation $(R - R_i)/(R_t - R_i)$ and mass asymmetry η for the double magic nucleus $^{304}120$. $^{304}120_{184}$.

The potential energy surfaces of superheavy elements, as they emerge from selfconsistent calculations within mean-field models in axial symmetry, exhibit some interesting features,[15] see Fig. 8 for the RMF force NL-Z2 and the Skyrme force SLy6. Nuclei in the vicinity of the nucleus with $Z = 108$ protons and $Z = 162$ have prolate ground-states and barriers. Going upward in proton and neutron numbers, one encounters transitional systems with two shallow minima, one on the oblate, one on the prolate side. Nuclei with proton numbers $Z = 120$ and neutron numbers $N = 178...184$ exhibit no pronounced deformation. Mean-field forces predict either a clear spherical shape or a rather soft potential energy surface around zero deformation with small wiggles. For these nuclei, however, triaxial degrees of freedom might become important and change the picture considerably.

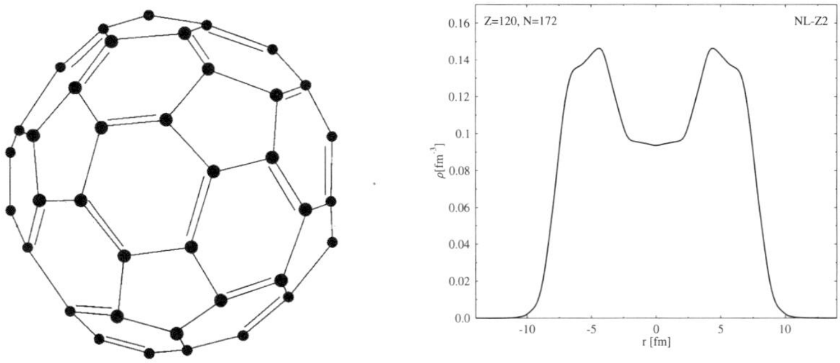

Fig. 6. Typical structure of the fullerene C_{60}. The double bindings are illustrated by double lines. In the nuclear case the carbon atoms are replaced by α particles and the double bindings by the additional neutrons. Such a structure would immediately explain the semi–hollowness of that superheavy nucleus, which is revealed in the mean–field calculations within meson–field theories. The radial density of the nucleus with 120 protons and 172 neutrons, as emerging from a meson-field calculation with the force NL-Z2 is shown on the right side. Note that the semi-bubble structure is mostly pronounced for this nucleus. When going to higher neutron numbers, this structures becomes less and less pronounced.

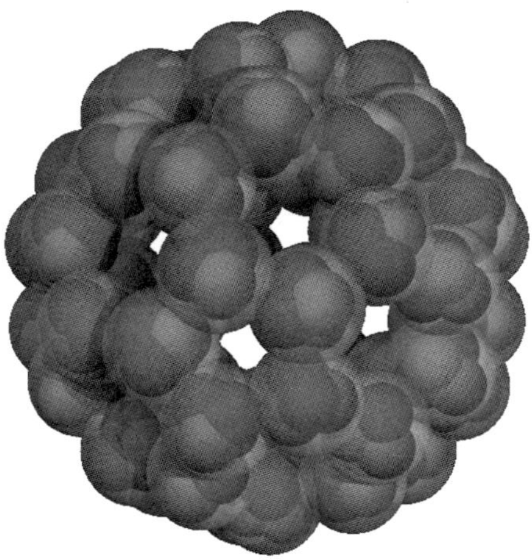

Fig. 7. An artists view of the nuclear fullerene structure that might occur for the superheavy nucleus $^{300}120_{180}$ (picture courtesy of Henning Weber).

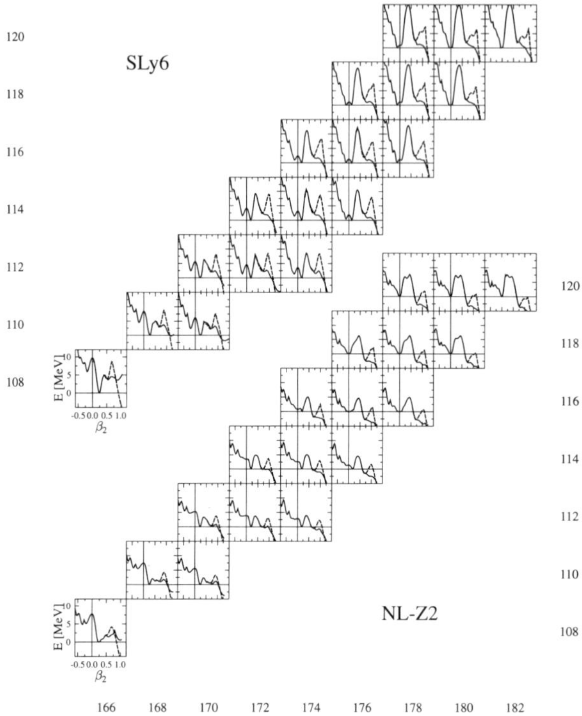

Fig. 8. Axial fission barriers for the Skyrme force SLy6 (top) and the relativistic force NL-Z2 (bottom). Solid (dashed) lines denote the reflection-asymmetric (reflection-symmetric) path.

The barriers correspond to a simple-humped structure for almost all forces. Isomeric states appear in the reflection-symmetric solutions but disappear when allowing for shapes including odd multipole moments. Globally, barriers calculated with Skyrme-forces appear to be up as twice as heigh as the ones emerging from RMF calculations (see Fig. 9 for a comparison). These trends become visible already in actinide nuclei, though they are much stronger in the extrapolations to superheavy elements. This effect has already been seen in former studies.[16] It indicates the need for a deeper understanding of these self -consistent approaches.

One might further ask how collective motions of these spherical superheavy elements might look like. We will take a first look at these aspects in the following section.

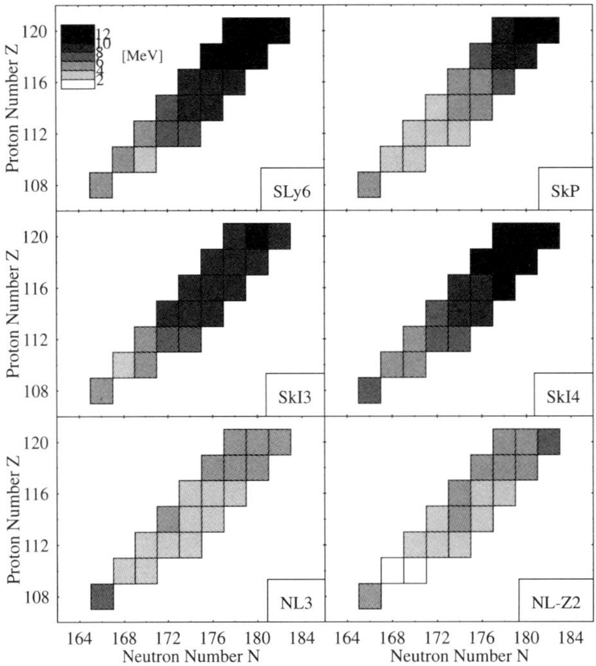

Fig. 9. The height of the symmetric (first) barrier calculated in axial symmetry for the models and forces as indicated.

4. Vibrational Modes in Spherical Superheavy Nuclei

We consider vibrational collective properties of the putative double magic SH nucleus $^{292}120$ as predicted by the RMF axial-symmetric model and compare them with those of the well-known double magic heavy nucleus ^{208}Pb.[17] As one can see in Fig. 10, the nucleus ^{208}Pb has a pronounced harmonic behaviour, at least for the three vibrational states, i.e. 0^+, 2^+ and the triplet $0^+, 2^+, 4^+$. In contrast, the SHE $^{292}120$, computed also with the force NL-Z2, exhibits a clear prolate-oblate asymmetry and consequently the sequence of states follows a non-equidistant behaviour. This result was expected because the SHE are less stable (calculations give barriers up to 5 times smaller when the first symmetric barrier of $^{292}120$ is compared with that of ^{208}Pb). Therefore the departure of the deformation energy curve from the harmonic oscillator well will be larger.

It is important to stress that in view of the width and height of the potential well in the β_2-coordinate, no more than two phonon states exist.

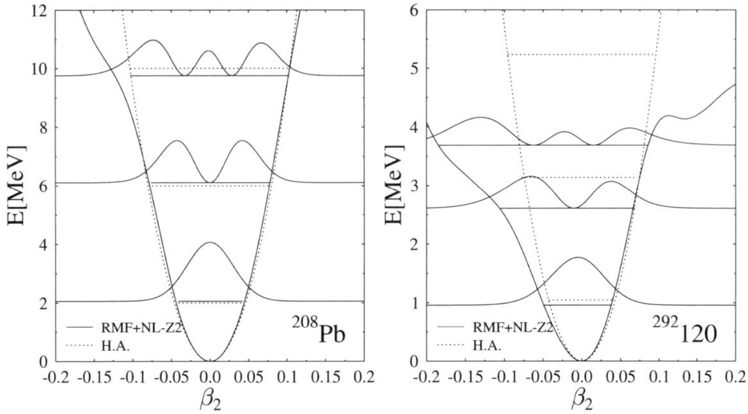

Fig. 10. Potential well and first three vibrational states of the potential, calculated in the frame of the RMF model with NL-Z2 force (RMF+NL-Z2) and in the Harmonic approximation (HA) for two nuclei. The wave functions of the states are also shown. The left panel represents the case of ^{208}Pb where the harmonic approximation works quite well. The right panel shows the putative double-magic nucleus $^{292}120$ for which the anharmonic distortions in the potential are inducing a sensitive departure of the collective level spacing from the equidistant harmonic behaviour.

Clearly, the future observation of such β-vibrational states will yield further useful information about the structure of these nuclei. Also the sensitivity of this structure to the underlying effective forces is interesting.

5. Asymmetric and Superasymmetric Fission–cluster Radioactivity

The potential energy surfaces, which are shown prototypically for $Z = 114$ in Fig 11, contain even more remarkable information: if a given nucleus, e. g. Uranium, undergoes fission, it moves in its potential mountains from the interior to the outside. Of course, this happens quantum mechanically. The wave function of such a nucleus, which decays by tunneling through the barrier, has maxima where the potential is minimal and minima where it has maxima.

The probability for finding a certain mass asymmetry $\eta = \dfrac{A_1 - A_2}{A_1 + A_2}$ of the fission is proportional to $\psi^*(\eta)\psi(\eta)d\eta$. Generally, this is complemented by a coordinate dependent scale factor for the volume element in this (curved) space. Now it becomes clear how the so-called asymmetric and superasymmetric fission processes come into being. They result from

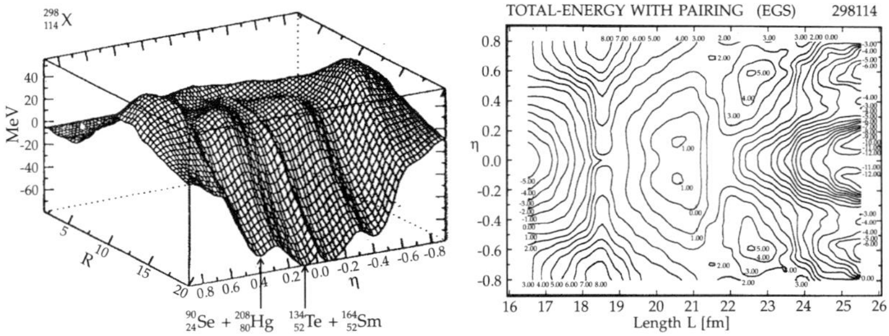

Fig. 11. The collective potential energy surface of $^{184}114$, calculated within the two center shell model by J. Maruhn et al., shows clearly the cold valleys which reach up to the barrier and beyond. Here R is the distance between the fragments and $\eta = \dfrac{A_1 - A_2}{A_1 + A_2}$ denotes the mass asymmetry: $\eta = 0$ corresponds to a symmetric, $\eta = \pm 1$ to an extremely asymmetric division of the nucleus into projectile and target. If projectile and target approach through a cold valley, they do not "constantly slide off" as it would be the case if they approach along the slopes at the sides of the valley. Constant sliding causes heating, so that the compound nucleus heats up and becomes unstable. In the cold valley, on the other hand, the created heat is minimized. Colleagues from Freiburg should be familiar with that: they approach Titisee (in the Black Forest) most elegantly through the Höllental and not by climbing its slopes along the sides.

the enhancement of the collective wave function in the cold valleys. And that is indeed what one observes. For large mass asymmetry ($\eta \approx 0.8,\ 0.9$) there exist very narrow valleys. They are not as clearly visible in Fig. 11, but they have interesting consequences. Through these narrow valleys nuclei can emit spontaneously not only α-particles (Helium nuclei) but also ^{14}C, ^{20}O, ^{24}Ne, ^{28}Mg, and other nuclei. Thus, we are lead to the cluster radioactivity (Poenaru, Sandulescu, Greiner[18]).

By now this process has been verified experimentally by research groups in Oxford, Moscow, Berkeley, Milan and other places. Accordingly, one has to revise what is learned in school: there are not only 3 types of radioactivity (α-, β-, γ-radioactivity), but many more. Atomic nuclei can also decay through spontaneous cluster emission (that is the "spitting out" of smaller nuclei like carbon, oxygen,...). Figure 12 depicts some examples of these processes.

The knowledge of the collective potential energy surface and the collective masses $B_{ij}(R,\eta)$, all calculated within the Two-Center-Shell-Modell (TCSM), allowed H. Klein, D. Schnabel and J. A. Maruhn to calculate lifetimes against fission in an "*ab initio*" way.[19] The discussion of much more very interesting new physics cannot be persued here. We refer to Refs.

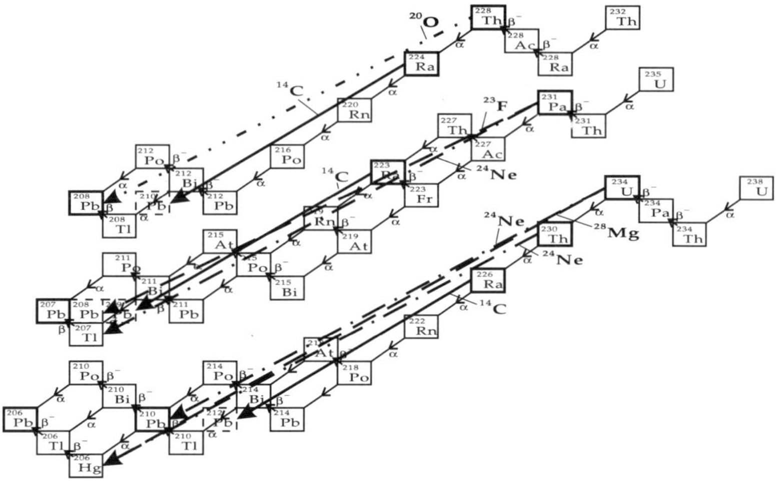

Fig. 12. Cluster radioactivity of actinide nuclei. By emission of ^{14}C, ^{20}O,... "big leaps" in the periodic system can occur, just contrary to the known α, β, γ radioactivities, which are also partly shown in the figure.

20, 24–26.

The "cold valleys" in the collective potential energy surface are essential for understanding this exciting area of nuclear physics! It is a master example for understanding the structure of elementary matter, which is so important for other fields, especially astrophysics, but even more so for enriching our "Weltbild", i.e. the status of our understanding of the world around us.

6. Concluding Remarks

For the Gesellschaft für Schwerionenforschung (GSI), which one of the authors (W.G.) helped initiate in the sixties, the questions raised here could point to the way ahead. Working groups have been instructed by the board of directors of GSI, to think about the future of the laboratory. On that occasion, very concrete (almost too concrete) suggestions are discussed - as far as it has been presented to the public. What is necessary, as it seems, is a vision on a long term basis. The ideas proposed here, the verification of which will need the commitment for 2–4 decades of research, could be such a vision with considerable attraction for the best young physicists. The new dimensions of the periodic system made of hyper- and antimatter cannot be examined in the "stand-by" mode at CERN (Geneva); a dedicated facility is necessary for this field of research, which can in future serve as a home for the universities. The GSI - which has unfortunately become much too self-sufficient - could be such a home for new generations of physicists, who are interested in the structure of elementary matter. GSI would then not develop only into a detector laboratory for CERN, and as such become obsolete. I can already see the enthusiasm in the eyes of young scientists, when I unfold these ideas to them - similarly to 30 years ago, when the nuclear physicists in the state of Hessen initiated the construction of GSI.

References

1. S.G. Nilsson et al., *Phys. Lett.* B 28, 458 (1969); *Nucl. Phys.* A 131, 1 (1969); *Nucl. Phys.* A 115, 545 (1968).
2. U. Mosel, B. Fink and W. Greiner, *"Contribution to Memorandum Hessischer Kernphysiker"* Darmstadt, Frankfurt, Marburg (1966).
3. U. Mosel and W. Greiner, *Z. f. Physik* 217, 256 (1968); 222, 261 (1968).
4. a) J. Grumann, U. Mosel, B. Fink and W. Greiner, *Z. f. Physik* 228, 371 (1969).
   b) J. Grumann, Th. Morovic, W. Greiner, *Z. f. Naturforschung* 26 A, 643 (1971).
5. W. Greiner, *Int. Journal of Modern Physics E, Vol. 5 , No. 1* (1995) 1-90. This review article contains many of the subjects discussed here in an extended version, see also for a more complete list of references.
6. A. Sandulescu, R.K. Gupta, W. Scheid, W. Greiner, *Phys. Lett.* B 60, 225 (1976);
   R.K. Gupta, A. Sandulescu, W. Greiner, *Z. f. Naturforschung* 32A, 704 (1977);
   R.K. Gupta, A.Sandulescu and W. Greiner, *Phys. Lett.* B 64, 257 (1977);
   R.K. Gupta, C. Parrulescu, A. Sandulescu, W. Greiner, *Z. f. Physik* 283A, 217 (1977).
7. G.M. Ter-Akopian et al., *Nucl. Phys.* A 255, 509 (1975);
   Yu.Ts. Oganessian et al., *Nucl. Phys.* A 239, 353 and 157 (1975).
8. D. Scharnweber, U. Mosel and W. Greiner, *Phys. Rev. Lett.* 24, 601 (1970);
   U. Mosel, J. Maruhn and W. Greiner, *Phys. Lett.* B 34, 587 (1971).
9. G. Münzenberg et al., *Z. Physik* 309A, 89 (1992);
   S.Hofmann et al., *Z. Phys* A 350, 277 and 288 (1995).
10. R. K. Gupta, A. Sandulescu and Walter Greiner, *Z. für Naturforschung* 32A, 704 (1977).
11. A. Sandulescu and Walter Greiner, *Rep. Prog. Phys* 55, 1423 (1992); A. Sandulescu, R. K. Gupta, W. Greiner, F. Carstoin and H. Horoi, *Int. J. Mod. Phys.* E 1, 379 (1992).
12. A. Sobiczewski, *Phys. of Part. and Nucl.* 25, 295 (1994).
13. R. K. Gupta, G. Münzenberg and W. Greiner, *J. Phys. G: Nucl. Part. Phys.* 23, L13 (1997).

14. K. Rutz, M. Bender, T. Bürvenich, T. Schilling, P.-G. Reinhard, J.A. Maruhn, W. Greiner, *Phys. Rev.* C 56, 238 (1997).
15. T. Bürvenich, M. Bender, J. A. Maruhn, P.–G. Reinhard, *accepted for publication in Phys. Rev.* C, nucl-th/0302056
16. M. Bender, K. Rutz, P.-G. Reinhard, J. A. Maruhn, W. Greiner, *Phys. Rev.* C 58, 2126 (1998).
17. Ş. Mişicu, T. Bürvenich. T. Cornelius, and W. Greiner, *J. Phys. G: Nucl. Part. Phys.* 28, 1441 (2002).
18. A. Sandulescu, D.N. Poenaru, W. Greiner, *Sov. J. Part. Nucl.* 11(6), 528 (1980).
19. Harold Klein, *thesis, Inst. für Theoret. Physik*, J.W. Goethe-Univ. Frankfurt a. M., (1992);
Dietmar Schnabel, *thesis, Inst. für Theoret. Physik*, J.W. Goethe-Univ. Frankfurt a.M., (1992).
20. D. Poenaru, J.A. Maruhn, W. Greiner, M. Ivascu, D. Mazilu and R. Gherghescu, *Z. Physik* 328A, 309 (1987), *Z. Physik* 332A, 291 (1989).
21. P. Papazoglou, D. Zschiesche, S. Schramm, H. Stöcker, W. Greiner, *J. Phys. G* 23, 2081 (1997); P. Papazoglou, S. Schramm, J. Schaffner-Bielich, H. Stöcker, W. Greiner, *Phys. Rev.* C 57, 2576 (1998).
22. P. Papazoglou, D. Zschiesche, S. Schramm, J. Schaffner–Bielich, H. Stöcker, W. Greiner, *accepted for publication in Phys. Rev.* C, nucl–th/9806087.
23. P. Papazoglou, *PhD thesis*, University of Frankfurt, 1998; C. Beckmann et al., nucl-th/0002046
24. E. K. Hulet, J. F. Wild, R. J. Dougan, R. W.Longheed, J. H. Landrum, A. D. Dougan, M. Schädel, R. L. Hahn, P. A. Baisden, C. M. Henderson, R. J. Dupzyk, K. Sümmerer, G. R. Bethune, *Phys. Rev. Lett.* 56, 313 (1986).
25. K. Depta, W. Greiner, J. Maruhn, H.J. Wang, A. Sandulescu and R. Hermann, *Intern. Journal of Modern Phys.* A 5, 3901 (1990);
K. Depta, R. Hermann, J.A. Maruhn and W. Greiner, in *"Dynamics of Collective Phenomena"*, ed. P. David, World Scientific, Singapore, 29 (1987);
S. Cwiok, P. Rozmej, A. Sobiczewski, Z. Patyk, *Nucl. Phys.* A 491, 281 (1989).
26. A. Sandulescu and W. Greiner in discussions at Frankfurt with J. Hamilton (1992/1993)

EFFECTS OF IONIC CORES IN SMALL RARE GAS CLUSTERS: POSITIVE AND NEGATIVE CHARGES

C. Di Paola, I. Pino, E. Scifoni, F. Sebastianelli and F.A. Gianturco

Department of Chemistry and INFM, University of Rome "La Sapienza", Piazzale A. Moro 5, 00185 Rome, Italy

E-mail: fa.gianturco@caspur.it

The structural properties and the energetics of some of the smaller He_nH^-, Ne_nH^-, Ne_n^+ and He_n^+ clusters are examined both with classical and quantum treatments. The results of the calculations, the physical reliability of the employed interaction modeling, and the comparison with previous results are discussed. The emerging picture shows very different features when comparing the positively ionized pure rare gas clusters with respect to those in which the negative impurity H^- is present.

1. Introduction

The interest in small ionic clusters involving rare-gas (Rg) atoms has markedly increased over the years on both the theoretical and experimental sides. A number of papers have focused their attention on neat and doped neon and helium clusters. From the experimental and theoretical standpoints, both positively ionized neon and helium clusters $[Rg_n]^+$, and the corresponding protonated aggregates $[He_nH]^+$, are known to be constituted by an arrangement of neutral, or almost neutral Rg atoms which are bound by polarization forces and, to a lesser degree, by dispersion forces to a moiety over which the majority of the charge resides, $[Rg_k]^+$ or $[He_kH]^+$ with k ranging from 2 to 4, depending on the cluster size.[1–7] They show very different structural features with respect to the neutral Rg_n clusters because the interaction potential between the ionic core and the other Rg atoms is usually more than three orders of magnitude greater than that between Rg atoms in the neutral aggregates. On the other hand, much less attention has been paid to negative clusters containing the H^- ion.[8] These species are of particular interest because, as we shall see below, much weaker forces are at play. Such complexes are governed by comparable strengths of interactions between the 'solute' with the 'solvent' molecules and the latter species among themselves. This is also what usually occurs in chemical species undergoing possible solvation in solution and therefore their study offers a realistic, but yet simpler, analogy that can help us to gain insight into anionic solvation at the microscopic level. Recently high accuracy Potential Energy Curves (PEC's) for the neon-H^- and for the helium-H^- interactions have become available[9] and therefore we decided to study the structure and the energetics of the Rg_nH^- systems (Rg=Ne, He). From a comparison of the electron affinity between the neon and the helium (atoms on one hand, and the hydrogen atom on the other hand), we should expect that these clusters are largely simple complexes in which the excess electron is chiefly localized on the added H^- within the cluster. In the following we describe the modeling of the interaction forces within each cluster and we carry out the analysis for these systems, comparing them with the results obtained for the singly ionized Rg_n^+ clusters.

2. Interaction Potentials

Previous studies converge on the finding that in the $(Ne)_n^+$ clusters there is a dimeric core over which the majority of the charge is spread.[10,11] On the other hand, for the $(He)_n^+$ system the number of helium atoms in the core, k, seems to depend on the size of the cluster, varying from 4 to 2 with the increasing of n.[5,6] Given these facts, we can model the global interaction potential within these positively charged clusters as sums of pairwise potentials, i.e. to approximate the full set of forces as the sum of the individual interactions between a Rg_2^+ and the relevant number of rare gas atoms, namely for a generic Rg_n^+ cluster, write

$$V_{TOT}^{dimer} = \sum_{i=3}^{n} V^{i}_{(Rg_2^+ - Rg)} + \sum_{\substack{i<j \\ i\geq 3}}^{n} V^{ij}_{(Rg-Rg)} \tag{1}$$

in which the first term is the sum of the interactions in the Rg_3^+ system considered as a rigid rotor, and the second term is the sum of the interactions between two neutral neon or helium atoms. We therefore need to set up the relevant $V^{i}_{(Rg_2^+ - Rg)}$ and the $V^{ij}_{(Rg-Rg)}$ interactions. For the calculation of the interaction which views Ne_2^+ as a rigid rotor system, we have used the DFT approach known as the Half & Half method of Becke to compute the PES[11] for the Ne_3^+ system using the familiar Jacobi coordinates, hold-

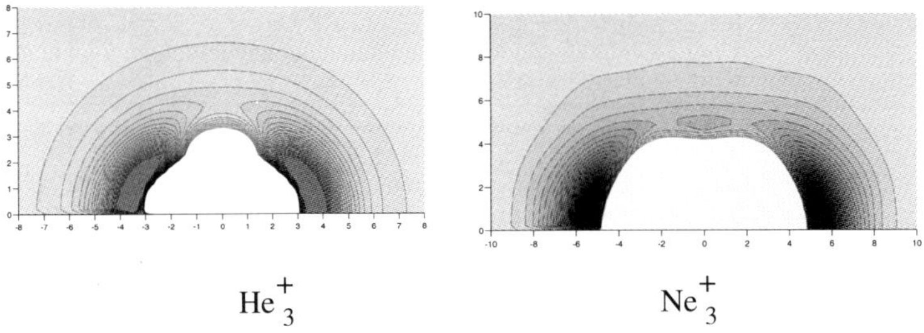

Fig. 1. Contour maps for the He_3^+, left panel, and Ne_3^+, right panel, energies in meV and distances in atomic units.

ing the molecular $r_{Rg_2^+}$ coordinate fixed at the optimized distance of the Ne_2^+ isolated molecule. Analogously we calculate the PES's for the He_3^+ system, in this case taking also into account the variation of the molecular coordinate r, knowing that in this case the core may not be the same for all the He_n^+ as n varies. The calculations were carried out employing the CCSD(T) method with the AUG-cc-pVQZ basis set.[12]

In Fig. 1 we report the contour plots for the two triatomic system. On the left part we show the PES for the He_3^+, in which the He_2^+ distance is fixed at 2.34 a.u., while on the right part the PES for the Ne_3^+, in which the Ne_2^+ distance is fixed to 3.28 a.u.

For the H^- doped Rg_n clusters we approximate the full set of forces as the individual interactions between H^- and the relevant number of Rg atoms, writing for a generic Rg_nH^- cluster

$$V_{TOT} = \sum_{i=1}^{n} V^i_{(Rg-H)^-} + \sum_{\substack{i,j=1 \\ i<j}}^{n} V^{ij}_{(Rg-Rg)}, \qquad (2)$$

in which the first term is the sum of pair-interactions from the RgH^- system for which we employ the PECs from Ref. 9, and the second term is the sum of the interactions between two Rg atoms for which we use the results of Refs. 13 and 14. These curves are shown in Fig. 2.

3. Classical and Quantum Structures for $(Ne)_nH^-$ and $(He)_nH^-$

Having set up all the necessary interaction potentials, our next task is to employ them for geometry optimizations. We use cubic splines to fit

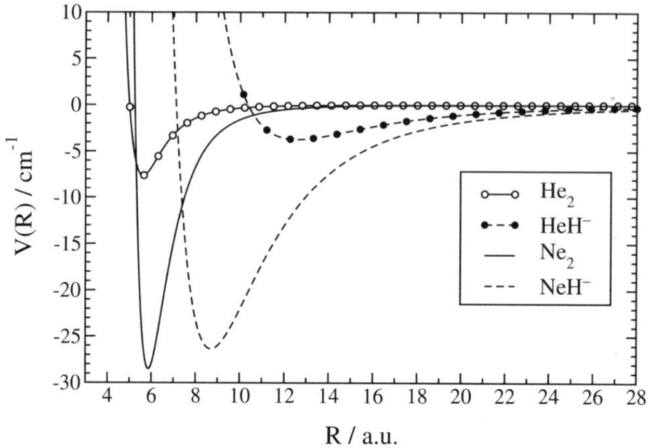

Fig. 2. Comparison between the potential energy curves for HeH^-, NeH^-, He_2 and Ne_2. The potential values are in cm^{-1} and the R values in a.u.

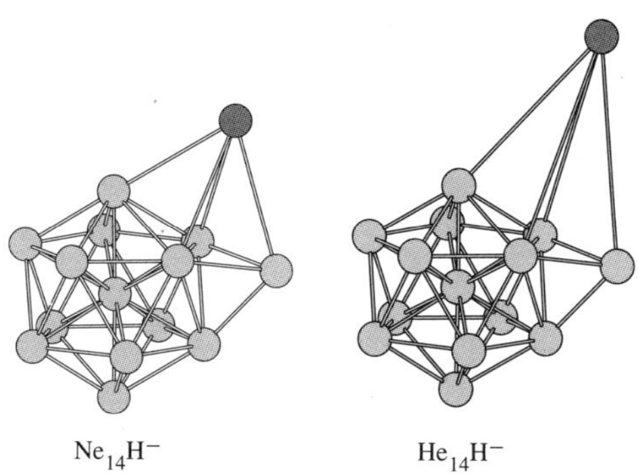

Fig. 3. Lowest energy structures found for $Ne_{14}H^-$ and $He_{14}H^-$.

the terms of Eqs. (1)–(2) in order to have an analytical representation of V_{TOT} and then write down its first and second derivatives. The total potential in each cluster is described by the sum of pairwise potentials and searching for the global minimum on this hypersurface will give us the lowest energy structure for each aggregate using a classical picture for its atom locations. All the classical minimizations were carried out using the OPTIM code[15] implemented by us for our potentials and generating the potential's first and second derivatives in Cartesian coordinates to yield the analytical expressions of the Hessian.

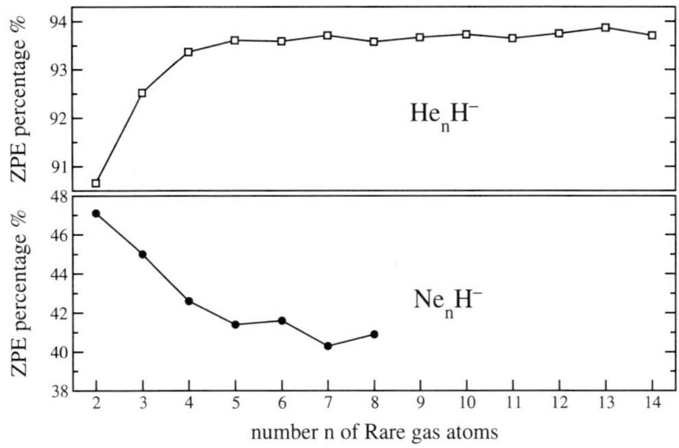

Fig. 4. ZPE percentage for He_nH^- and Ne_nH^-.

In Fig. 3 we draw two of the lowest energy minima found for the doped species: as one can see, in both cases the impurity locates itself far away from the Rg moiety and this is true for all the complexes under inspection (i.e. n from 2 up to 14). Furthermore the two structures show similar shape, indicating that in these anionic clusters the H^- does not perturb much the corresponding geometries of the neutral $(Rg)_n$ counterparts. The main difference between the He and Ne case for doped anionic clusters remains in the Zero-Point-Energy (ZPE) effects: we carried out quantum Diffusion Monte Carlo (DMC) calculations for $(Rg)_nH^-$[16] using the same potential modelling and found the ground states for these systems in order to estimate the importance of the ZPE effects. In Fig. 4 we report the percentage of ZPE (calculated as $\{(E_{cl}-E_Q)/E_{cl}\} \cdot 100)$ for the two sets of clusters, where E_{cl} is the classical minimum of the well depth for the potential of Eq. (1) and E_Q is the energy obtained with our DMC calculations.[16]

We can see that for the helium case more than 90 % of the total potential well depth is taken up by the ZPE effects, while for the neon case the percentage is about 40 %.

4. The $(He)_n^+$ and $(Ne)_n^+$ Structures

When considering the Ne_n^+ clusters the situation changes dramatically:[11,17,18] for all the structures we consider (n up to 25) the presence of a dimer core is observed, around which axis all other almost neutral neon atoms locate themselves in successive planes of three, four or five sides (see

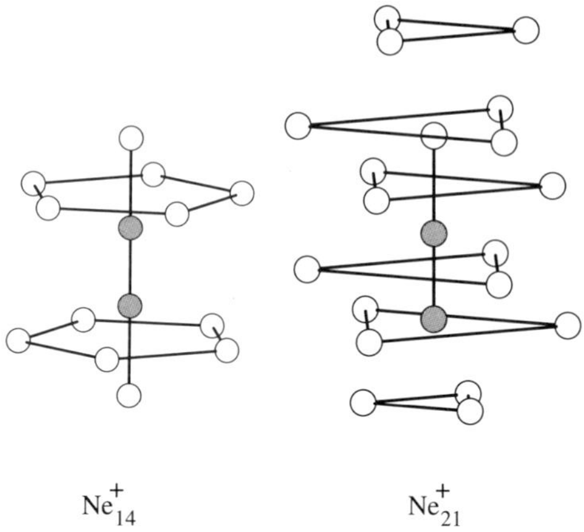

Fig. 5. Lowest energy structures for Ne^+_{14} and Ne^+_{21}.

Fig. 5), in which we draw two of the lowest energy minima found employed for the description of the global potential within each cluster in Eq. (1). We clearly see there the formation of a definite, ionic dimeric core. For the He^+_n systems, previous study[5,6] found that for very small clusters the ionic core is a tetramer moiety, while from He^+_7 structures a dimer core begin to appear, becoming the most stable structure upon increasing the size of the cluster. Hence we started a study on these systems looking at the most stable geometries using the MP4(SDQ) method with AUG-cc-pQTZ basis set. In Fig. 6 we sketch the results we obtained: we can see that up to He^+_5 the aggregates show a symmetrical trimeric core (with distances of 2.34 a.u.) in which the majority of the charge is shared, while from He^+_6 the dimeric core (now with a distance of 2.00 a.u., as in the isolated He^+_2 molecule) appear, confirming the picture that sees the larger clusters formed by this dimeric moiety around which all the other helium atoms build in. Then we can approximate the interaction forces within each cluster as in Eq. (1) where the $V^i_{(Rg^+_2-Rg)}$ term is given by the PES for the rigid rotor with He^+_2 distance fixed at 2.34 a.u. (see Fig. 1). In the upper part of Fig. 7 we show the lowest energy structures obtained holding the r distance fixed at 2.34 a.u., while in the lower part we also report, for comparison, the minima we found when the r distance is fixed at 2.00 a.u., i.e. at the distance in the isolated He^+_2 molecule.

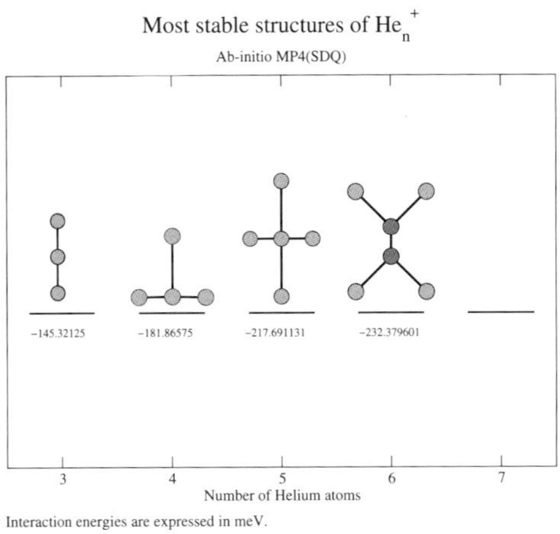

Fig. 6. Lowest energy structures for He_n^+. Interaction energy in meV.

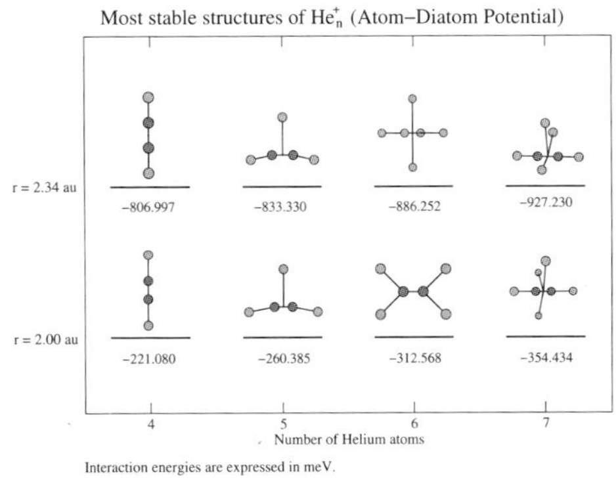

Fig. 7. Lowest energy structures for He_n^+ using Atom-Diatom potential model of Eq. (1). Interaction energy in meV.

5. Conclusions

We have considered the structural behaviour of H^- doped helium and neon clusters, using the sum of pairwise potentials for the global interactions within each cluster and carrying out both classical and quantal calculations. The overall picture emerging from our calculations indicates that the H^- dopant always remains outside the Rg atoms moiety, i.e. for clusters of such size the H^- dopant is not solvated. We also obtained the lowest energy geometries for the Rg_n^+, where Rg=He and Ne, finding that in this case the presence of a ionic core is the driving force in the building up of the structures and that such a core is invariably larger than the simpler atomic ion that dominates the anionic clusters.

Acknowledgments

The financial support of the Ministry for University and Research (MUIR), of the University of Rome I Research Committee, of the INFM institute, are gratefully acknowledged.

References

1. B. Balta, F.A. Gianturco, and F. Paesani, *Chem. Phys.* **254**, 215 (2000).
2. B. Balta and F.A. Gianturco, *Chem. Phys.* **254**, 203 (2000).
3. F. Filippone and F.A. Gianturco, *Europhys. Lett.* **44**, 585 (1998).
4. F.A. Gianturco and F. Filippone, *Chem. Phys.* **241**, 203 (1999).
5. F.A. Gianturco and M.P. De Lara-Castells, *J. Quantum Chem.* **60**, 593 (1996).
6. P. Knowles and J.N. Murrell, *Mol. Phys* **87**, 827 (1996).
7. C.Y. Ng, T. Baer, and I. Powis (Ed.s), *Cluster ions* John Wiley & Sons, New York, (1993).
8. J. Kalcher and A. F. Sax, *Chem. Rev.* **94**, 2291 (1994).
9. V. Vallet, G.L. Bendazzoli, and S. Evangelisti, *Chem. Phys.* **263**, 33 (2001).
10. F. A. Gianturco and F. Sebastianelli, *Eur. Phys. J.* D **10**, 399 (2000).
11. F. Sebastianelli, E. Yurtsever, and F. A. Gianturco, *Int. J. Mass Spect.* **220**, 193 (2002).
12. E. Scifoni and F.A. Gianturco, *Eur. Phys. J.* D **21**, 323 (2002).
13. R.A. Aziz and M.J. Slaman, *J. Chem. Phys.* **94**, 8047 (1991).
14. U. Kleinekathöfer, K.T. Tang, J.P. Toennies, and C.L. Yiu, *Chem. Phys. Lett.* **249**, 257 (1996).
15. D.J. Wales, *J. Chem. Phys.* **101**, 3750 (1994).
16. F. Sebastianelli, I. Baccarelli, C. Di Paola, and F.A. Gianturco, *J. Chem. Phys.* (to be published).
17. F. Sebastianelli, F. A. Gianturco, and E. Yurtsever, *Chem. Phys.* **290**, 279 (2002).
18. F.Y. Naumkin, D.J. Wales, *Mol. Phys.* **93**, 633 (1998)

METAL CLUSTER FISSION: JELLIUM MODEL AND MOLECULAR DYNAMICS SIMULATIONS

Andrey Lyalin

Institute of Physics, St Petersburg State University,
198504 St Petersburg, Petrodvorez, Russia
and
Institut für Theoretische Physik der Johann-Wolfgang Goethe Universität,
Robert-Mayer Str. 8-10, D-60054 Frankfurt am Main, Germany
E-mail: lyalin@th.physik.uni-frankfurt.de

Oleg Obolensky

A. F. Ioffe Physical-Technical Institute, 194021 St. Petersburg, Russia

Ilia A. Solov'yov, Andrey V. Solov'yov

A. F. Ioffe Physical-Technical Institute, 194021 St. Petersburg, Russia
and
Institut für Theoretische Physik der Johann-Wolfgang Goethe Universität,
Robert-Mayer Str. 8-10, D-60054 Frankfurt am Main, Germany

Walter Greiner

Institut für Theoretische Physik der Johann-Wolfgang Goethe Universität,
Robert-Mayer Str. 8-10, D-60054 Frankfurt am Main, Germany

Fission of doubly charged sodium clusters is studied using the open-shell two-center deformed jellium model approximation and *ab initio* molecular dynamic approach accounting for all electrons in the system. Results of calculations of fission reactions $Na_{10}^{2+} \to Na_7^+ + Na_3^+$ and $Na_{18}^{2+} \to 2Na_9^+$ are presented. Dependence of the fission barriers on isomer structure of the parent cluster is analyzed. Importance of rearrangement of the cluster structure during the fission process is elucidated. This rearrangement may include transition to another isomer state of the parent cluster before actual separation on the daughter fragments begins and/or forming a "neck" between the separating fragments.

1. Introduction

Fission of charged atomic clusters occurs when repulsive Coulomb forces, arising due to the excessive charge, overcome the electronic binding energy of the cluster.[1–3] This mechanism of cluster fission is very similar to the nuclear fission phenomena. Experimentally, multiply charged metal clusters can be observed in the mass spectra when their size exceeds the critical size of stability, which depends on the metal species and cluster charge.[4–6]

In the present work we report the results of calculations of the fission barriers for the symmetric and asymmetric fission processes $Na_{10}^{2+} \to Na_7^+ + Na_3^+$ and $Na_{18}^{2+} \to 2Na_9^+$. Fission of doubly charged sodium clusters is studied using the open-shell two-center deformed jellium model approximation and *ab initio* molecular dynamics (MD) approach accounting for all electrons in the system.

We have investigated the parent cluster isomer dependence of the fission barrier for the reaction $Na_{10}^{2+} \to Na_7^+ + Na_3^+$. To the best of our knowledge, a comparative study of fission barriers for various isomers by means of quantum chemistry methods has not been carried out before. Note that such a study is beyond the scope of simpler approaches which do not account for ionic structure of the cluster. We found that the direct separation barrier for the reaction $Na_{10}^{2+} \to Na_7^+ + Na_3^+$ has a weak dependence on the isomeric structure of the parent cluster. We note, however, that the groups of atoms to be removed from the parent cluster isomers must be chosen with care: one has to identify homothetic groups of atoms in each fissioning isomer. The weak dependence on the isomeric state of the parent Na_{10}^{2+} cluster implies that the particular ionic structure of the cluster is largely insignificant for the shape and height of the fission barrier. This is due to the fact that the maximum fission barriers in considered cases are located at distances comparable or exceeding the sum of the resulting fragments radii. At such distances the interaction between the fragments, apart from the Coulombic repulsion, is mainly determined by the electronic properties rather than by the details of the ionic structure of the fragments. This is an important argument for justification of the jellium model approach to the description of the fission process of multiply charged metal clusters.

We have demonstrated the importance of rearrangement of the cluster ionic structure during the fission process. The possibility of rearrangement of the cluster structure leads to the fact that direct fission of a cluster isomer in some cases may not be the energetically optimum path for the fission reaction. Alternatively, the reaction can go via transition to another isomer state of the parent cluster. This transition can occur in the first phase of the fission process, before separation of the fragments actually begins. We show that this is the case for the fission of C_{4v} and D_{4d} isomers of Na_{10}^{2+} cluster.

The rearrangement of ionic structure may be important also after the fragments began to separate. For $Na_{18}^{2+} \rightarrow 2Na_9^+$ reaction, two magic fragments Na_9^+ form a metastable transitional state in which the fragments are connected by a "neck". This "necking" results in significant reduction of the height of the fission barrier. The similar necking phenomenon is known for the nuclear fission process.[7]

2. Theoretical Methods

According to the main postulate of the jellium model, the electron motion in a metallic cluster takes place in the field of the uniform positive charge distribution of the ionic background. For the parameterization of the ionic background we consider the model in which the initial parent cluster, having the form of the ellipsoid of revolution (spheroid), splits into two independently deformed spheroids of smaller size.[8,9] The two principal diameters a_k and b_k of the spheroids can be expressed via the deformation parameter δ_k as

$$a_k = \left(\frac{2+\delta_k}{2-\delta_k}\right)^{2/3} R_k, \qquad b_k = \left(\frac{2-\delta_k}{2+\delta_k}\right)^{1/3} R_k. \tag{1}$$

Here partial indexes $k = 0, 1, 2$ correspond to the parent cluster ($k = 0$) and the two daughter fragments ($k = 1, 2$), R_k ($k = 0, 1, 2$) are the radii of the corresponding undeformed spherical cluster, which are equal to $R_k = r_s N_k^{1/3}$, where N_k is the number of atoms in the k-th cluster, and r_s is the Wigner-Seitz radius. For sodium clusters, $r_s = 4.0$, which corresponds to the density of the bulk sodium. The deformation parameters δ_k characterize the families of the prolate ($\delta_k > 0$) and the oblate ($\delta_k < 0$) spheroids of equal volume $V_k = 4\pi a_k b_k^2/3 = 4\pi R_k^3/3$. For overlapping region the radii $R_1(d)$ and $R_2(d)$ are functions of the distance d between the centers of mass of the two fragments. They are determined so that the total volume inside the two spheroids is equal to the volume of the parent cluster $4\pi R_0^3/3$.

The Hartree-Fock and LDA equations have been solved in the system of the prolate spheroidal coordinates as a system of coupled two-dimensional second order partial differential equations.[10–12] In the present work we use the Gunnarsson and Lundqvist parameterization of the density of electron exchange-correlation energy.[13]

The MD calculations have been carried out with the use of the GAUSSIAN 98 software package.[14] We have utilized the density functional theory based on the hybrid Becke-type three-parameter exchange functional paired with the gradient-corrected Lee, Yang and Parr correlation functional (B3LYP).[14] The B3LYP functional has proved to be a reliable tool for studying the structure and properties of small metal clusters. It provides high accuracy at comparatively low computational costs. For a discussion and a comparison with other approaches, see our recent works.[15,16] Note that the density of the parent cluster and two daughter fragments (including the overlapping region before scission point) almost does not change during the fission process (by analogy with the deformed jellium model[8,9]). This means that the B3LYP method works adequately for any fragment separation distances, d, during the fission process.

The 6-311G(d) and LANL2DZ basis sets of primitive Gaussian functions have been used to expand the cluster orbitals.[14] The 6-311G(d) basis has been used for simulations involving Na_{10}^{2+} cluster. This basis set takes into account electrons from all atomic orbitals, so that the dynamics of all particles in the system are taken into account. For Na_{18}^{2+} cluster we have used a more numerically efficient LANL2DZ basis, for which valent atomic electrons move in an effective core potential.[14]

3. $Na_{10}^{2+} \rightarrow Na_7^+ + Na_3^+$ Process

3.1. *Deformed jellium model calculations*

Let us present and discuss the results of calculations performed within the models described above. Figure 1 shows fission barriers calculated within the jellium model for the asymmetric channel $Na_{10}^{2+} \rightarrow Na_7^+ + Na_3^+$ as a function of the fragments separation distance d. We have minimized the total energy of the system over the parent and daughter fragments spheroidal deformations with the aim finding the fission pathway corresponding to the minimum of the fission barrier. We have also used the assumption of continuous shape deformation during the fission process. The evolution of cluster shape during the fission process is shown on top of Fig. 1. Solid and dashed lines in Fig. 1 are the result of the two-center jellium HF and LDA calculations respectively. The zero of energy put at $d = 0$.

Within the framework of the two-center deformed jellium Hartree-Fock approximation, the parent cluster Na_{10}^{2+} is unstable towards the asymmetric channel $Na_{10}^{2+} \rightarrow Na_7^+ + Na_3^+$. Accounting for many-electron correlations within the LDA theory leads to the formation of the fission barrier and

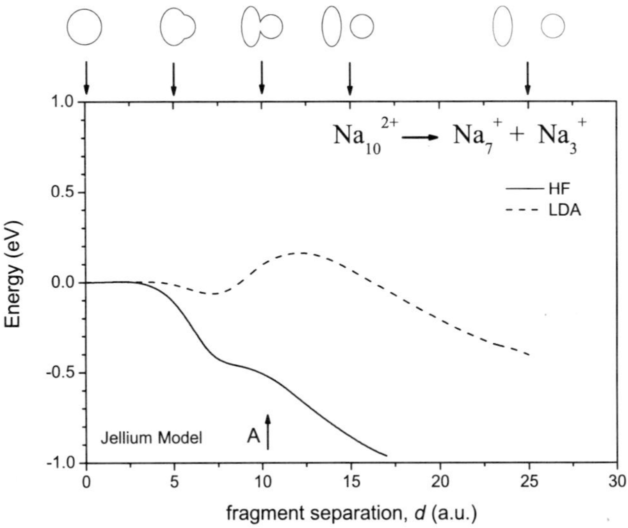

Fig. 1. Fission barriers in the two-center deformed jellium Hartree-Fock (solid lines) and LDA (dashed lines) approaches for the asymmetric channel $Na_{10}^{2+} \rightarrow Na_7^+ + Na_3^+$ as a function of fragments separation distance d (for details see our recent works[8,9]). The evolution of jellium cluster shape during the fission process is shown on top of figure.

the appearance of a local minimum on the energy curve at $d = 7.2$ a.u., corresponding to the super deformed asymmetric prolate state of the parent Na_{10}^{2+} cluster before the scission point A. The latter is located at $d = 10.4$ a.u.

3.2. *Isomer dependence of the fission barrier*

The results of jellium model calculations are compared with the results of *ab initio* MD simulations accounting for all electrons in the system. To simulate the fission process we start from the optimized geometry of a cluster (for details of the geometry optimization procedure see Refs.[15,16]) and choose the atoms that the resulting fragments would consist of. The atoms chosen for a smaller fragment are shifted from their locations in the parent cluster to a certain distance. Then, the multidimensional potential energy surface, its gradient and forces with respect to the molecular coordinates are calculated. These quantities specify the direction along the surface in which the energy decreases the most rapidly and provide information for the determination of the next step for the moving atoms. If the fragments were not removed far enough from each other then the attractive forces prevailed over the repulsive ones and the fragments stuck together forming the unified cluster again. In the opposite situation the repulsive forces dominate and the fragments drift away from each other. The dependence of the total energy of the system on the fragment separation distance forms the fission barrier. The aim of our simulations is to find the fission pathway corresponding to the minimum of the fission barrier.

There are usually many stable isomers of a cluster, with energies slightly exceeding the energy of the ground state isomer. In order to analyze the isomer dependence of the fission barrier in the reaction $Na_{10}^{2+} \rightarrow Na_7^+ + Na_3^+$ we have picked two energetically low-lying isomers with the point symmetry groups C_{4v} and D_{4d} differing from the distorted T_d point symmetry group of the ground state parent Na_{10}^{2+} cluster. Three isomer states of the Na_{10}^{2+} cluster are shown in Fig. 2.

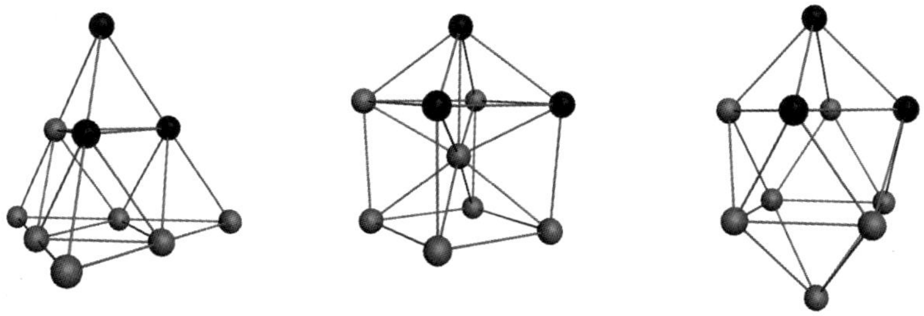

Fig. 2. Three isomers of Na_{10}^{2+} cluster. From left to right: the ground state isomer of distorted T_d point symmetry group (total energy is -1622.7063 a.u.); an isomer of C_{4v} point symmetry group (total energy is -1622.6888 a.u., that exceeds the lowest energy state by 0.476 eV); an isomer of D_{4d} point symmetry group (total energy is -1622.6860 a.u., that exceeds the lowest energy state by 0.553 eV). The homothetic group of three atoms marked by black color.

In Fig. 3 we show fission barriers for separation of three atoms from the C_{4v}, D_{4d}, and T_d isomers of the Na_{10}^{2+} cluster. In this figure zero level of energy is chosen for each parent isomer separately and corresponds to the minimum of total energy of that isomer. The initial distances between the centers of mass of two (future) fragments are finite so that the barriers do not start at the origin.

The barriers for all three channels are close. The weak sensitivity of

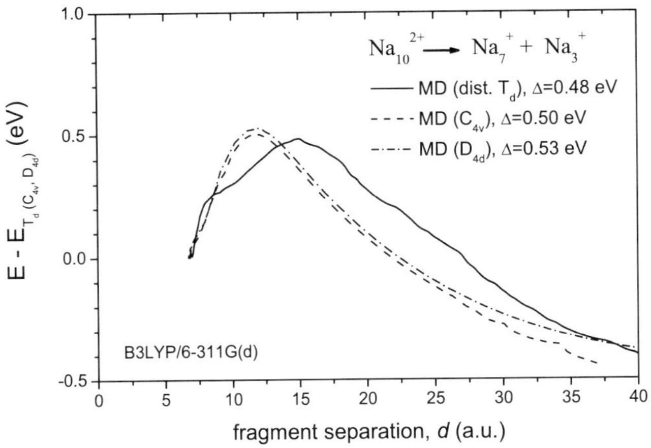

Fig. 3. Fission barriers for separating the homothetic group of three atoms (marked by black color in Fig. 2) from three isomers of Na_{10}^{2+} cluster derived from molecular dynamics simulations (direct $Na_{10}^{2+} \rightarrow Na_7^+ + Na_3^+$ fission channel). The barriers plotted versus distance between the centers of mass of the fragments. Solid, dashed, and dashed-dotted lines correspond to distorted T_d, C_{4v}, and D_{4d} point symmetry groups isomers of the parent cluster, respectively. Energies are measured from the energy of the ground state of the corresponding isomers, i.e. we plot $E - E_{T_d(C_{4v},D_{4d})}$, where E is the total energy of the system and $E_{T_d(C_{4v},D_{4d})}$ are the ground energies of the T_d, C_{4v} and D_{4d} isomer states of the parent Na_{10}^{2+} cluster, respectively.

the fission barrier on the isomeric states of the reactants can be explained if one notices that the barrier maxima are located at distances comparable to or exceeding the sum of the resulting fragments radii, that is not far from the scission point. At such distances the interaction between the fragments, apart from Coulombic repulsion, is mainly determined by the electronic properties rather than by the details of the ionic structure of the fragments. This is an important argument for justification of the jellium model approach to the description of the fission process of multiply charged metal clusters.

It is important to note that the barriers presented in Fig. 3 are calculated in assumption that fission occurs for the fixed (given) isomers. However, since C_{4v} and D_{4d} isomers are not the lowest energy states of Na_{10}^{2+} system, there could be other processes competing with fission. One of such processes is rearrangement of the cluster structure.

3.3. *Rearrangement of the cluster structure during the fission process*

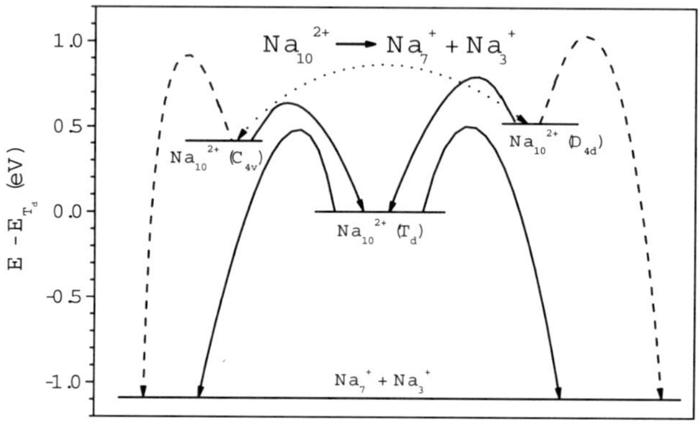

Fig. 4. Energy levels of the selected isomer states of the Na_{10}^{2+} system and schematic barriers for transitions between these states. Energies are measured from the energy of the ground T_d state of the Na_{10}^{2+} cluster.

Rearrangement of the cluster structure during the fission process may significantly reduce the fission barrier. Such rearrangement may occur before the actual separation of the daughter fragments begins or after that.

Fission of C_{4v} and D_{4d} isomers of Na_{10}^{2+} cluster is an example of a situation where rearrangement of the cluster structure takes place before the fragments start to separate. In Fig. 4 we show schematically the total energies of the Na_{10}^{2+} T_d, C_{4v} and D_{4d} isomers and barriers for the transitions between those states. It is seen from the figure 4 that transition to the ground (T_d) state with subsequent fission into the Na_3^+ and Na_7^+ fragments, $Na_{10}^{2+}(C_{4v}$ or $D_{4d}) \to Na_{10}^{2+}(T_d) \to Na_7^+ + Na_3^+$, (shown by solid lines) is the preferred path for fission of both C_{4v} and D_{4d} isomers of the Na_{10}^{2+} cluster and requires only about 0.2 eV for the C_{4v} isomer and 0.26 eV for the D_{4d} isomer. In contrast, the direct fission process, $Na_{10}^{2+}(C_{4v}$ or $\mathrm{D_{4d}}) \to \mathrm{Na}_7^+ + \mathrm{Na}_3^+$, (shown by dashed lines) requires about 0.5 eV. We also show the barrier for the transition between the C_{4v} and D_{4d} isomers.

4. $Na_{18}^{2+} \rightarrow 2Na_{9}^{+}$ Process

4.1. *Deformed jellium model calculations*

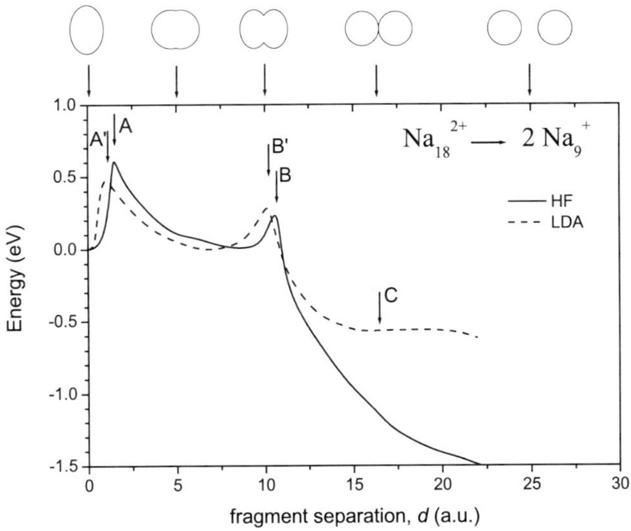

Fig. 5. The same as in figure 1 but for for the symmetric channel $Na_{18}^{2+} \rightarrow 2Na_{9}^{+}$

Figure 5 shows the dependence of the fission barrier on separation distance d for the symmetric channel $Na_{18}^{2+} \rightarrow 2Na_{9}^{+}$. The parent cluster changes its shape from an oblate to a prolate one in the initial stage of the fission process ($d \approx 1$ a.u.). This transition is accompanied by the first rearrangement[9] of the electronic configuration (marked by vertical arrow A for HF and A$'$ for LDA). The total fission barrier for the symmetric channel $Na_{18}^{2+} \rightarrow 2Na_{9}^{+}$ is equal to $\Delta_{HF} = 0.63$ eV and $\Delta_{LDA} = 0.48$ eV in the two-center jellium Hartree-Fock and LDA models respectively. On the next stage of the reaction the prolate deformation develops, resulting in the highly deformed cluster shape, as it is shown on top of Fig. 5. At the distance $d \approx 11$ (marked by vertical arrow B for HF, and B$'$ for LDA) the electronic configuration reaches its final form being the same as in the spherical Na_{9}^{+} products.[9]

4.2. *MD simulations and rearrangement of the cluster structure*

Figure 6 shows a fission barrier for the symmetric channel $Na_{18}^{2+} \to 2Na_9^+$ derived from molecular dynamics simulations.

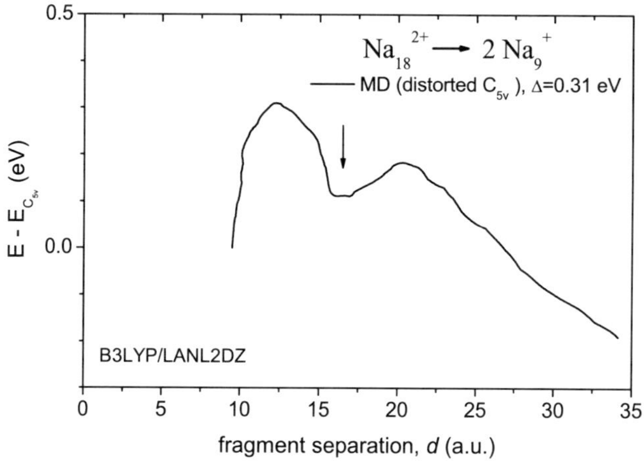

Fig. 6. Fission barrier for $Na_{18}^{2+} \to 2Na_9^+$ channel derived from molecular dynamics simulations as a function of distance between the centers of mass of the fragments. Energy is measured from the energy of the ground C_{5v} state of the Na_{18}^{2+} cluster. The arrow shows the position of the metastable transitional state, see also Fig. 7.

The reaction $Na_{18}^{2+} \to 2Na_9^+$ is another example of the importance of the cluster structure rearrangement in the fission process. If two fragments of the parent cluster were not allowed to adjust their ionic structure the fission barrier goes up to about 1 eV. Rearrangement of the cluster structure allows reduction of the fission barrier down to 0.31 eV. These results are in a reasonable agreement with the results of the jellium model.[8,9] During the fission process the daughter fragments start to drift away from each other and a "neck" forms between the fragments. Formation of the "neck" results in a metastable transitional state. The geometry of this state, as well as the geometry of the parent cluster, are shown in Fig. 7.

In Tab. 1 we have summarized our results for the fission barrier heights and compared them with the results of other molecular dynamics simula-

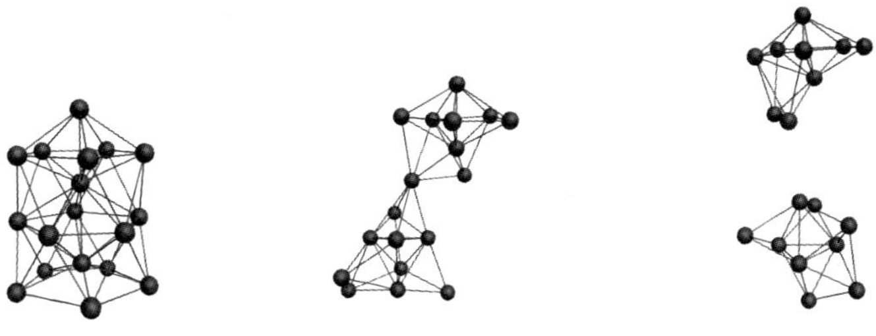

Fig. 7. Rearrangement of the cluster structure during the fission process $Na_{18}^{2+} \rightarrow 2Na_9^+$. From left to right: ground state of the parent cluster; "necking" between the two fragments leads to a meta stable intermediate state and significantly reduces the fission barrier height; two Na_9^+ fragments drifting away from each other.

tions and with the predictions of the jellium model.

Table 1. Summary of the fission barrier heights (eV).

	$Na_{10}^{2+} \rightarrow Na_7^+ + Na_3^+$	$Na_{18}^{2+} \rightarrow 2Na_9^+$
MD (this work)	0.49 (distorted T_d)	0.31
MD[17]	0.67	0.52
MD[18]	0.54	–
Jellium model[8, 9]	0.16	0.48

5. Conclusions

We have investigated two aspects of the charged metal cluster fission process: dependence of the fission barrier on the isomer state of the parent cluster and the importance of rearrangement of the cluster ionic structure during the fission process.

We found that for a consistent choice of the atoms removed from the cluster the fission barrier for the reaction $Na_{10}^{2+} \rightarrow Na_7^+ + Na_3^+$ has a weak dependence on the initial isomer structure of the parent cluster. This implies that the particular ionic structure of the cluster is largely insignificant for the height of the fission barrier.

We have demonstrated the importance of rearrangement of the cluster ionic structure during the fission process. The fission reaction can go through transition to another isomer state of the parent cluster. This transition can occur before actual separation of the fragments begins and/or a "neck" between the separating fragments is formed. In any case the resulting fission barrier can be significantly lower compared to the one for the direct fission path.

Acknowledgments

The authors acknowledge support of this work by the Alexander von Humboldt Foundation, the Studienstiftung des deutschen Volkes, the INTAS (grant No 03-51-6170), Russian Foundation for Basic Research (grant No 03-02-16415-a), and the Russian Academy of Sciences (grant No 44).

References

1. K. Sattler, J. Mühlbach, O. Echt, P. Pfau, and, E. Recknagel, *Phys. Rev. Lett.* **47**, 160 (1981).
2. U. Näher, S. Bjornholm, F. Frauendorf, and C. Guet, *Phys. Rep.* **285**, 245 (1997).
3. C. Yannouleas, U. Landman, and R.N. Barnett, in *Metal Clusters*, p.145, edited by W. Ekardt, Wiley, New York, (1999).
4. C. Bréchignac, Ph. Cahuzac, F. Carlier, and J. Leygnier, *Phys. Rev. Lett.* **63**, 1368 (1989).
5. C. Bréchignac, Ph. Cahuzac, F. Carlier, *et al*, *Phys. Rev.* B **49**, 2825 (1994).
6. T.P. Martin, *J. Chem. Phys.* **81**, 4426 (1984).
7. J.M. Eisenberg, and W. Greiner, *Nuclear Theory. Vol.1. Collective and Particle Models*, North Holland, Amsterdam, (1985).
8. A. Lyalin, A.V. Solov'yov, W. Greiner and S. Semenov, *Phys. Rev.* A **65**, 023201 (2002).
9. A. Lyalin, A.V. Solov'yov and W. Greiner, *Phys. Rev.* A **65**, 043202 (2002).
10. A.G. Lyalin, S.K. Semenov, A.V. Solov'yov, N.A. Cherepkov and W. Greiner, *J. Phys.* B **33**, 3653 (2000).
11. A.G. Lyalin, S.K. Semenov, A.V. Solov'yov, N.A. Cherepkov, J.-P. Connerade, and W. Greiner, *J. Chin. Chem. Soc.* (Taipei) **48**, 419 (2001).
12. A. Matveentsev, A. Lyalin, Il.A. Solov'yov, A.V. Solov'yov and W. Greiner, *Int. J. Mod. Phys.* E **12**, 81 (2003).
13. O. Gunnarsson and B.I. Lundqvist, *Phys. Rev.* B **13**, 4274 (1976).
14. M.J. Frisch *et al*, computer code GAUSSIAN 98, Rev. A. 9, Gaussian Inc., Pittsburgh, PA, 1998; James B. Foresman and Æleen Frisch *Exploring Chemistry with Electronic Structure Methods*, Pittsburgh, PA: Gaussian Inc, (1996)
15. Il.A. Solov'yov, A.V. Solov'yov and W. Greiner, *Phys. Rev.* A **65**, 053203 (2002).
16. A. Lyalin, Il.A. Solov'yov, A.V. Solov'yov and W. Greiner, *Phys. Rev.* A **67**, 063203 (2003).
17. B. Montag and P.-G. Reinhard, *Phys. Rev.* B **52**, 16365 (1995).
18. P. Blaise, S.A. Blundell, C. Guet and R.R. Zopa *Phys. Rev. Lett.* **87**, 063401 (2001).

MULTIFRAGMENTATION, CLUSTERING, AND COALESCENCE IN NUCLEAR COLLISIONS

Stefan Scherer and Horst Stöcker

Institut für Theoretische Physik, Johann Wolfgang Goethe Universität, Robert Mayer-Str. 10, D-60054 Frankfurt am Main, Germany
E-mail: scherer@th.physik.uni-frankfurt.de, stoecker@uni-frankfurt.de

Nuclear collisions at intermediate, relativistic, and ultra-relativistic energies offer unique opportunities to study in detail manifold fragmentation and clustering phenomena in dense nuclear matter. At intermediate energies, the well known processes of nuclear multifragmentation – the disintegration of bulk nuclear matter in clusters of a wide range of sizes and masses – allow the study of the critical point of the equation of state of nuclear matter. At very high energies, ultra-relativistic heavy-ion collisions offer a glimpse at the substructure of hadronic matter by crossing the phase boundary to the quark-gluon plasma. The hadronization of the quark-gluon plasma created in the fireball of a ultra-relativistic heavy-ion collision can be considered, again, as a clustering process. We will present two models which allow the simulation of nuclear multifragmentation and the hadronization via the formation of clusters in an interacting gas of quarks, and will discuss the importance of clustering to our understanding of hadronization in ultra-relativistic heavy-ion collisions.

While most experimental studies concerning clustering and fragmentation of matter focus on the scale of atoms and molecules, there are prominent examples of these phenomena on the more fundamental scale of nuclear matter. In this note, we want to briefly present two of them: the multifragmentation transition for heated, diluted nuclear matter, and the clustering of quarks and hadrons at the transition from a quark-gluon-plasma to a gas of hadrons. The theoretical models we will use to study the relevant physics are the Quantum Molecular Dynamics (QMD) for nuclear matter, and the quark Molecular Dynamics (qMD) for the subnuclear degrees of freedom, respectively. We will further discuss how clustering helps to un-

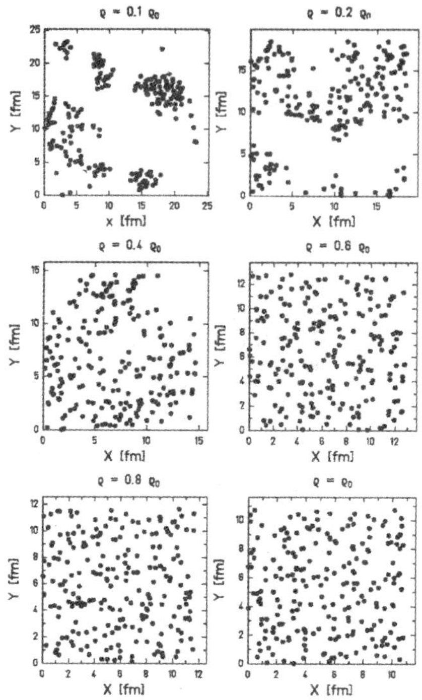

Fig. 1. Results of a Quantum Molecular Dynamics simulation of bulk nuclear matter at different densities. While at normal nuclear densities (lower row), nuclear matter is distributed homogeneously, prominent clustering builds up at diluted densities, $\rho \sim 0.1\rho_0$ (upper row).

derstand data from ultra-relativistic heavy-ion collisions at the Relativistic Heavy Ion Collider (RHIC), on the level of clustering both of partons and of hadrons.

1. Multifragmentation in Nuclear Matter

At the heart of all matter, nearly all the mass of every atom is concentrated in the tiny atomic nucleus, taking roughly $1/10^{-15}$ of the volume of the atom. The atomic nucleus is build up of protons and neutrons, which are bound together by nuclear forces, effective remnants of the fundamental strong interaction between quarks and gluons. Understanding the nuclear forces is essential in order to understand, for example, which nuclei can be stable, and to gain a complete overview of the chart of isotopes. From the theoreticians point of view, a possible way to study nuclear forces is to incorporate them in a model which is then solved numerically on a computer. Such a model is, e. g. the Quantum Molecular Dynamics (QMD) of nuclear

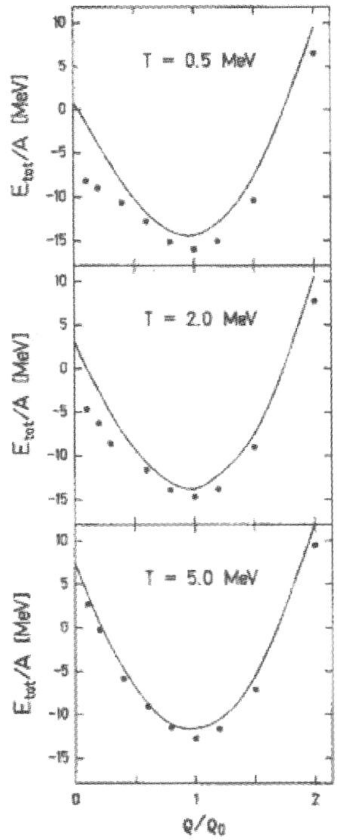

Fig. 2. The Equation of State (EoS) of nuclear matter at different temperatures. Solid lines show the energy per nucleon for infinite, homogeneous nuclear matter at different temperatures. This energy will be lowered significantly if the clustering of nucleons is taken into account (dotted marks).

matter. Here, nucleons are modelled as Gaussian wavepackets, with realistic potential interactions corresponding to the nuclear forces, and Fermi statistics is mimicked by a Pauli potential.[1] Results of QMD calculations of bulk nuclear matter at different nuclear densities are shown in Fig. 1. While at normal nuclear densities, bulk nuclear matter is homogeneous, a strong clustering is observed at low densities ($\rho \approx 0.1\rho_0$).

What are the physical consequences of this clustering? While it is difficult to access nuclear forces directly by experiment, a lot of information can be gained by the study of the Equation of State (EoS) of nuclear matter, which gives the energy per nucleon as a function of nuclear density. The EoS can be probed, for example, in nuclear collisions. In a first approximation, looking at homogeneous, infinite nuclear matter, the energy per nucleon depends on bulk nuclear density and temperature. These relations are plotted for different temperatures in Fig. 2 as solid lines. However, calculations with QMD show that allowing for clustering will lower the energy per nucleon. These energy shifts are most prominent at low densities and temperatures, where clustering is strongest.

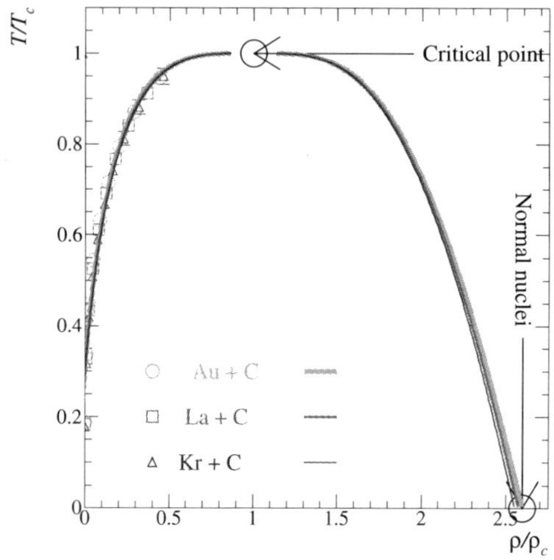

Fig. 3. The Guggenheim plot for finite nuclear matter in different nuclear collisions: nuclear fragments populate the low density (vapour) branch of the coexistence curve of finite nuclear matter (from Elliott *et al.*[3])

How can these calculations be checked by experiment in the laboratory? One possibility is the analysis of nuclear collisions at intermediate energies (about 100–500 MeV/N). Such collisions yield in a first stage compressed nuclear matter, which subsequently expands, thereby running through a stage of diluted nuclear matter, which fragments in clusters of different sizes. Of course, the systems studied in such collisions are far from representing infinite nuclear matter which exists only in neutron stars, so it is essential to take into account finite size effects.[2] It emerges from of the study of the cluster size distribution that the fragmentation of nuclear matter in these collisions can be understood in terms of a liquid-gas phase transition: diluted and heated nuclear matter fragments and evaporates like a Van der Waals fluid! Figure 3 shows the corresponding Guggenheim plot, representing the results of this fragmentation analysis.[3]

2. Clustering and the Transition to the Quark-Gluon Plasma

At the liquid-gas transition of nuclear matter, the substructure of the nucleons does not matter. However, as is well known since the 1970s, all hadrons, such as protons and neutrons, do have such a substructure: they are composed of quarks and gluons. Hadrons consisting of three quarks are called baryons, hadrons made up of a quark and an antiquark are called mesons. (Very recently, a short living state consisting of five quarks – a so called pentaquark state, with 4 quarks and 1 anti-s-quark and the electric charge of the proton – has been found in nuclear reactions,[4] but we will not discuss this topic further.) One may ask whether it is possible to separate single quarks from nucleons by suitable scattering experiments. This, however, can not happen, a consequence of Quantum Chromodynamics (QCD), the gauge theory describing the interaction between quarks.

In QCD, quarks carry a so-called colour charge, which comes in three types (red, green, blue), corresponding to the fundamental representation of the gauge group $SU(3)$. This colour charge should not be confused with the quark flavour, which can be up, down, strange, charm, top, and bottom, where only the first four are relevant in current nuclear collision experiments. The gauge bosons mediating the interactions between quarks are called gluons. Since the gauge group $SU(3)$ is non-abelian, gluons also interact among themselves. As a consequence, the colour field created by two quarks of opposing colour does not spread over all space as in electrodynamics, but is confined to a so-called flux tube. This means that the interaction energy between two quarks increases linearly with distance. A large enough increase of the distance between two quarks hence deposits enough energy in the flux tube that a new quark-antiquark pair will be created in the flux tube, not allowing a single quark to escape. For the same reason, all hadrons are colour neutral, hence internally carrying colour and anticolour (mesons) or three different colours (baryons). This property of QCD is called colour confinement.

2.1. *Experimental studies of the quark-gluon plasma*

Colour confinement does not mean, however, that quarks must always be bound to hadrons. It means that there can be no single, free colour charges. Larger chunks of nuclear matter consisting of hadrons can indeed undergo transition to a dense system of free quarks and gluons – this is the transition to the quark-gluon plasma (QGP). Figure 4 shows a simplified, schematic

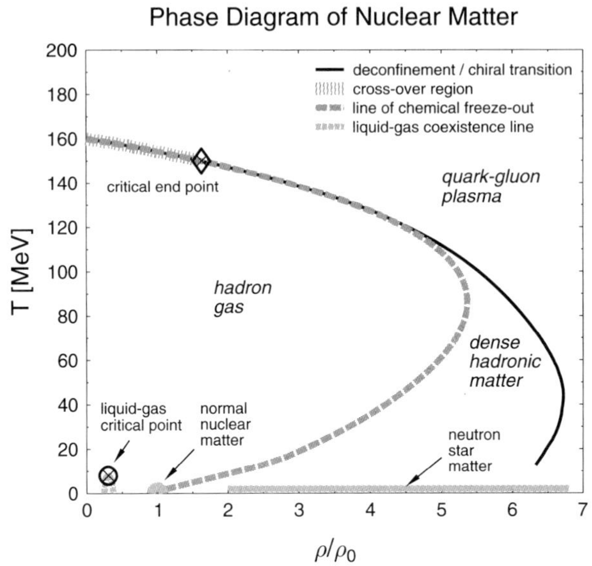

Fig. 4. A schematic view of the phase diagram of nuclear matter. For diluted, cool systems, we find the liquid-gas transition. For hot or dense systems, hadronic matter undergoes the transition to the quark-gluon plasma.

version of the corresponding phase diagram of nuclear matter. Normal nuclear matter from atomic nuclei is at $T = 0$ MeV and at ρ_0 =112 MeV/fm^3. At lower densities and slightly higher temperatures, we find the liquid-gas transition with its critical point, which manifests itself in the multifragmentation of nuclear matter. At zero temperature and higher densities, we find the bulk nuclear matter which is found in neutron stars. At higher temperatures there is the deconfinement transition, above which quarks and gluons can move freely in the hot and dense system. At the transition to the quark-gluon plasma, quark masses drop to their current masses and chiral symmetry is restored, which is why the QGP transition is also called chiral transition.

Probably the only place in nature where the quark-gluon plasma transition has ever occurred is the early universe. Nevertheless, it is possible to study this transition in the laboratory – this is the scientific aim of the ultra-relativistic heavy ion programs at GSI in Darmstadt, CERN, and the RHIC at BNL. In these experiments, heavy nuclei such as Au or Pb are brought to collisions at energies of $\sqrt{s}_{NN} \approx 7 - 18$ GeV (CERN-SPS) or even $\sqrt{s}_{NN} \approx 130 - 200$ GeV (BNL-RHIC). In such a collision, nuclear matter is compressed, and a fireball – a zone of very hot and dense nuclear matter – is created, where the transition to the QGP state occurs. In the subsequent expansion and cooling of the fireball, the quark-gluon matter condenses again to a dense, interacting system of hadrons, which further expands and undergoes the chemical freeze-out after which there are no more changes in the composition of the system. The final state hadrons are the particles that can be measured in detectors. Temperatures and chemical potentials which can be extracted from the measured hadrons at different experiments yield the curve of chemical freeze-out shown in Fig. 4.

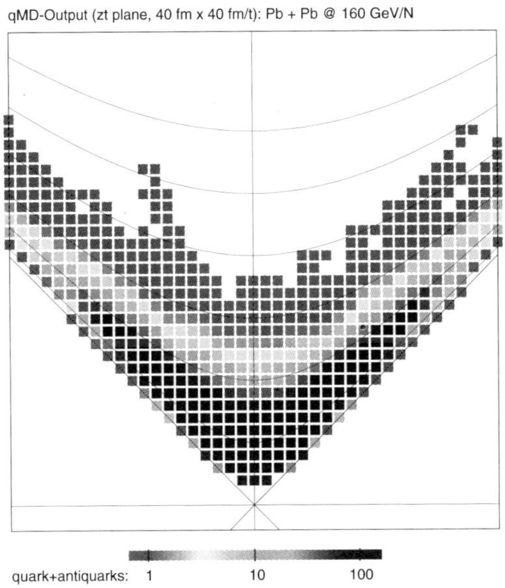

Fig. 5. Time evolution of the number of quarks and anti-quarks in a Pb+Pb collision at SPS energies ($\sqrt{s}_{NN} = 17.3$ GeV), as calculated from qMD. Quarks form clusters of three quarks or of a quark and an antiquark, which are mapped to baryons and mesons, respectively.

2.2. *Modelling the quark-gluon plasma: qMD*

Since QCD is a very complex theory which is not yet solved analytically, theoretical studies of the quark-gluon plasma always involve the construction

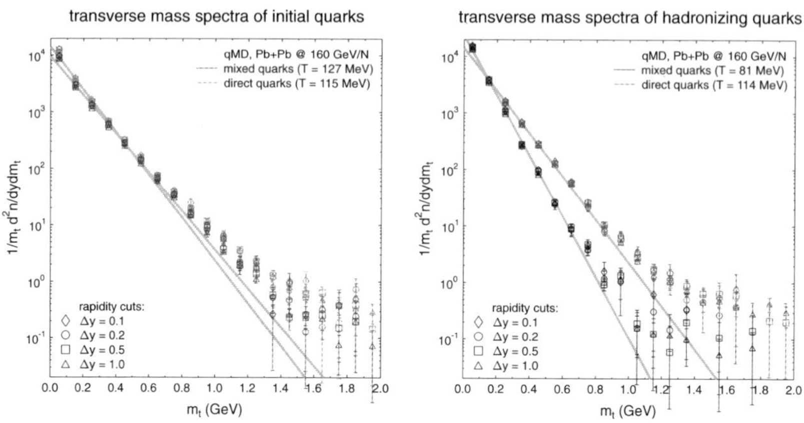

Fig. 6. Transverse mass spectra of quarks in the initial phase of the qMD calculation (left) and at hadronization (right). The spectra can be fitted quite reasonably with a thermal model. Temperatures from the slope of the spectra show cooling during the time evolution, which is much stronger for mixed quarks than for direct quarks.

of models. One such model which can be used to examine the hadronization of an expanding quark-gluon plasma is the quark molecular dynamics (qMD).

The idea of this model is to treat quarks (and antiquarks) as classical particles carrying a colour charge and interacting via a potential which increases linearly with distance and thus mimics the confining properties of colour flux tubes. The relative strength of the coupling depends on the colours of the quarks involved, and it can be both attractive and repulsive. Thus, the Hamiltonian of the model reads

$$\mathcal{H} = \sum_{i=1}^{N} \sqrt{\vec{\mathbf{p}}_i^2 + m_i^2} + \frac{1}{2} \sum_{i,j} C_{ij} V(|\vec{\mathbf{r}}_i - \vec{\mathbf{r}}_j|) , \quad V(r) = -\frac{3}{4}\frac{\alpha_s}{r} + \kappa\, r. \quad (1)$$

The time evolution of a system of quarks described by this Hamiltonian yields the formation of clusters of two quarks (quark and antiquark with colour and anticolour) and of three quarks (or three antiquarks) of three different colours. This is due to the colour-dependency of the interaction, which favours a redistribution of a homogeneous system in colour neutral clusters. In qMD, these clusters are mapped on hadronic states according to their masses and quantum numbers such as spin and isospin.

Starting from this Hamiltonian, Monte Carlo calculations show a tran-

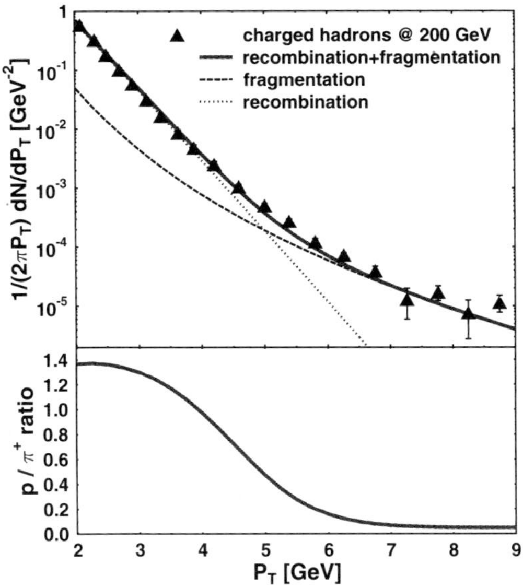

Fig. 7. Transverse momentum spectrum of charged hadrons in central Au+Au collisions at RHIC energies ($\sqrt{s}_{NN} = 200$ GeV). Data from the PHENIX experiment can be understood as showing the sum of two contributions from parton clustering and parton fragmentation, where parton fragmentation dominates at $p_\perp > 4$ GeV (top). The transition from clustering to fragmentation is also seen in the ratio of p/π^+ (bottom). Note different $p_\perp$ ranges of this plot and the plots in Fig. 6 (from Fries *et al.*.[10])

sition between two very distinct phases,[5] from one dominated by clusters at low temperatures, to a phase of free quarks at high energies. This can be seen as a simple model of the quark-gluon plasma transition.

qMD can thus be used to simulate the hadronization of the expanding fire ball in a heavy ion collision.[6] Figure 5 shows the time evolution of the number of quarks and antiquarks in a Pb+Pb collision at SPS ($\sqrt{s}_{NN} = 17.3$ GeV$/N$). One sees that after an eigenzeit ≈ 15 fm/c, hadronization is over, what is a very reasonable result. While qMD allows to calculate experimental observables like particle numbers and momentum spectra, it offers also the opportunity to look into the microscopic dynamics of hadronization not directly accessible to experiment. Figure 6 shows the transverse momentum spectra (a measure of temperature) of "direct" quarks (quark correlations from one initial hadron forming again the "same" hadron) and "mixed quarks", which regroup to form new hadronic clusters. While the initial temperatures of the two populations of quarks are essentially the same, mixed quarks show a much stronger cooling in the expansion than direct quarks. This means that, in this model, the quark system is made up of two distinct subsystems: quarks which interact and interchange their role in between the hadronic correlations, and quarks which escape, essentially without interaction, from the system, forming hadrons which can be traced back to the initial stage of the collision.

2.3. *Partonic clustering at RHIC*

One of the remarkable discoveries in the experiments at RHIC in Brookhaven has been the suppression of pion yields at transverse momenta $p_\perp > 2$ GeV in central Au+Au collisions in comparison to p+p collisions. This is generally understood as a consequence of jet quenching and interpreted as a strong signal of the creation of a quark-gluon plasma. At the collision energies of RHIC, quarks and gluons are considered as partons which may scatter with large exchange of momentum. The scattered partons then fragment into hadrons (as in the string picture mentioned before), which carry away the transverse momentum. If the scattered partons have to cross a colour-charged medium (as if produced within a QGP) before fragmentation, they lose energy by processes such as gluon bremsstrahlung, thus depositing less transverse momentum in the final hadrons. This is the simply physical picture of the processes yielding jet quenching.

However, data from RHIC showed a puzzle, which was called the proton/pion anomaly:[7] whereas jet quenching was observed in the transverse momentum spectra of pions as expected, it was found to be much smaller for protons and antiprotons. This would mean that the partons fragmenting into baryons would suffer less energy loss in the medium than those producing pions, which is hard to understand. This missing suppression for baryons is seen best in the ratio of protons to pions at transverse momenta of 2–3 GeV, where it surmounts 1 – a very unusual result. A similar riddle showed up in the analysis of elliptic flow.[8,9]

These problems can be solved by considering not only parton fragmentation, but also parton recombination:[10–12] in the colour-charged medium, scattered partons can recombine and cluster to form colour-neutral hadrons. This is the same idea as in the qMD model. Thus, partons with relatively small transverse momentum can cluster to build up protons with $p_\perp \sim 2$–3 GeV without the need of parton fragmentation. Figure 7 shows how the combination of both parton clustering and fragmentation yields an excellent description of the transverse mass spectra of charged hadrons. It also shows how the observed high ratio of p/π can be understood – and makes the prediction that this ratio should drop at higher transverse momenta.

It should be noted that the parton recombination involved to explain the RHIC observables do not include microscopic parton dynamics, which is carried out in qMD for quarks. Instead, they work by coalescence in phase space. It would be tempting to apply qMD to look at these questions for RHIC events. However, beside the problem of applying the instantaneous potential interaction at RHIC energies, at the moment it is hard to obtain enough statistics with qMD simulations to get reliable data for the high $p_\perp$ region which is most interesting. Note that the $m_\perp$ spectra in Fig. 6 (which are essentially $p_\perp$ spectra for massless quarks) end in statistical noise at $m_\perp = 2$ GeV, which is the lower $p_\perp$ offset of Fig. 7.

2.4. *Nucleonic clustering at RHIC: antimatter*

Once partons in a heavy ion collision have fragmented or recombined to hadrons, this is not the end of the story as far as clustering phenomena are concerned. The dense hadronic medium in the expanding fireball after the transition from the QGP to hadrons allows for many rescattering processes. During rescattering, the formation of nuclei and even anti-nuclei by coalescence of nucleons is possible.[13]

The STAR collaboration has been looking for anti-deuteron and anti-helium in the final particle yields of Au+Au collisions at RHIC.[14] Production rates of $\bar{\mathrm{d}}$ and $^3\overline{\mathrm{He}}$ were found which are larger than in nucleus-nucleus collisions with lower energies – this can be understood by the much more copiously produced anti-nucleons in the nearly net-baryon free fireball of a RHIC event, as compared to the net-baryon rich events at lower energies. In fact, the production of light anti-nuclei fits very well the expectations from anti-nucleon coalescence models: anti-nucleons cluster together to form anti-deuteron and anti-helium.

3. Conclusion

We have presented two examples of fragmentation and clustering phenomena from nuclear physics at mediate and high energies. There are many other cases in nuclear physics where clustering and fragmentation are important – strange nuclear matter with such objects as the pentaquark states, strangelets and MEMOs are among them. The examples presented here can only give a scarce impression of this very rich and interesting field.

Acknowledgements

The authors thank Steffen Bass and Marcus Bleicher for helpful hints and fruitful discussions.

References

1. G. Peilert *et al*, *Phys. Rev.* C **39**, 1402 (1989).
2. M. Kleine Berkenbusch *et al.*, *Phys. Rev. Lett.* **88**, 022701 (2002).
3. J. B. Elliott *et al.* [EOS Collaboration], *Phys. Rev.* C **67**, 024609 (2003).
4. T. Nakano *et al.* [LEPS Collaboration], *Phys. Rev. Lett.* **91**, 012002 (2003).
5. M. Hofmann *et al. Phys. Lett.* B **478**, 161 (2003).
6. S. Scherer *et al. New J. Physics* **3**, 8 (2001).
7. I. Vitev, M. Gyulassy, *Phys. Rev.* C **65**, 041902 (2002).
8. Z. W. Lin, C. M. Ko, *Phys. Rev. Lett.* **89**, 202302 (2002).
9. S. A. Voloshin, *Nucl. Phys.* A **715**, 379c (2003).
10. R. J. Fries *et al. Phys. Rev. Lett.* **90**, 202303 (2003).
11. V. Greco *et al. Phys. Rev. Lett.* **90**, 202302 (2003).
12. D. Molnar, S. A. Voloshin, *Phys. Rev. Lett.* **91**, 092301 (2003).
13. C. Spieles *et al. Phys. Rev.* C **53**, 2011 (1996).
14. C. Adler *et al.* [STAR Collaboration], *Phys. Rev. Lett.* **87**, 279902 (2001).

DYNAMICS OF MULTIPLE EVAPORATION IN THE MIXED ATOMIC Ar_6Ne_7 CLUSTER

P. Parneix and Ph. Bréchignac

Laboratoire de Photophysique Moléculaire, C.N.R.S.,
Bât. 210, Université Paris-Sud, F 91405 ORSAY CEDEX, France
E-mail: pascal.parneix@ppm.u-psud.fr

The dynamics of multiple evaporation in the mixed Lennard-Jones atomic cluster Ar_6Ne_7 has been studied from classical molecular dynamics simulations. A relationship between the liquid to gas like phase transition and pertinent observables has been explored. In particular the mean kinetic energy of the atomic fragments and the ratio between successive times of evaporation have been carefully analyzed as a function of energy to find such a link between thermodynamics and multi-evaporation dynamics.

1. Introduction

Very recently the liquid to gas like phase transition has been experimentally characterized in free clusters.[1–3] As the "boiling" of a free cluster is intrinsically linked to the occurrence of multiple evaporation events during the experimental time-scale of interest, the study of the dynamics and energetics of this process is important. Theoretical studies have already shown that the melting of a free cluster could be probed from the analysis of the kinetic energy release following the evaporation of non-rotating[4–6] and also of rotating clusters.[7] This effect is a direct consequence of the change in the evolution of the vibrational entropy as a function of energy near the melting transition in the product cluster. Pursuing this idea, we want to analyze the evolution of this observable at higher energy, i.e. near the boiling of the atomic cluster. Moreover, a careful analysis of the evaporation times characterizing the sequential loss of monomers could also yield information on the "boiling" of small clusters.

In a previous study[6] concerning the vibrational dynamics of mixed Lennard-Jones (LJ) Ar_pX_m clusters, it was shown that the use of Neon atoms for X has the property to enhance the overall dynamics, so that evaporation occurs on a shorter timescale for a given energy. It is then an adequate case to be able to observe successive evaporation events along the same trajectory without the need to extend its duration in an unreasonable way. For the purpose of the present study we have chosen to focus on the Ar_6Ne_7 cluster, for which the melting and boiling temperatures have been previously calculated.[8,9]

2. Theory and Computational Details

2.1. *Statistical theory*

We note $k^{(n)}(E)$ the evaporation rate for the dissociation of the n-atom cluster. In the phase space theory framework, $k^{(n)}$ is given by[10]

$$k^{(n)}(E) = C_n \frac{\int_0^{E-E_{0,n}} \Gamma(\epsilon; J=0)\, \omega_{n-1}(E - E_{0,n} - \epsilon)\, d\epsilon}{\omega_n(E)} \,. \tag{1}$$

C_n is a constant which does not depend on the energy E. ω_n and ω_{n-1} are respectively the vibrational densities of states for the parent and product clusters. $\Gamma(\epsilon; J = 0)$ is the rotational density of states (RDOS). Finally $E_{0,n}$ corresponds to the energy difference between the most stable isomers of the parent n-atom cluster and the product (n-1)-atom cluster. When we neglect the centrifugal barrier in the exit channel, the RDOS can be considered as a linear function with respect to the kinetic energy release ϵ. As the kinetic energy release (KER) distributions are relatively peaked, we will not consider the integral in the numerator of the previous equation but only the ensemble-averaged value of ϵ, noted $\overline{\epsilon}_n$. Thus we obtain

$$k^{(n)}(E) = C'_n \frac{\omega_{n-1}(E - E_{0,n} - \overline{\epsilon}_n)}{\omega_n(E)} \,, \tag{2}$$

in which $C'_n = \overline{\epsilon}_n \times C_n$.

As we are interested in the statistical description of the sequential evaporation, the previous equation can be written for two successive events. The ratio between these two evaporation rates is given by

$$\frac{k^{(n-1)}(E - E_{0,n} - \overline{\epsilon}_n)}{k^{(n)}(E)} = \frac{C'_{n-1}}{C'_n} \times \frac{\omega_{n-2}(E - E_{0,n} - E_{0,n-1} - \overline{\epsilon}_n - \overline{\epsilon}_{n-1})}{\omega_{n-1}(E - E_{0,n} - \overline{\epsilon}_n)} \times \frac{\omega_n(E)}{\omega_{n-1}(E - E_{0,n} - \overline{\epsilon}_n)} \,. \tag{3}$$

We note $< t_{p+1} - t_p >$ $(=1/k^{(n-p)})$ the ensemble-averaged evaporation time of the (n-p)-atom cluster, n being the initial cluster size. From this definition, $< t_1 >$ and $< t_2 >$ correspond respectively to the ensemble-averaged evaporation times for the loss of one and two atoms. The previous equation can be rewritten as

$$\ln[\frac{< t_1 >}{< t_2 - t_1 >}] = \ln[\frac{C'_{n-1}}{C'_n}] + \frac{1}{k_B}[S_n(E) \\ +S_{n-2}(E - E_{0,n} - E_{0,n-1} - \overline{\epsilon}_n - \overline{\epsilon}_{n-1}) \\ -2S_{n-1}(E - E_{0,n} - \overline{\epsilon}_n)] \ , \tag{4}$$

in which S_n is the vibrational entropy of the n-atom cluster and k_B is the Boltzmann's constant.

If we make the approximation $E_{0,n} = E_{0,n-1} = E_0$, we obtain a more compact equation

$$\ln[\frac{< t_1 >}{< t_2 - t_1 >}] = \ln[\frac{C'_{n-1}}{C'_n}] + \frac{1}{k_B}[S_n(E) \\ +S_{n-2}(E - 2E_0 - \overline{\epsilon}_n - \overline{\epsilon}_{n-1}) \\ -2S_{n-1}(E - E_0 - \overline{\epsilon}_n)] \ . \tag{5}$$

It is interesting to transform this expression in the harmonic limit. We obtain

$$\ln[\frac{< t_1 >}{< t_2 - t_1 >}] = \ln[\frac{C'_{n-1}}{C'_n}] + (3n - 7)\ln(E) \\ +(3n - 13)\ln(E - 2E_0 - \overline{\epsilon}_n - \overline{\epsilon}_{n-1}) \\ -(6n - 20)\ln(E - E_0 - \overline{\epsilon}_n) \ , \tag{6}$$

with $\overline{\epsilon}_n = \frac{2(E-E_0)}{3n-7}$ and $\overline{\epsilon}_{n-1} = \frac{2(E-2E_0-\overline{\epsilon}_n)}{3n-10}$, expressions of the mean kinetic energy release in the harmonic approximation with the RDOS taken as a linear function with respect to ϵ.

When we consider the distribution of kinetic energy release in this harmonic limit, it can be shown analytically that the expression becomes

$$\ln[\frac{< t_1 >}{< t_2 - t_1 >}] = \ln[\frac{C'_{n-1}}{C'_n}] + \ln[\frac{\overline{\epsilon}_n}{\overline{\epsilon}_{n-1}}] + \ln[\frac{(3n-8)^2(3n-9)}{(3n-10)(3n-11)(3n-12)}] \\ +(3n - 7)\ln(E) \\ +(3n - 9)\ln(E - 2E_0) \\ -(6n - 16)\ln(E - E_0) \ . \tag{7}$$

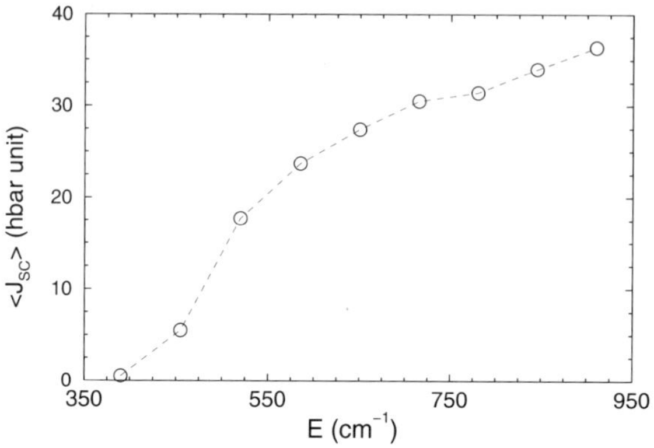

Fig. 1. Mean angular momentum of the sub-cluster as a function of E.

This expression (7), valid in the harmonic limit, will be later used in order to compare with the results of MD calculations.

2.2. *MD simulation*

The multi-evaporation process has been studied from classical Molecular Dynamics simulations. A fifth-order Adams-Moulton predictor-corrector algorithm was used to integrate Hamilton's equations. The integration time step was equal to 4 fs. For each energy, 4000 independent trajectories were propagated during 4 ns. Each 2.5 ps the cluster was interrogated in order to determine the number of evaporated atoms and their kinetic energy, calculated in the laboratory frame. One atom was considered as evaporated when its distance to all the other atoms was larger than 12 Å. The angular momentum of the sub-cluster was finally deduced. Times associated to the successive evaporation events were also recorded in order to obtain additional dynamical information on the processes. The potential energy surface was built as a sum of pairwise atom-atom LJ potentials [σ_{Ar-Ar}=3.405 Å, σ_{Ne-Ne}=2.749 Å, ϵ_{Ar-Ar}= 83.26 cm^{-1} and ϵ_{Ne-Ne}= 24.74 cm^{-1}]. The LJ potential parameters for the Ar-Ne interaction were deduced from the empirical Lorentz-Berthelot combination rules.

3. Results and Discussion

First of all, we recall the main features of the mixed Ar_6Ne_7 cluster. Its lowest energy isomer is icosahedral and the corresponding binding energy

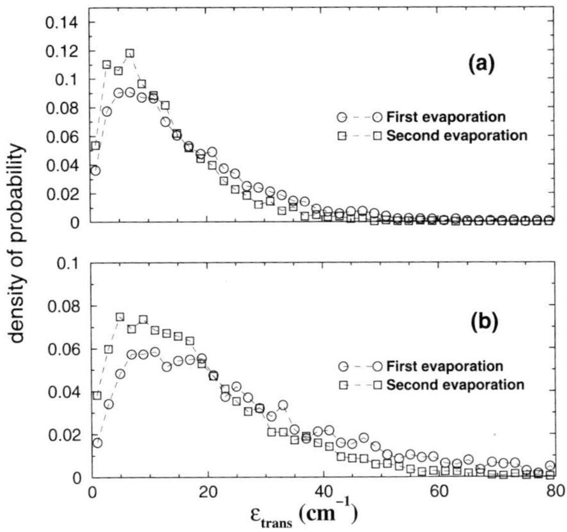

Fig. 2. Distributions of the translational kinetic energy of the first two ejected Ne atoms for 2 different energies: (a) $E = 650$ cm^{-1}; (b) $E = 910$ cm^{-1}.

is equal to -2142.7 cm^{-1}. Thermodynamical informations have been previously collected from the calculation of the vibrational density of states.[8,9] The melting temperature is equal to 8.5 K and the liquid-gas like transition occurs near 25 K. As demonstrated previously,[9] only Ne atoms are ejected from the mixed Ar/Ne clusters in the energy range studied here. The mean number of Ne atoms, $< Ne >$, is a linear function of E and $< Ne >$ becomes larger than 1 in the range of energy which corresponds to the increase of the heat capacity, which signals the liquid-to-gas like phase transition, *i.e.* around E=520 cm^{-1} (here n=13). Moreover, it is important to note that only very few dimers are observed in the dissociation process. Consequently only the Ar_6Ne_p ($p \leq 7$) will be considered in the statistical description of the sequential evaporations. The energy difference between Ar_6Ne_p and Ar_6Ne_{p-1} has been found always very close to $E_0 = 160$ cm^{-1}, whatever the value of p is, which is in accord with the linear behavior of $< Ne >= f(E)$.

First information on the energetics can be derived from the behavior of the rotational excitation of the sub-cluster which is not broken after 5 ns. The mean angular momentum $< J_{sc} >$ is plotted in Fig. 1 as a function of E. A saturation in the increase of this quantity clearly appears around E=520 cm^{-1}, which is a direct consequence of the multi-evaporation regime. A

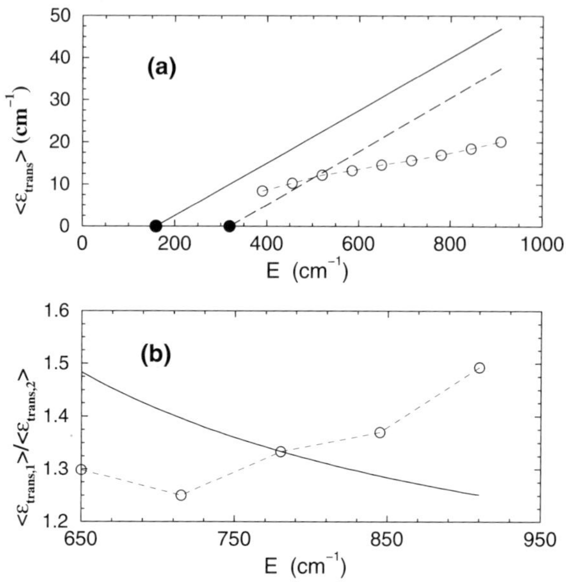

Fig. 3. (a) Mean translational kinetic energy of the ejected Ne atoms as a function of E, the lines correspond to the harmonic values (solid = first evaporation, dash = second evaporation); (b) ratio between the mean translational kinetic energies of the first and second evaporation. The solid line corresponds to the harmonic prediction.

first evaporation of the initially non-rotating cluster automatically induces a rotational heating. When the rotational angular momentum of this first product sub-cluster is larger than a critical value, the second evaporation will induce a decrease of the angular momentum. Following a succession of Neon atom losses, the mean angular momentum will reach a limiting value, only slightly dependent on E. This competition between rotational heating and cooling has been recently described from the PST formalism in the case of evaporation of rotating clusters.[11] In the multi-evaporation regime, the rotational energy of the sub-cluster will thus also be only very slightly dependent on E. Consequently, the evolution of the translational contribution versus E can be significantly compared with the evolution of KER (rotation + translation).

The distribution of the translational kinetic energy release has been plotted in Fig. 2 for $E = 650$ and 910 cm^{-1}, both for the first and second evaporations. The characteristic width of these distributions, noted $\Delta\epsilon$, is equal to about 20 cm^{-1}, which is very small with respect to $E - E_0$ (equal to 490 and 750 cm^{-1} respectively for E= 650 and 910 cm^{-1}). The condition $\Delta\epsilon << E - E_0$ makes the approximation undertaken for calculating $\ln[\frac{<t_1>}{<t_2-t_1>}]$ in the previous section justified. In Fig. 3a, the mean kinetic energy of all the ejected Ne atoms (calculated at the end of the evaporative trajectories, i.e. after 5 ns) is plotted as a function of energy E. It appears that $< \epsilon_{trans} >$ increases linearly as a function of E but with a change of the slope occurring around E=520 cm^{-1}, obviously linked to the influence of the liquid-to-gas like phase transition. The solid line and the long-dashed line respectively correspond to the harmonic approximations for the evaporation of the first and second atom. When compared to the expectations from these harmonic approximations, it is striking that ϵ_{trans} is only weakly dependent on E, which is mainly due to the effect of the liquid-to-gas like phase transition. It is important to note that this effect could be much more important if the duration of the simulation trajectory was much larger than 5 ns. Indeed more evaporation events could be detected with low translational kinetic energies.

In Fig. 3b, the ratio between the mean translational kinetic energy releases in the first and second evaporation events, obtained from MD simulations, has been plotted and compared to the harmonic prediction. Although this ratio is a monotonic decreasing function versus E in the harmonic approximation, the ratio obtained from MD simulations is now increasing. This effect can be explained in the following way: the kinetic energy released in the second evaporation is much lower than predicted by the harmonic approximation, and this becomes more and more true as the energy increases. This reflects directly that a significant portion of the excess energy is devoted to induce the liquid-like to gas-like transition and consequently less energy is kept within the cluster for the successive evaporations.

To confirm this interpretation we have plotted in Fig. 4 the quantity $\ln[\frac{<t_1>}{<t_2-t_1>}]$ as a function of E. By comparing the MD results with the harmonic PST prediction (Eq. (7)), it is clear that the evaporation time ($< t_2 - t_1 >$) devoted to the loss of the second Ne atom is larger than expected in the harmonic description. Again this feature is more pronounced at high energy.

4. Conclusion

We have explored in this work the thermodynamical behavior of the model atomic cluster Ar_6Ne_7, in the regime where multiple evaporation of atoms takes place. Very large deviations from the predictions of phase space theory in the harmonic limit have been found in the results of Molecular Dynamics

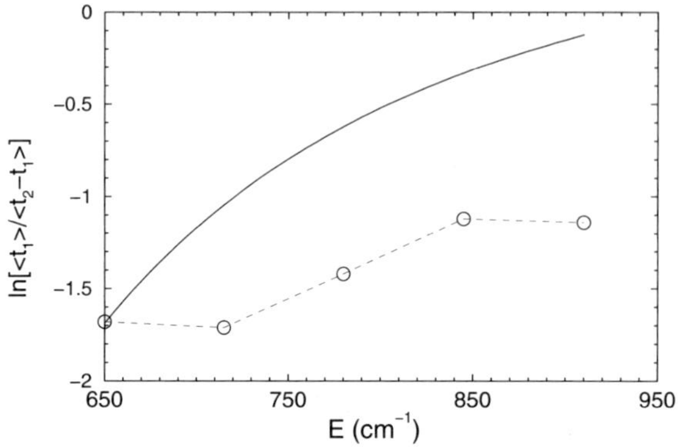

Fig. 4. $\ln[\frac{<t_1>}{<t_2-t_1>}]$ as a function of E from MD simulations (open circles), from harmonic PST with KER distribution (solid line).

simulations. It is shown that observables, such as the kinetic energy release and the evaporation rates associated with successive atom losses, which are potentially accessible to experiment, can be sensitive probes of the dynamics of fundamental importance accompanying the liquid-like to gas-like phase transition in isolated clusters.

References

1. M. Schmidt, T. Hippler, J. Donger, W. Kronmüller, B. von Issendorff, H. Haberland, and P. Labastie, *Phys. Rev. Lett.* **87**, 203402 (2001).
2. C. Bréchignac, Ph. Cahuzac, B. Concina, J. Leygnier, *Phys. Rev. Lett.* **89**, 203401 (2002).
3. F. Gobet, B. Farizon, M. Farizon, M.J. Gaillard, J.P. Buchet, M. Carré, P. Scheier, and T.D. Märk, *Phys. Rev. Lett.* **89**, 183403 (2002).
4. S. Weerasinghe, F.G. Amar, *J. Chem. Phys.* **98**, 4967 (1993).
5. P. Parneix, F.G. Amar, Ph. Bréchignac, *Chem. Phys.* **239**, 121 (1998).
6. P. Parneix and Ph. Bréchignac, *J. Chem. Phys.* **118**, 8234 (2003).
7. F. Calvo, P. Parneix, *J. Chem. Phys.* **119**, 256 (2003).
8. D.D. Frantz, *J. Chem. Phys.* **107**, 1992 (1997).
9. G.S. Fanourgakis, P. Parneix and Ph. Bréchignac, *Eur. Phys. J.* D **24**, 207 (2003).
10. M.F. Jarrold, "Introduction to statistical reaction theories", in *Clusters of Atoms and Molecules I*, edited by H. Haberland, Springer-Verlag, Berlin, (1991).
11. P. Parneix, F. Calvo, *J. Chem. Phys.* **119**, 0000 (2003).

Electron Scattering on Clusters

LOW-ENERGY ELECTRON ATTACHMENT TO VAN DER WAALS CLUSTERS

I. I. Fabrikant

Department of Physics and Astronomy, University of Nebraska, Lincoln, NE 68588, USA E-mail: ifabrikant1@unl.edu

H. Hotop

Fachbereich Physik, Universität Kaiserslautern, 67663 Kaiserslautern, Germany E-mail: hotop@physik.uni-kl.de

We review experimental data on electron attachment to CO_2 and N_2O clusters showing very narrow vibrational Feshbach resonances of the type $[(XY)_{N-1}{\cdot}XY(\nu \geq 1)]^-$ which occur at energies below those of neutral cluster $[(XY)_{N-1}{\cdot}XY(\nu \geq 1)]$. These resonances appear due to the stabilization of the cluster anion by the polarization interaction between the electron and the cluster. Based on this result, we develop an R-matrix model describing nondissociative electron attachment to CO_2 clusters. The results of calculations describe major features in electron attachment: very narrow vibrational Feshbach resonances and the weak dependence of their widths on the cluster size.

1. Introduction

One of the interesting features of Van der Waals clusters is their role as nanoscale prototypes for studying the effects of solvation on the characteristics of both solvent and solvated particle, due to the interaction between a solvated molecule or ion and its surrounding solvent environment. Solvation effects also play a key role in the formation of negative ions by attachment of slow electrons to clusters. Since the first pioneering work of Klots and Compton,[1,2] many interesting features have been observed in electron attachment to molecular clusters of the type $(XY)_N$, including a prominent resonance at zero energy in cases where such a feature is absent in monomers XY.[3–5] Using the laser photoelectron attachment (LPA) method at energy width around 1 meV, it was recently demonstrated[6,7] that the anion yield in electron collisions with N_2O and CO_2 clusters within the energy range 0-180 meV is mediated by narrow vibrational Feshbach resonances (VFR), i.e. temporary negative ion states of the type $[(XY)_{N-1}\cdot XY(\nu_i \geq 1)]^-$, which occur at energies below those of neutral cluster $[(XY)_{N-1}\cdot XY(\nu_i \geq 1)]$. It was thus shown that the 'zero energy resonance' in N_2O and CO_2 clusters, as observed in the early work, is due to the combined influence of the previously non-resolved, overlapping VFRs.

Two very different types of VFRs were detected by the experimental group at the University of Kaiserslautern. The first type, observed in electron attachment to methyl iodide dimers and trimers,[8] has its origin in dissociative attachment to the monomer. With increasing number of monomers in the cluster, the energy of VFR is rapidly shifting away from the vibrational excitation threshold and the resonance width is rapidly growing. Essentially no structure is left in the attachment spectrum for $(CH_3I)_2\cdot I^-$. This phenomenon was explained by the effects of solvation and increased electron-target long-range polarization interaction in dissociative attachment.

In contrast, the VFRs observed in electron attachment to CO_2 and N_2O clusters remain sharp with increasing N. The position of VFRs in these systems can be explained by simple model calculations[7,9] for the binding energy of the captured electron in the VFR state relative to the energy of the neutral cluster which carries the same amount of intramolecular vibrational energy.

In the present paper we develop this model further and apply it to calculation of nondissociative attachment cross sections. We present the major features observed in electron attachment allowing us to formulate quantitative theory which is applied then to electron attachment to CO_2 clusters.

2. Basic Experimental Features

Before formulating the theoretical model, we summarize the major experimental results.

Low-energy electron attachment to N_2O clusters produces heterogeneous $(N_2O)_qO^-$ and homogeneous $(N_2O)_p^-$ cluster anions in a highly size selective way, with the dominant anion species to be heterogeneous cluster anions with $q = 5, 6$ and the homogeneous cluster anions with $p = 7, 8$.[6,9] In all attachment spectra, astoundingly narrow peaks are observed at energies close to, but not identical with, the excitation energies for the bending ($\nu_2 = 1, 2$) and the N–O stretching ($\nu_1 = 1$) vibrational mode of free N_2O molecules. The width of these peaks is very narrow (in the few meV range), but substantially broader than the experimental resolution, and is essentially independent of cluster ion size.

Similar observations were made for electron attachment to $(CO_2)_N$ clusters[7] except that only homogeneous ions $(CO_2)_q^-$ ($q \geq 4$) were observed in the anion mass spectra. The narrow peaks in the attachment spectra were interpreted as VFRs of the type $[(CO_2)_{N-1}CO_2(\nu_i)]^-$ with the vibrational excitation ν_i corresponding to the bending mode (010) and the Fermi resonance combining the bending mode (020) and the symmetric stretch (100). For each vibrational series, the resonance position of the VFR mirrors the binding energy of the captured electron in the $[(CO_2)_{N-1}CO_2(\nu_i)]^-$ anion state relative to the energy of the neutral $[(CO_2)_{N-1}CO_2(\nu_i)]$ cluster which carries the same quanta of intramolecular energy. The binding energy can be estimated[7,9] using a simple model potential including the long-range attraction between the electron and the cluster and a short-range interaction V_0 at distances smaller than the cluster radius R_N. We only take into account the polarization attraction $V_{pol} = -Ne^2\alpha/2r^4$, where α is the polarizability of the monomer, and cut it off at the cluster radius; moreover, we set V_0 constant at electron-cluster distances smaller than $R_N = R_0(1.5N)^{1/3}$ (R_0 = effective radius of a monomer) and treat V_0 as a parameter. Figure 1 shows the results of these calculations using six constant values of the short range potential V_0. The experimental results for attachment to N_2O clusters are well described by this model with $V_0 = 0.2$ eV. For CO_2 clusters, a better fit is obtained if we assume that V_0 depends on the cluster size. Calculations of electron affinities to clusters[10] suggest that this dependence can be described by the equation

$$V_0 = aN^{-1/3} + b. \tag{1}$$

The dashed curve in Fig. 1 represents the results for the binding energy obtained by using Eq. (1) with $a = 0.7$ eV and $b = -0.866$ eV. The difference between N_2O and CO_2 in N-dependence of the binding energies lies in the average short range interaction between the electron and the respective molecular constituents (the polarizabilities of N_2O and CO_2 agree to within 5%). This conclusion is confirmed by analysis[7] of experimental results and theoretical calculations of low-energy electron scattering by CO_2 and N_2O molecules.

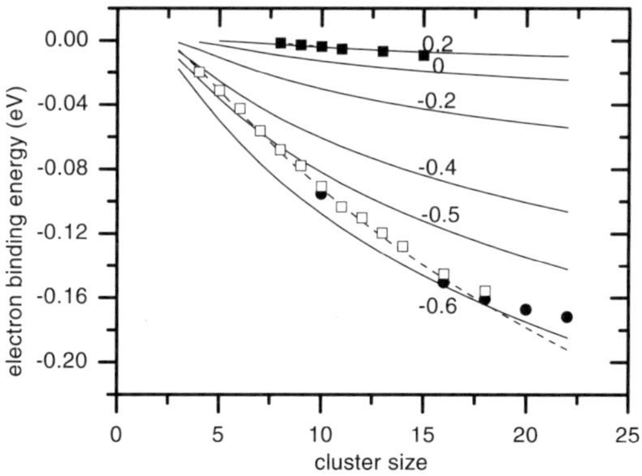

Fig. 1. Binding energies of electron in a cluster as a function of the cluster size N for different values of the short-range potential V_0 (in eV). Dashed curve presents calculations with a variable depth. Experimental data: open squares, CO_2, average of the two (020) and (100) series; full circles, CO_2, average of the two (030) and (110) series; full squares, N_2O, (100) series.

3. Theoretical Model

Since only homogeneous ions $(CO_2)_q^-(q \geq 4)$ are observed in electron attachment to CO_2 clusters, this case seems to be simpler for our first theoretical attempt to calculate attachment cross sections. We neglect the dissociative attachment channels and include explicitly vibrational channels of CO_2 and CO_2^-. Each vibrational channel of CO_2^- is also coupled with the phonon modes of the cluster that provides the path to nondissociative attachment. This coupling is included phenomenologically by adding an imaginary part to the resonance energy of the intermediate negative ion state. Details of this approach were given by Thoss and Domcke.[11] They worked out an effective Hamiltonian describing the interaction of the system mode (in our case, a vibrational mode of a single molecule) with residual bath modes (in our case, phonon modes of the cluster). In the Markov approximation the effective Hamiltonian is reduced to the Hamiltonian of a damped harmonic oscillator with a damping rate and a frequency shift related to the bath spectral density. In the absence of information on the spectral density of the phonon modes in CO_2 clusters, we consider the damping rate as an empirical parameter, and for the frequency shift we use the polarization shift obtained from model calculations as described above.

Since the long-range polarization interaction plays the crucial role in formation of VFRs, we use the resonance R-matrix approach[12] in which long-range effects can be easily incorporated. We choose the origin of our coordinate system at the center of mass of a particular monomer which undergoes vibrational excitation by the electron impact and divide the whole space into two regions. Outside a sphere of radius r_0 we include only the long-range interaction between the electron and the cluster. The matrix of the radial solutions for the electron wave functions in this region has the following form

$$\mathbf{u} = \mathbf{u}^- - \mathbf{u}^+\mathbf{S} \tag{2}$$

where $\mathbf{S}$ is the scattering matrix, and $\mathbf{u}^\pm$ are ingoing and outgoing wave solutions. Matrix (2) is matched with the internal wave functions in the form

$$\mathbf{u}(\mathbf{r_0}) = \mathbf{R}\frac{d\mathbf{u}}{dr}|_{r=r_0} \tag{3}$$

where the R matrix has the following form in the fixed-nuclei approximation

$$\mathbf{R}(q) = \frac{\gamma(q) \times \gamma(q)}{W(q) - E_e} + \mathbf{R}_b, \tag{4}$$

where q represents the totality of all vibrational coordinate of the molecule, $\gamma(q)$ is the vector of the surface amplitudes, E_e is the electron energy, and $\mathbf{R}_b$ is the background (nonresonant) term. The function $W(q)$ is the R-matrix pole as a function of vibrational coordinates. Quite often it can be associated with the position of the resonance in electron-molecule scattering. However, in the present application we are interested in the near-threshold electron scattering by the CO_2 molecule which is dominated by a virtual state transformed into a bound state due to polarization interaction between the electron and the cluster environment. Therefore the function $W(q)$ can be connected to the position of the virtual or a bound state in the complex energy plane, but it does not correspond to a resonance.

To describe properly threshold effects, we take into account the vibrational dynamics according to Schneider *et al.*[13] The denominator in Eq. (4) is replaced by the operator $T + U(q) - E$, where T is the kinetic energy operator for the nuclear motion, E is the total energy of the system, and $U(q) = W(q) + U_0(q)$, where $U_0(q)$ is the potential surface of the neutral molecule. Introducing the explicit dependence of the surface amplitudes on the electron orbital angular momentum l, we can rewrite the dynamical R matrix in the form

$$R_{vv'}^{ll'} = \sum_\lambda \frac{\langle v|\gamma_l|\lambda\rangle\langle\lambda|\gamma_{l'}|v'\rangle}{\epsilon_\lambda - E} + R_b^{ll'}\delta_{vv'}. \tag{5}$$

where ϵ_λ and $|\lambda\rangle$ are eigenvalues and eigenstates of the negative-ion Hamiltonian $H_I = T+U(q)$, and $|v\rangle$ are eigenstates of the molecular Hamiltonian $H_0 = T + U_0(q)$.

As was outlined above, the coupling of vibrational modes of an individual molecule with the phonon modes of the cluster is introduced by the substitution

$$\epsilon_\lambda \to \epsilon_\lambda + \Delta - \frac{i}{2}\Gamma \tag{6}$$

where we assume that neither the polarization shift Δ nor the damping rate Γ depend on specific vibrational mode. Substitution (6) makes the Hamiltonian of the problem non-Hermitian and the S matrix nonunitary. The attachment cross section to the vth vibrational state of the neutral molecule can be determined as

$$\sigma_v = \frac{\pi}{k^2} \sum_{ll'v'} [\delta_{ll'}\delta_{vv'} - |S^{ll'}_{vv'}|^2]. \tag{7}$$

For specific calculations of electron attachment to CO_2 clusters we have made several simplifying assumptions. First, we assume that only one vibrational mode of the monomer is involved in the attachment process. More specifically, we include only symmetric stretch vibrations in CO_2. It should be emphasized, that in general bending vibrations in CO_2 are important for two reasons: first, the bent CO_2 molecule acquires a dipole moment[14] and the virtual state supported by the linear configuration turns into bound state even without the polarization due to the cluster environment. The related threshold structures were recently observed[15] in vibrational excitation of CO_2. Secondly, symmetric stretch and bending vibrations in CO_2 interact due to the Fermi resonance: two quanta of the bending vibrations correspond, almost exactly, to the one quantum of the symmetric stretch vibrations, and in order to find the true eigenstates of the vibrational Hamiltonian, anharmonic terms should be included. In particular, a pronounced selectivity was observed[16] in the excitation of the Fermi-coupled vibrations (100)/(020) in the virtual-state range. Therefore our calculations, employing one-mode approximation, have a model character. Nevertheless, they represent the important physics of the process by incorporating the stabilization of the negative ion due to polarization interaction between the molecule and the cluster.

In the linear-configuration approximation, the dipole moment of CO_2 can be neglected, and the electron interaction with the cluster becomes nearly isotropic. We will assume that the molecule which undergoes vibra-

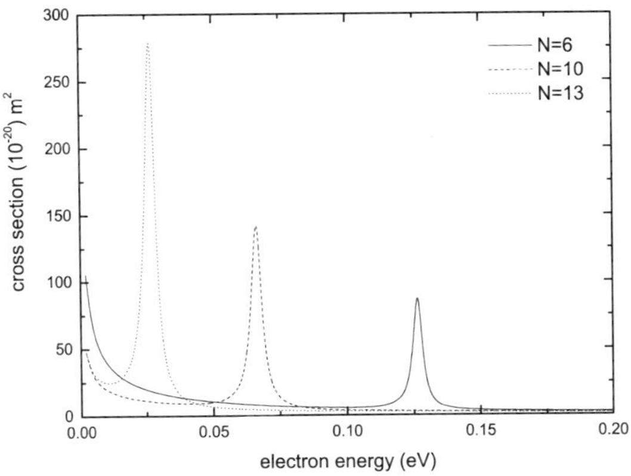

Fig. 2. Cross sections for electron attachment to CO_2 clusters for different cluster sizes. The damping width is 105 meV.

tional excitation is located near the center of the cluster. Then the electron-cluster interaction potential remains isotropic even in the reference frame with the origin at the molecular center of mass, and the interaction outside the R-matrix sphere can be described by the model potential discussed in Sec. 2.

Our next simplification is employing the displaced harmonic oscillator model[17] according to which both neutral potential curve $U_0(q)$ and the negative-ion curve $U(q)$ are described by the harmonic potential with the same frequency. In this case all Franck-Condon factors entering Eq. (4) are conveniently expressed in terms of Laguerre polynomials. This model is useful for description of vibrational motion which does not lead to dissociation (in our case, to dissociative attachment) and has been employed in several calculations of vibrational excitation of molecules[17] and attachment to clusters.[8]

4. Results

The R-matrix parameters of our model were chosen to reproduce major features of low-energy electron scattering by CO_2 molecules: very large elastic cross section at zero energy corresponding to the scattering length $A = -7.2$ a.u.[18] and a sharp threshold peak in excitation of symmetric stretch vibrations.[19] Both features appear due to a virtual state which becomes bound due to the electron-cluster polarization interaction.

In Fig. 2 we present the energy dependence of an electron attachment cross section for different cluster sizes. The model incorporates vibrational excitation of a single mode whose energy, 166 meV, is equal to the average energy of the Fermi-coupled pair (020)/(100). The damping width Γ entering Eq. (6) was varied in a broad range between 13 and 105 meV. The corresponding VFR width is varying in a narrow range between 2 and 5 meV in accord with experimental observations. The width of VFR does not change significantly with N. The resonance amplitude is growing with N, however. In the zero-energy regions the cross section is proportional to $1/E^{1/2}$. However, its absolute value strongly depends on the cluster size.

Acknowledgments

This work was supported by Forschergruppe Niederenergetische Elektronenstreuprozesse and U.S. National Science Foundation, Grant No. PHY-0098459.

References

1. C.E. Klots and R.N. Compton, *J. Chem. Phys.* **67**, 1779 (1977).
2. C.E. Klots and R.N. Compton, *J. Chem. Phys.* **69**, 1636 (1978).
3. T.D. Märk *et al*, *Phys. Rev. Lett.* **55**, 2559 (1985).
4. A. Stamatovic *et al*, *J. Chem. Phys.* **83**, 2942 (1985).
5. M. Knapp *et al*, *J. Chem. Phys.* **85**, 636 (1986).
6. J.M. Weber *et al*, *Phys. Rev. Lett.* **82**, 516 (1999).
7. E. Leber *et al*, *Eur. Phys. J. D* **12**, 125 (2000).
8. J.M. Weber *et al*, *Eur. Phys. J. D* **11**, 247 (2000).
9. E. Leber *et al*, *Chem. Phys. Lett.* **325**, 345 (2000).
10. P. Stampfli, *Phys. Rep.* **255**, 1 (1995).
11. M. Thoss and W. Domcke, *J. Chem. Phys.* **109**, 6577 (1998).
12. I.I. Fabrikant, *Phys. Rev. A* **43**, 3478 (1991).
13. B.I. Schneider *et al*, *J. Phys. B* **12**, L365 (1979).
14. G.L. Gutsev *et al*, *J. Chem. Phys.* **108**, 6756 (1998).
15. M. Allan, *J. Phys. B* **35**, L387 (2002).
16. M. Allan, *Phys. Rev. Lett.* **87**, 033201 (2001).
17. W. Domcke and L.S. Cederbaum, *Phys. Rev. A* **16**, 1465 (1977).
18. S. Mazevet *et al*, *Phys. Rev. A* **64**, 040701 (2001).
19. B.L. Whitten and N.F. Lane, *Phys. Rev. A* **26**, 3170 (1982).

PLASMON EXCITATIONS IN ELECTRON COLLISIONS WITH METAL CLUSTERS AND FULLERENES

Andrey V. Solov'yov

A. F. Ioffe Physical-Technical Institute, Russian Academy of Sciences, Polytechnicheskaya 26, St. Petersburg 194021, Russia

E-mail: solovyov@th.physik.uni-frankfurt.de

This paper gives a survey of physical phenomena manifesting themselves in electron scattering on atomic clusters. The main emphasis is made on electron scattering on fullerenes and metal clusters, however some results are applicable to other types of clusters as well. This work is addressed to theoretical aspects of electron-cluster scattering; however some experimental results are also discussed. It is demonstrated that the electron diffraction plays an important role in the formation of both elastic and inelastic electron scattering cross sections. The essential role of the multipole surface and volume plasmon excitations is elucidated in the formation of electron energy loss spectra on clusters (differential and total, above and below ionization potential) as well as the total inelastic scattering cross sections. Particular attention is paid to the elucidation of the role of the polarization interaction in low energy electron-cluster collisions. This problem is considered for electron attachment to metallic clusters and the plasmon enhanced photon emission. Finally, mechanisms of electron excitation widths formation and relaxation of electron excitations in metal clusters and fullerenes are discussed.

1. Introduction

Clusters have been recognized as new physical objects with their own properties relatively recently. This became clear after such experimental successes as the discovery of electron shell structure in metal clusters, observation of plasmon resonances in metal clusters and fullerenes, formation of singly and doubly charged negative cluster ions and many more. Complete review of the field can be found in review papers and books.[1–6]

Properties of clusters can be studied by means of photon, electron and ion scattering. These methods are the traditional tools for probing properties and internal structure of various physical objects.

In this paper we consider electron collisions with metal clusters and fullerenes, being in a gas phase, and focus on the following physical problems: manifestation of electron diffraction both in elastic and inelastic collisions,[7–10] the role of surface and volume plasmon excitations in the formation of electron energy loss spectra (differential and total, above and below ionization potential) as well as the total inelastic scattering cross sections,[7–11] the importance of the polarization effects in electron attachment and photon emission processes.[12–22] We also discuss briefly mechanisms of electron excitation width formation and relaxation of electron excitations in metal clusters.[23–25] The choice of these problems is partially made because of their links with experimental efforts performed in the field.

In this paper the atomic system of units, $\hbar = |e| = m_e = 1$, is used.

2. Theoretical Methods

Metallic clusters are characterized by the property that their valence electrons are fully delocalized. To some extent this feature is also valid for fullerenes, where the delocalization of electrons takes place on the surface in the vicinity of the fullerene's cage. When considering electron collisions involving metal clusters and fullerenes, it is often the valence delocalized electrons that play the most important role in the formation of the cross sections of various collision processes. Therefore, it is possible to achieve an adequate description of such processes on the basis of the jellium model (see Refs. 1, 2, 5 and 6 for a review).

The jellium model of metal clusters and fullerenes can be examined in electron elastic scattering of fast electrons on metal clusters and fullerenes. Indeed, the jellium model implies that there is a rigid border in the ionic density distribution of a cluster. The presence of a surface in a cluster results in the specific oscillatory behaviour of the electron elastic scattering cross sections, which can be interpreted in terms of electron diffraction of the cluster surface.[7,10]

The detailed theoretical treatment of the diffraction phenomena arising in electron scattering on metal clusters and fullerenes has been given in Refs. 7–9. Experimentally, diffraction behaviour of electron elastic scattering cross sections on fullerenes in the gas phase has been observed for the first time in Ref. 10. The angular dependence of the electron elastic scattering cross section is shown in Fig. 1.

Let us explain the physical nature of the diffraction phenomena arising in elastic electron-cluster scattering on the example of fast electron scattering on the fullerene C_{60}. Due to the spherical-like shape of the C_{60} molecule, the charge densities of electrons and ions near the surface of the fullerene are much higher than in the outer region. These densities are characterized by the radius of the fullerene R and the width of the fullerene shell, $a \ll R$. The de Broglie wavelength of a fast electron in collision with a fullerene is small compared to R. At the same time the characteristic scattering length, $1/q$, where q is the transferred momentum, can be larger than the average inter-atomic distance in C_{60}, which is approximately as large as a. Under these conditions the electron scattering amplitude becomes determined by the charge density of the entire fullerene rather than the charge density of a single carbon atom. In this limit, the scattering amplitude and the corresponding cross section possess a series of diffraction maxima and minima. The positions of these minima and maxima are mainly determined by the radius of the molecule. In this case the process of electron elastic scattering on C_{60} becomes qualitatively similar to the diffraction of an electron at a shell. Similar behaviour occurs for other type of clusters as well.

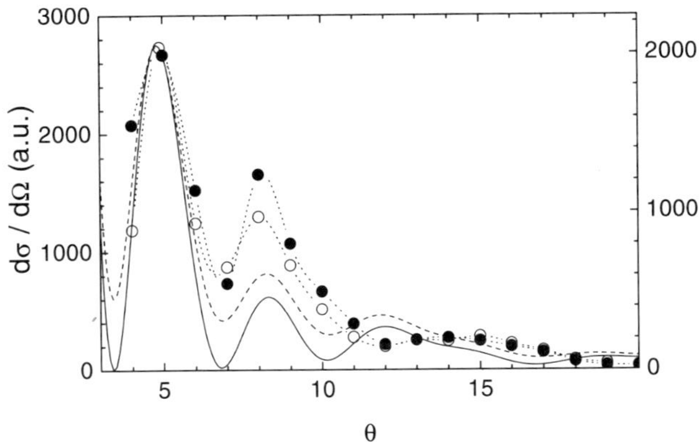

Fig. 1. Experimental (full and open circles) and theoretical (solid curve) angular dependencies of the differential elastic scattering cross section in collision of $809eV$ electron with the C_{60} molecule.[10] Full and open circles correspond to the two independent sets of measurements. Dashed line is the differential cross section for the mixture containing 60% of C_{60} and 40% of equivalent isolated carbon atoms. The scale for this curve is given in the right hand side of the figure.

Metal clusters and fullerenes possess prominent dynamic properties. Due to the presence of the highly movable delocalized valence electrons in the system, these clusters are highly polarizable. Collective oscillations of the delocalized electrons can be excited in a cluster by a photon or a projectile charged particle (electron or ion). These collective excitations are known as plasmons. Polarization effects are especially strong when considering low-energy electron cluster collisions, when the velocity of the projectile electron is comparable with the energy of the delocalized cluster electrons. Multipole plasmon excitations are easier to study in fast-electron cluster collisions. Note that the dipole plasmon excitation mode can also be effectively probed in the photoabsorption or photoionization process.

The dynamic properties of metal clusters are mainly determined by the delocalized cluster electrons and thus can be very well described on the basis of the jellium model. The transition amplitude M_{fi} is defined as follows:

$$M_{fi} = \langle \Psi_f(\mathbf{r}')\Psi_n(\mathbf{r}) \left| \frac{1}{|\mathbf{r} - \mathbf{r}'|} \right| \Psi_i(\mathbf{r}')\Psi_m(\mathbf{r}) \rangle. \tag{1}$$

This amplitude describes the transition of the projectile electron from the state Ψ_i to the state Ψ_f, with simultaneous excitation of the target electron from the state Ψ_m to Ψ_n. This transition is caused by the Coulomb interaction between the electrons. We assume that the final states Ψ_n and Ψ_f of the electrons can either belong to continuous or discrete spectrum.

A similar amplitude arises, when considering electron-cluster collisions on the basis of the perturbation theory. Depending on the final states of the particles this amplitude can describe either inelastic electron-cluster collision or electron attachment processes.

The bound states for the extra electron in the field of the positive cluster can be calculated within the frozen-core Hartree-Fock approximation and by accounting independently for the cluster dynamic polarization potential. Bound states in the system of extra electron plus neutral target cluster do not exist if considering the system in the HF approximation. They appear however, when accounting for the polarization interaction between the electron and the cluster. The negative ion wave functions Ψ_n and the energies ε_n can be obtained as Dyson's equation with the non-local polarization potential:[26]

$$\hat{H}^{(0)}\Psi_n(\mathbf{r}) + \int \Sigma_{\varepsilon_n}(\mathbf{r}, \mathbf{r}')\, \Psi_n(\mathbf{r}')\, d\mathbf{r}' = \varepsilon_n \Psi_n(\mathbf{r}). \tag{2}$$

Here, $\hat{H}^{(0)}$ is the static single-particle Hamiltonian of the cluster and $\Sigma_E(\mathbf{r}, \mathbf{r}')$ is the energy-dependent non-local potential, which is equal to the irreducible self-energy part of the single-electron Green's function of the system cluster + electron. $\Sigma_E(\mathbf{r}, \mathbf{r}')$ can be represented diagrammatically as a series on the inter-electron correlation interaction.[18,26] It is natural to calculate $\Sigma_E(\mathbf{r},\mathbf{r}')$ and solve Eq. (2) by using the eigen single-particle wave functions ψ_i of the Hartree-Fock $\hat{H}^{(0)}$ Hamiltonian:

$$\hat{H}^{(0)}\psi_i(\mathbf{r}) = \epsilon_i\psi_i(\mathbf{r}) \tag{3}$$

where

$$\begin{aligned}\hat{H}^{(0)}\psi_i(\mathbf{r}) = {} & \left(-\frac{\Delta}{2} - U(\mathbf{r}) + \sum_j^{occ}\int \psi_j(\mathbf{r}')\frac{1}{|\mathbf{r}-\mathbf{r}'|}\psi_j(\mathbf{r}')\,d\mathbf{r}'\right)\psi_i(\mathbf{r}) \\ & - \sum_j^{occ}\int \psi_j(\mathbf{r}')\frac{1}{|\mathbf{r}-\mathbf{r}'|}\psi_i(\mathbf{r}')\,d\mathbf{r}'\psi_j(\mathbf{r}).\end{aligned} \tag{4}$$

Here, $U(\mathbf{r})$ is the potential of the positive cluster core. The exchange interaction in Eq. (4) is taken into account explicitly, which makes the potential in Eq. (4) non-local, contrary to the local density approximation in which the exchange correlation interaction is always local.

Note that when the collective electron excitations in a cluster become important, one should treat the Coulomb many-electron correlations properly in order to calculate the matrix element (1) or similar correctly. For this purpose, we treat the matrix element (1) and excitation energies ω_{fi} in the RPAE scheme, using the Hartree-Fock wave functions calculated within the jellium model as a basis. This method, similar to the one used in the dipole case[28] for photoabsorption by metal clusters.

3. Inelastic Scattering of Fast Electrons on Metal Clusters and Fullerenes

We now consider the inelastic scattering of fast electrons on metal clusters and fullerenes, using approaches and methods described in the previous section. This process is of interest because the many-electron collective excitations of various multipolarity provide significant contribution to the cross section as demonstrated in Refs. 7–10.

Plasmon excitations in metal clusters and fullerenes have been intensively studied during last years.[1–6] They were observed in photoabsorption experiments with metal clusters and in photoionization studies with the fullerenes. In photoionization experiments with metal clusters and fullerenes only dipole collective excitations have been investigated.

Electron collective modes with higher angular momenta can be excited in metal clusters and fullerenes by electron impact if the scattering angle of the electron is large enough.[7–10] The plasmon excitations manifest themselves as resonances in the electron energy loss spectra. Dipole plasmon resonances of the same physical nature as in the case of the photoabsorption, dominate the electron energy loss spectrum if the scattering angle of the electron, and thus its transferred momentum, is sufficiently small. With increasing scattering, angle plasmon excitations with higher angular momenta become more probable. The actual number of multipoles coming into play depends on the cluster size.

3.1. *Electron inelastic scattering cross section*

The triply differential cross section of fast electron inelastic scattering on a cluster reads as

$$\frac{d^3\sigma}{d\varepsilon' d\Omega} = \frac{16\pi p'}{pq^4} \sum_{lm} (2l+1) \sum_{f} \left| \langle \Psi_f \left| \sum_a j_l(qr_a) Y_{lm}(\mathbf{n_a}) \right| \Psi_i \rangle \right|^2$$

$$\delta(\frac{\mathbf{p}'^2}{2} + \varepsilon_f - \frac{\mathbf{p}^2}{2} - \varepsilon_i). \qquad (5)$$

Here, $d\Omega$ is the solid angle of the scattered electron. The summation over f implies the summation over the discrete spectrum and the integration over the continuous spectrum of the final states of the cluster. We calculate the many electron wave functions Ψ_i, Ψ_f and the excitation energies ε_f and ε_i, using the Hartree-Fock jelium model. As soon as collective electron excitations in a cluster play the significant role, then in order to obtain the correct result when calculating the matrix elements in Eq. (5), one should properly take into account many-electron correlations. This problem can be solved in the RPAE described in the previous section.

Integrating the triply differential cross section Eq. (5) over $d\Omega$, we derive the total differential energy loss spectrum

$$\frac{d\sigma}{d\varepsilon'} = \frac{32\pi^2}{p^2} \sum_f \int_{q_{min}}^{q_{max}} \frac{dq}{q^3} \sum_{lm} (2l+1) \left| \langle \Psi_f \left| \sum_a j_l(qr_a) Y_{lm}(\mathbf{n_a}) \right| \Psi_i \rangle \right|^2$$

$$\delta(\frac{\mathbf{p}'^2}{2} + \varepsilon_f - \frac{\mathbf{p}^2}{2} - \varepsilon_i). \qquad (6)$$

Integrating this equation over the transferred energies of the electron, we come to the expression for the total cross section of inelastic scattering

$$\sigma = \frac{32\pi^2}{p^2} \sum_f \int_{q_{min}}^{q_{max}} \frac{dq}{q^3} \sum_{lm} (2l+1) \sum_f \left| \langle \Psi_f \left| \sum_a j_l(qr_a) Y_{lm}(\mathbf{n_a}) \right| \Psi_i \rangle \right|^2 . \qquad (7)$$

The minimum and the maximum transferred momenta in Eqs. (6) and (7) are equal to $q_{min} = p(1-\sqrt{1-\omega_{fi}/\varepsilon})$ and $q_{max} = p(1+\sqrt{1-\omega_{fi}/\varepsilon})$. Here, ε_f and ε_i are the energies of the levels i and f respectively; ε is the energy of the projectile electron. The contributions of different multipolarity do not interfere in Eqs. (5-7), which is the result of spherical symmetry of the target cluster.

3.2. *Plasmon resonance approximation: diffraction phenomena, comparison with experiment and RPAE*

Besides complex numerical calculations of the cross section (5-7) one can derive rather simple approximate analytical results giving the distinct physical picture of the process.[7] Indeed, let us consider the behaviour of the inelastic cross sections in the vicinity of the giant collective resonance when surface plasmon excitations give the main contribution. In this case the interaction of the projectile with electrons in the surface layer of the width a near the surface of the cluster should play the most significant role in the inelastic scattering process. The width a of this layer is determined by the width of the region near the surface of the cluster where oscillations of electron density mainly occur. This width is of the same order of magnitude as the size of a single atom. Oscillations of electron density take place mainly near the surface, because the electron density inside the cluster is well compensated by the oppositely charged density of the ionic background. Mathematically this means that theory has a small parameter, namely $a/R \ll 1$, which allows one to make the simplification of final results. Indeed, in the case of the collective excitation the main contribution to the matrix elements in Eqs. (5-7) arises from the $r_a \approx R$.[7,9]

The condition $r_a \approx R$ allows us to simplify the matrix elements in Eq. (5) and express them via the multipole matrix elements Q_{fi}^{lm} as

$$\langle \Psi_f \left| \sum_a j_l(qr_a) Y_{lm}(\mathbf{n_a}) \right| \Psi_i \rangle \simeq \sqrt{\frac{2l+1}{4\pi}} \frac{j_l(qR)}{R^l} Q_{fi}^{lm}, \tag{8}$$

where

$$Q_{fi}^{lm} = \sqrt{\frac{4\pi}{2l+1}} \langle \Psi_f \left| \sum_a r_a^l Y_{lm}(\mathbf{n_a}) \right| \Psi_i \rangle.$$

Substituting these equations to Eq. (5) and using the relationship

$$\int \left| Q_{fi}^{lm} \right|^2 \delta(\omega - \varepsilon_i - \varepsilon_f) df = \frac{1}{\pi} Im\alpha_l(\omega), \tag{9}$$

where $\alpha_l(\omega)$ is the multipole dynamic polarizability of the cluster, we finally obtain the expression for the cross section via the imaginary parts of the multipole dynamic polarizabilities

$$\frac{d^3\sigma}{d\varepsilon' d\Omega} = \frac{4p'}{\pi p q^4} \sum_l (2l+1)^2 \frac{j_l^2(qR)}{R^{2l-2}} Im\alpha_l(\Delta\varepsilon). \quad (10)$$

Here, $\Delta\varepsilon = \mathbf{p}^2/2 - \mathbf{p}'^2/2$.

Integrating the cross section (10) over $d\Omega$, we derive the expression for the total spectrum of the electron energy loss

$$\frac{d\sigma}{d\varepsilon'} = \frac{8}{p^2} \sum_l \frac{(2l+1)^2}{R^{2l-2}} S_l(\Delta\varepsilon) Im\alpha_l(\Delta\varepsilon). \quad (11)$$

Here, $\Delta\varepsilon = \varepsilon - \varepsilon'$. We have also introduced functions $S_l(\Delta\varepsilon)$ as follows

$$S_l(\Delta\varepsilon) = \int_{q_{min}R}^{q_{max}R} \frac{dx}{x^3} j_l^2(x). \quad (12)$$

The limits q_{min} and q_{max} are equal to $q_{min} = p(1 - \sqrt{1 - \Delta\varepsilon/\varepsilon})$, $q_{max} = p(1 + \sqrt{1 - \Delta\varepsilon/\varepsilon})$. In the most interesting region for our consideration, one derives $\Delta\varepsilon \sim \omega_p$ and $\varepsilon \sim \omega_p^2 R^2$. Therefore, $\Delta\varepsilon/\varepsilon \ll 1$ and the limits of integration in Eq. (12) are approximately equal to $q_{min} \approx \Delta\varepsilon/v$ and $q_{max} \approx 2p$. In this case $q_{max}R \gg 1$, the upper limit in Eq. (12) can be replaced by the infinity and S_l becomes a function of $\Delta\varepsilon R/v \ll 1$.

Equations (11-12) establish the connection between the inelastic scattering cross section and the imaginary parts of the multipole dynamic polarizabilities. These polarizabilities have a resonance behaviour in the region of frequencies where collective electron modes in a cluster can be excited. In the plasmon resonance approximation the polarizability $\alpha_l(\omega)$ can be written as

$$\alpha_l(\omega) = R^{2l+1} \frac{\omega_l^2}{\omega_l^2 - \omega^2 - i\omega\Gamma_l}, \quad (13)$$

where ω_l is the resonance frequency of the plasmon excitation with the angular momentum l, which for metal clusters according to the Mie theory is equal to $\omega_l = \sqrt{\frac{3lN_e}{(2l+1)R^3}}$ and $\omega_l = \sqrt{\frac{l(l+1)N_e}{(2l+1)R^3}}$ for fullerenes.[7] Here, N_e is the number of delocalized electrons in the cluster. The parameter Γ_l in Eq. (13) is the width of the plasmon resonance with the angular momentum l.

These formulae demonstrate that due to the resonance behaviour of the polarizability, the differential inelastic scattering cross section should also exhibit resonances, if the transferred energy lies in the range characteristic for plasmon excitations.

Integrating Eq. (11) over ε', we derive the plasmon contribution to the total inelastic cross section. Assuming that the plasmon resonance is narrow enough and using the pole approximation Eq. (13), we obtain the following expression for the plasmon contribution to the inelastic cross section

$$\sigma = \frac{4\pi e^2 R^3}{v^2} \sum_l (2l+1)^2 \omega_l S_l(\omega_l). \tag{14}$$

Note that excitations with large enough angular momenta l have a single particle nature rather than collective character. It follows, for instance, from the fact that with increasing l the wave length of the surface plasmon mode, $2\pi R/l$, becomes smaller than the characteristic wave length of the delocalized electrons on the surface of the cluster, $2\pi/\sqrt{2\varepsilon}$, where ε is the characteristic kinetic electron energy in the cluster. In other words, excitations with the angular momenta comparable or larger than the characteristic electron angular momenta of the ground state should have a single particle character rather than a collective nature. Therefore, analyzing contribution of the collective modes, which have prominent resonance character, we should restrict the consideration only by relatively low angular momenta. For example, according to the jellium model the maximum angular momentum of the delocalized electrons in the Na_{40} cluster is equal to 4. Therefore, only the dipole, the quadrupole and the octupole collective modes can be expected in this case. This means that only the first three terms in Eq. (13) should be considered. With increasing cluster size the number of collective modes grows as R.

The plasmon resonance approximation and its classical and quasiclassical modifications can be used for the estimation of the cross sections and investigation of their dependence on various parameters even for large clusters having hundreds and thousands of atoms, when *ab initio* calculations are hardly possible. For the small cluster systems its validity can be verified by performing the *ab initio* quantum mechanical calculation of cross sections. Such calculations for the electron collision with the Na_{20}, Na_{40}, Na_{58} and Na_{92} clusters have been performed in Refs. 8 and 9. In both approaches we have described the collision process in the Born approximation, which is applicable when the energy of the projectile electron surpasses typical energy of the delocalized electrons. Comparison of the two approaches demonstrates that collective excitations of electrons in a cluster provide dominating contribution to inelastic scattering cross sections and explain their most important features.

The plasmon resonance approximation provides a simple criterion for the estimation of the relative importance of various plasmon modes. Indeed, according to the plasmon resonance approximation the minima and the maxima of the contribution of the plasmon mode with the angular momentum l are determined by the minima and the maxima of the diffraction factor $j_l^2(qR)$ as it follows from Eq. (7). This factor shows that diffraction phenomena arise also in electron inelastic scattering on clusters. The main maximum of the partial contribution with the angular momentum l arises when $qR \sim l$. This condition reflects a simple fact that the probability of the excitation of the collective plasmon mode is maximum, when the characteristic collision distance is about the wave length of plasmon. This results in the significant dependence of the profile of the spectrum on the angle of the scattered electron.

In Fig. 2, we consider the behaviour of the cross sections for the electron-fullerene C_{60} collision (left panel) and the electron-Na_{40} collision (right panel). Figures 2a and b (left panel) show quite reasonable agreement of theoretical results with the experimental data. At low momentum transfer ($\theta = 1.5^o$) the dipole excitation dominates in the energy loss spectrum (fig. 2a, left panel), while for $\theta = 10^o$ the quadrupole contribution (Fig. 2b, left panel) provides the main contribution. As a result the position of the maximum of the energy loss spectrum shifts from the dipole plasmon frequency ω_1 at $\theta = 1.5^o$ to the quadrupole plasmon frequency ω_2 at $\theta = 10^o$. At $\theta = 5^o$, the position of the maximum is close to the octupole plasmon resonance frequency ω_3 (Fig. 2c, left panel).

To check the validity of the pure resonance treatment of plasmon excitations in a cluster and establish the relative role of the collective modes in the formation of the inelastic scattering cross sections, we compare the results obtained from the direct quantum calculations with those derived from the resonance treatment of plasmon excitations in a cluster. Figure 2a (right panel) shows that at 1° the dipole plasmon excitation dominates in the electron energy loss spectrum. The quadrupole excitation provides relatively small contribution to the spectrum. The contributions of the monopole and all higher multipole excitations are almost negligible at this scattering angle. Figure 2b (right panel) demonstrates that at 6° the quadrupole excitation becomes the leading excitation in the electron energy loss spectrum, shifting the maximum of the spectrum towards higher energies and also changing the profile of the resonance. The dipole and the octupole excitations also provide considerable contributions in a wide range of transferred energies broadening the spectrum. The dominance of quadrupole excitation is not as large as for the dipole excitation at 1°. At 9° the picture (Fig.

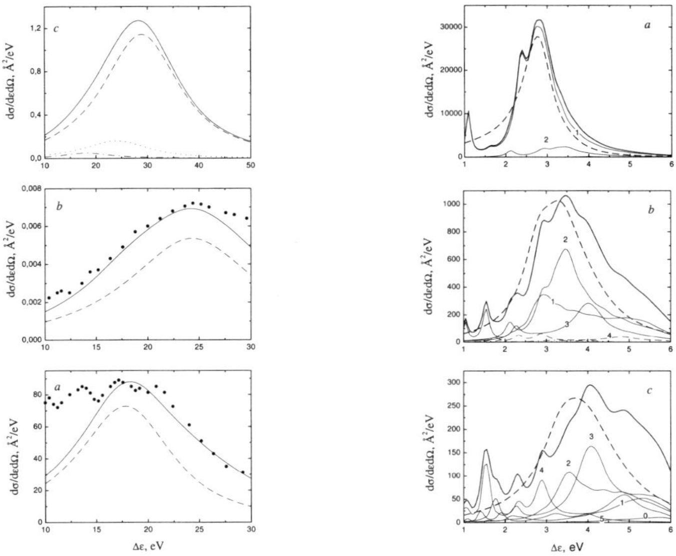

Fig. 2. Left panel: differential cross section $d\sigma/d\varepsilon' d\Omega$ (5) as a function of transferred energy $\Delta\varepsilon$ calculated for electron-fullerene C_{60} collision.[9] The impact electron energy is $\varepsilon = 1keV$. The scattering angle is $\theta = 1.5^o$ (a), $\theta = 9^o$ (b) and $\theta = 5^o$ (c). Results of the plasmon resonance approximation are shown by solid lines. Dots in figures (a) and (b) represent the experimental data from Ref. 29. Dashed lines show the leading multipole (dipole (a), quadrupole (b) and octupole (c)) plasmon contribution to the spectrum. In Fig. c, contributions of the qudrupole and the dipole plasmon modes are shown by dotted and dashed-dotted lines respectively. Right panel: differential cross section, $d\sigma/d\varepsilon' d\Omega$, (5) as a function of transferred energy $\Delta\varepsilon$ calculated for electron-Na_{40} collision.[9] The impact electron energy is $\varepsilon = 50eV$. The electron scattering angle is $\theta = 1^o$ (a), $\theta = 6^o$ (b) and $\theta = 8^o$ (c). Solid lines represent results of the RPAE calculation with the Hartree-Fock jellium model basis wave functions. Thick solid line is the total energy loss spectrum. Thin solid lines marked with the angular momentum number represent various multipole contributions to the energy loss spectrum. With the dashed line we plot the electron energy loss spectrum calculated in the plasmon resonance approximation.

2c, right panel) becomes more complex. In this case the octupole excitations provide the dominating contribution to the spectrum in the vicinity of the maximum of the energy loss spectrum at $\Delta\varepsilon \approx 4eV$. Besides this region, the dipole, quadrupole and even excitations with angular momentum 4 give comparable contributions to the energy loss spectrum and form rather broad structure. The monopole excitations and the excitations with angular momentum 5 and higher are almost negligible. With increasing scattering angle, excitations with the angular momentum 4 become more important. However, the corresponding spectrum does not possess a resonance behaviour, because it is mainly formed by single electron transitions.

Comparison of the results derived from the RPAE calculations with those obtained in the plasmon resonance approximation shows that, in spite of the simplicity, the plasmon resonance treatment is in quite good agreement with the consistent many-body quantum calculation. The main discrepancy between the two approaches arises from the single particle transitions omitted in the plasmon resonance approximation, but taken into account in the RPAE calculation. These transitions bring some structure to the final energy loss spectra manifesting themselves over the smooth resonance behaviour which is reproduced by the plasmon resonance approximation. At larger scattering angles plasmons with larger angular momenta can be excited. However, as we know, excitations with large enough angular momenta occur due to single particle transitions rather than due to collective excitations. Therefore, the agreement between the plasmon resonance approximation and the RPAE is better at small angles.

4. Surface and Volume Plasmon Excitations in the Formation of Electron Energy Loss Spectrum

We discuss now the formation of the widths of plasmon resonances. Damping of the plasmon oscillations is connected with the decay of the collective electron excitations to the single-particle ones similar to the mechanism of Landau damping in infinite electron gas. Frequencies of the surface plasmon excitations in neutral metal clusters usually lie below the ionization threshold. Therefore single-particle excitations in the vicinity of the surface plasmon resonance have the discrete spectrum. In this case the width of a surface plasmon excitation caused by the Landau damping should be treated as the width of the distribution of the oscillator strengths in the vicinity of the resonance. The problem of the formation of the surface plasmon resonance widths in clusters has been studied during the last decade in a number of works.[11,32,33]

A similar situation takes place for the volume plasmon excitations in metal clusters.[11] In metal clusters the resonance frequencies of volume plasmons are above the ionization threshold. This means that the volume plasmon excitations are quasi-stable. They have the real channel of the Landau damping leading to the ionization of the cluster. Thus, the process of inelastic scattering in the region of transferred energies above the ionization threshold can be described as follows: the projectile particle induces the oscillations of the electron density in the cluster. Oscillations of the electric field caused by the electron motion result in the ionization of the cluster. Note that the similar scenario takes place with damping of the surface plasmon resonances in fullerenes,[34] which also decay via the autoionization channel.

The differential cross section of the electron inelastic scattering on metal clusters obtained in the plasmon resonance approximation, accounting for both surface and volume excitations, reads as[11]

$$\frac{d^2\sigma}{d\varepsilon' d\Omega} = \frac{4p'R}{\pi p q^4}\sum_l (2l+1)^2 j_l^2(qR)\frac{\omega_l^2 \Delta\varepsilon \Gamma_{sl}}{(\Delta\varepsilon^2 - {\omega_l}^2)^2 + \Delta\varepsilon^2\ \Gamma_{sl}^2} + \\ + \frac{2p'R^3}{\pi p q^2}\sum_l (2l+1)\frac{\omega_p^2 \Delta\varepsilon \Gamma_{vl}}{(\Delta\varepsilon^2 -\ {\omega_p}^2)^2 + \Delta\varepsilon^2 \Gamma_{vl}^2} \times \\ \times \left(j_l^2(qR) - j_{l+1}(qR) j_{l-1}(qR) - \frac{2}{qR} j_{l+1}(qR) j_l(qR) \right). \tag{15}$$

Here $\omega_p = \sqrt{3N_e/\alpha}$ is the volume plasmon resonance frequency, $\omega_l = \sqrt{l/(2l+1)}\omega_p$ is the frequency of surface plasmon excitation with the angular momentum l, N_e is the number of delocalized electrons, α is the static polarizability of the cluster, Γ_{vl} and Γ_{sl} are the widths of the volume and surface plasmon resonances, which are defined below. Note that volume plasmon excitations with different angular momenta have the equal resonance frequency ω_p.

This cross section is totally determined by collective electron excitations in the cluster. The first and the second terms in Eq. (15) describe contributions of the surface and the volume plasmon excitations respectively. Note, that similar expressions for the cross section have been also obtained in Ref. 30 for electron scattering on small metal particles by means of classical electrodynamics.

According to Ref. 11 the width of the surface plasmon resonance in the plasmon resonance approximation is equal to:

$$\Gamma_{sl} = \frac{4\pi\omega_l}{(2l+1)R}\sum_{\nu\mu} \left|\langle\psi_\mu \left|\varphi_{sl}(\mathbf{r})\right| \psi_\nu\rangle\right|^2 \delta(\omega_l - \varepsilon_\mu + \varepsilon_\nu), \tag{16}$$

where $\varphi_{sl}(\mathbf{r}) = \left((r/R)^l \theta(R-r) + (R/r)^{(l+1)}\theta(r-R)\right) Y_{lm}(\mathbf{n})$. Note that the same expression was obtained earlier in Refs. 31 and 32 using other methods. Evaluation of the expression (16) for sufficiently large clusters leads to the well known result for the Landau damping of the surface plasmon oscillations $\Gamma_{sl} = 3lv_F/R$,[32] where v_F is the velocity of the cluster electron on the Fermi surface.

In the plasmon resonance approximation one can also determine the autoionization width of the volume plasmon resonance,[11] which is equal to:

$$\Gamma_{vl} = \frac{8\pi^2\omega_p}{q^2R^3}\frac{\sum_\nu \int |\langle\psi_\mu |\varphi_{vl}(\mathbf{r})|\,\psi_\nu\rangle|^2\,\delta(\omega_p-\varepsilon_\mu+\varepsilon_\nu)d\mu}{\left(j_l^2(qR)-j_{l+1}(qR)j_{l-1}(qR)-\frac{2}{qR}j_{l+1}(qR)j_l(qR)\right)}, \tag{17}$$

where $\varphi_{vl}(\mathbf{r}) = \left(j_l(qr) - j_l(qR)(r/R)^l\right)\theta(R-r)Y_{lm}(\mathbf{n})$. Here the summation is performed over the occupied single-electron states ν and the integration is performed over the electronic states μ of the continuous spectrum.

The projectile particle excites simultaneously numerous modes of the volume plasmon. The sum of the potentials of all the modes gives the resulting potential $\varphi_{vl}(\mathbf{r})$. It is essential that all normal modes of the volume plasmon have the same resonance frequency ω_p, but the excitation probability for these modes depends on the kinematics of collision. This leads to the dependence of the volume plasmon potential $\varphi_{vl}(\mathbf{r})$ upon the transferred momentum. The oscillations of the volume plasmon potential result in the ionization of the cluster, which probability and the volume plasmon resonance width depend on transferred momentum q. However, the numerical analysis performed in Ref. 11 shows that the dependence of Γ_{vl} on q is rather weak in the region of $q \ll 1$, where collective electron oscillations mainly take place. Therefore the volume plasmon resonance width with the given l can be approximated by the limiting value following from Eq. (17) at $q = 0$:

$$\Gamma_{vl} = (2l+5)\frac{\pi^2\omega_p}{R}\sum_\nu \int |\langle\psi_\mu |\varphi^0_{vl}(\mathbf{r})|\,\psi_\nu\rangle|^2\,\delta(\omega_p-\varepsilon_\mu+\varepsilon_\nu)d\mu, \tag{18}$$

where $\varphi^0_{vl}(\mathbf{r}) = (r/R)^l\left(1-(r/R)^2\right)\theta(R-r)Y_{lm}(\mathbf{n})$.

Figure 3 shows the dependence of the autoionization width on the transferred momentum q for the volume plasmon modes, which provide significant contribution to the EELS. The width of the dipole, the quadrupole and the octupole volume plasmon resonances has been calculated according to Eq. (17). The transferred momentum q plays the role of the wave vector for the volume plasmon excitations. All three plasmon modes have the similar dependence of Γ_{vl} upon q. The width grows slowly in the region of small q and it decreases rapidly at larger q. In the latter region the probability of volume plasmon excitation by the incoming electron is correspondingly reduced. Note that the wave length of a collective electron oscillation should be larger than the inter-electronic distance in the cluster, i.e. plasmon wave vector should be smaller than the Fermi momentum of

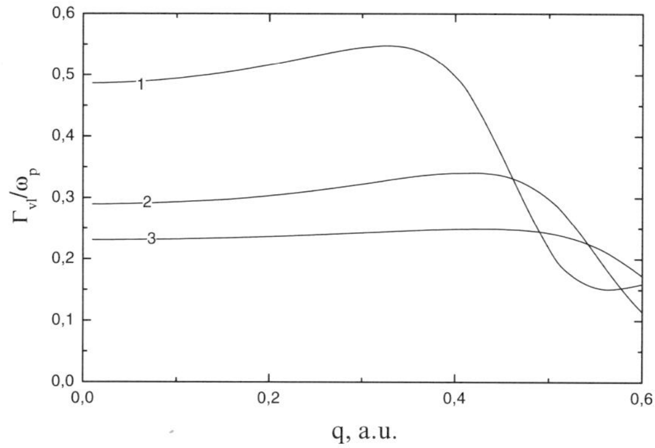

Fig. 3. Autoionization width Γ_{vl} of the dipole (1), the quadrupole (2) and the octupole (3) volume plasmon excitations as a function of transferred momentum q.[11]

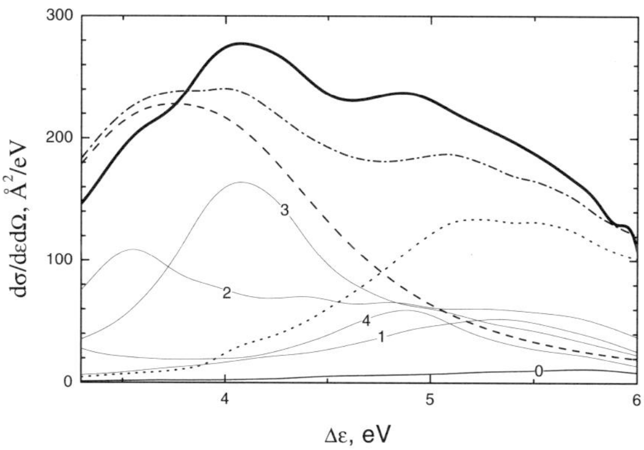

Fig. 4. Differential cross section, $d\sigma/d\varepsilon' d\Omega$ as a function of the transferred energy $\Delta\varepsilon$ calculated for the collision of $50eV$ electron with the Na_{40} clusters.[11] The electron scattering angle is $\theta = 9^o$. Solid lines represent the RPAE results. Thick solid line is the total energy loss spectrum. Thin solid lines marked by the corresponding angular momentum number represent various partial contributions to the energy loss spectrum. Contributions of the surface and the volume plasmons calculated in the plasmon resonance approximation (15) are shown by dashed and dotted lines respectively. Dashed-dotted line represents the sum of the surface and volume plasmon contributions to the EELS.

cluster electrons $q < 0.5$. In the region $q < 0.5$, where the latter condition is fulfilled, the dependence of Γ_{vl} upon q is rather weak. We can approximate the resonance width by following values $\Gamma_{v1} \simeq 0.5\omega_p$, $\Gamma_{v2} \simeq 0.3\omega_p$, $\Gamma_{v3} \simeq 0.23\omega_p$. Contrary to surface plasmons, the autoionization width of a volume plasmon decreases with the growth of the angular momentum.

The electron energy loss spectrum in the collision of an electron with the Na_{40} cluster in the region above the ionization potential, where volume plasmon modes become significant, is presented in Fig. 4.

The partial contributions to the EELS with $l < 3$ have the broad maximum in the vicinity of $\Delta\varepsilon \simeq 5.1eV$. Comparison of the EELS calculated in the two different approaches proves the assumption that the peculiarity in the EELS in the vicinity of $\Delta\varepsilon \sim 5eV$ is connected with the volume plasmon excitation. Figure 4 demonstrates that collective excitations provide dominating contribution to the total EELS determining its pattern.

5. Polarization Effects in Low Energy Electron – Cluster Collision and Photon Emission Processes

Our consideration in the previous sections has been mainly focused on the collisions of fast electrons with metal clusters and fullerenes, because the results of the fast electron-cluster collisions theory have the most straightforward and simple connection to the experiment.

However, the review of the electron-cluster collisions theory would not be complete if one said nothing about the low energy electron-cluster collisions. In the low energy electron-cluster collision, the electron collision velocity is lower or comparable with the characteristic velocities of the cluster delocalized electrons. This criterion can be traced from the Born theory of electron-cluster collisions and has very much in common with the Born theory of electron-atom collisions. In Ref. 7 it was shown that the electron collisions with metal clusters in the region of collision energies below 3-5 eV should be treated as slow, while for fullerenes, the region extends up to 30 eV.

In the low energy electron-cluster collisions the role of the cluster polarization and exchange correlation effects increases dramatically. The polarization potential of electron-cluster interaction sometimes changes completely the qualitative picture of the collision. This for example takes place, when considering low energy electron elastic scattering on metal clusters. In this case the resonant structures can appear in the energy dependence of the electron elastic scattering cross section due to the presence of the bound or quasi-bound states in the system.[35,36] The resonance structure turns out to be very sensitive to the choice of the approximations made for its description and has not been experimentally observed so far.

Collisions of low and intermediate energy electrons with metal clusters were experimentally investigated in Refs. 19–22. The response function of the fullerene and its relation to the inelastic scattering problem was considered in Ref. 37.

During the last years, considerable attention has been devoted both experimentally and theoretically to the problem of electron attachment to metal clusters and fullerenes. The electron attachment process is one of the mechanisms which leads to negative cluster ion formation in gases and plasmas and it therefore attracts the interest of numerous researchers. For fullerenes, the experimental observations of electron attachment have been performed in Refs. 38 and 39. For metal clusters, the electron attachment problem has been the subject of the intensive experimental[19–22] and theoretical investigations[12,13,16,18] and is not yet completely understood. Let us further discuss this problem in more detail.

The very simple picture of attachment is described in many textbooks. Let us assume that there exists a Langevin attraction potential of the form

$$V(r) = -\frac{1}{2}\frac{\alpha e^2}{r^4} \tag{19}$$

outside the cluster radius. The constant α is the static polarizability of the cluster. One can then show that there is an orbiting cross section

$$\sigma = \pi \left\{ \frac{2\alpha e^2}{\varepsilon} \right\}^{1/2}, \tag{20}$$

which sets an upper limit bound to the attachment cross section (so called the Langevin limit). Here ε is the kinetic energy of the projectile electron. This simple treatment, if valid, would explain the behaviour of the cross section in the vicinity of the threshold.

It is known that metal clusters possess a high polarizability, see e.g. Ref. 1. Hence, large capture cross sections are anticipated. However, simple attempts to account for attachment by using the static polarizability α are not in accordance with observation.[19] A recent review of electric polarizability effects in metal clusters is given in Ref. 21. The great weakness of the Langevin model is the treatment of α as an approximate constant. In fact, it possesses a complicated energy dependence, due to the dynamical polarizability of the metallic cluster.

The possibility of resonances in the capture cross section was first considered theoretically in Refs. 12 and 13. It was theoretically demonstrated that electrons of low energy can excite a collective plasmon resonance within the metal cluster in the electron attachment process as a result of a strong dipole deformation of the charge density of the cluster. Later this idea commented in the context of the measurements performed in Ref. 20, although no clear evidence of resonant behaviour has been found. In Ref. 20 there have been measured total inelastic scattering cross sections, which include attachment as only one of several possible contributing channels.

The resonant electron attachment mechanism was named in Refs. 12 and 13 as polarizational. An important consequence of the polarization mechanism is that the low energy electron falls into the target and the probability of this process is enhanced. Since the process as a whole is resonant, the enhancement is greatest for energies rather close to the plasmon resonance in the dynamic polarizability of the cluster. This fact explains another reason for the interest in this process. Indeed, the predictions of resonantly enhanced electron capture by metal clusters have been based on the jellium model. While there is convincing evidence for the jellium picture through the occurrence of magic numbers, ellipsoidal structures in studies of the stability of metallic clusters and through the energy splittings of plasmon resonances, the range of validity of the jellium model remains uncertain and is subject to current discussions.

In the attachment process the electron losses its excess energy. Polarizational bremsstrahlung is one of possible channels of the electron energy loss.[12–15,17] Energy of the electron can also be transferred to the excitations of the ionic background of the cluster,[24] which may lead to an increase in its vibrations and final fragmentation. In spite of the significant physical difference between various channels of the electron energy loss, they have one important common feature: they all go via the plasmon excitation. Therefore, calculating the total electron attachment cross section including all possible channels of the electron energy loss in the system, one obtains[16,18] qualitatively similar dependence of the cross section as it was obtained initially for the radiative channel of electron energy loss.[12]

In Ref. 12 the attachment cross section has been calculated within the jellium model in a scheme which holds best if the kinetic energy of the electrons is somewhat higher than the energy of the resonance. Also, it was assumed that the attached ion is created in the ground state. As a useful step in simplifying the calculation, a Kramers-Kronig transformation procedure was introduced to compute the polarizability from the absorption coefficient, thereby circumventing the need for full *ab initio* calculations.

Within this approximate scheme, it was found[12,13] that the resonant attachment cross section dominates over the non-resonant by a factor of about 10^3–10^4 near resonance, and is therefore a very significant pathway for electrons of low enough energy.

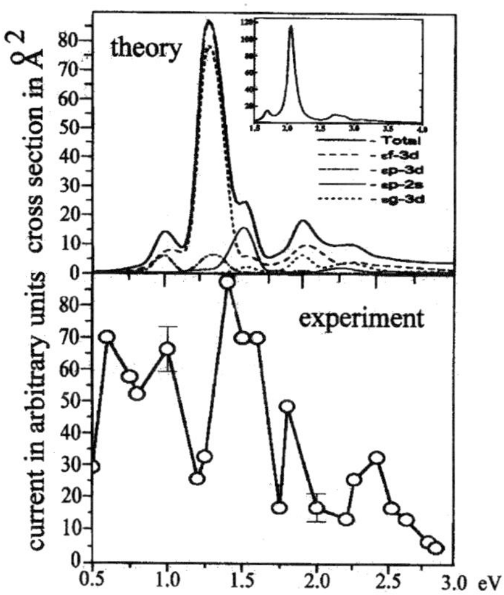

Fig. 5. A comparison between theory and experiment in the vicinity of plasmon resonance. Theoretical curves are from Ref. 18. The corresponding photoabsorption spectrum is shown in the inset. The experimental curves are from Ref. 22.

In Refs. 16 and 18 the earlier theoretical work investigating attachment was extended by including the following improvements: (a) all possible channels of the electron attachment were included and the total cross section of the process was calculated rather than analyzing a particular single channel; no assumption that the system can only return to its ground state had been made; (b) theoretical approximation was used to treat electron energies not only in the resonance region, but actually throughout the range of interest; (c) an RPAE calculation of the dynamical polarizability was performed along with the corresponding electron attachment cross sections on the basis of the consistent many-body theory with the use of the Hartree-Fock jellium model wave function; (d) calculations were performed for both neutral and charged cluster targets; (e) the polarization effect on the incoming particle has been taken into account and collective excitations of different multipolarity in the target electron system are taken into account; (f) Dyson's equation was used to reduce the problem of the interaction of an extra electron with a many-electron target system to a quasi-one-particle problem in a similar way as it was used for negative atomic ions calculations.[26]

An example of such a calculation is shown on the upper plot of Fig. 5. This plot represents the total and partial electron capture cross sections calculated for neutral potassium K_8 cluster.[18] The insert demonstrates the photoabsorption spectrum of K_8. It was found that the resonance pattern in the electron capture cross section for the K_8 cluster turns out to be similar in various approaches, although for some other sodium and potassium clusters it is more sensitive to the approximations made.[18] The plasmon resonance in the electron capture cross section is shifted on the value of energy of the attached electron as compared to the photoabsorption case shown in the insert.

Experimental evidence for the resonance enhancement of the cross sections of the electron attachment process has been recently obtained.[22] The experimental points from Ref. 22 are shown on the lower panel of Fig. 5. Comparison of the upper and lower parts of Fig. 5 demonstrates the reasonable agreement between theory predictions and experiment, although the more precise measurements would be desirable for the resolution of the more detailed structures in the electron attachment cross sections.

Finally, note that strong polarization effects arise also in the process of photon emission by an electron colliding with a cluster.[13–15,17] Such a process is known as the polarization bremsstrahlung or polarization radiation (see e.g. Ref. 15). In these papers it was demonstrated that the plasmon resonance structure manifests itself in the photon emission spectrum in collisions of electrons with metal clusters and fullerenes. This effect occurs because the cross section of this process is mainly determined by the dynamical polarizability of the cluster and is applicable to any polarizable system possessing a collective giant resonance.

6. How Electron Excitations in a Cluster Relax

In metal clusters, the plasmon resonances lie below the ionization thresholds, i.e. in the region of the discrete spectrum of electron excitations.[2] This fact raises an interesting physical problem regarding the eigen widths of the electron excitations, which possess large oscillator strengths and form the plasmon resonances. Knowledge of these widths is necessary for the complete description of the electron energy loss spectra, electron attachment, polarizational bremsstrahlung and photoabsorption cross sections in the vicinity of the plasmon resonances and the description of their dependence on the cluster temperature. Note, that experimentally the dependence of the plasmon resonance photoabsorption patterns of metal clusters on temperature has been studied in Ref. 40.

In metal clusters, the origination of the electron excitation widths is mainly connected with the dynamics of the ionic cluster core.[23, 24, 41–45] This is an example of effect, which has no analogy in atomic physics. Contraryly, in fullerenes the discrete transitions with the energies above the ionization threshold, which form mainly the plasmon resonance in the vicinity of 20 eV, possess the autoionization widths.[34] The latter mechanism of the line width broadening is well known in atomic physics and we do not discuss it here.

Instead, let us focus on the influence of the dynamics of ions on the motion of delocalized electrons in metal clusters and discuss it on the basis of the dynamic jellium model suggested in Ref. 23 and further developed in Refs. 24 and 25. This model generalizes the static jellium model, which treats the ionic background as frozen, by taking into account vibrations of the ionic background near the equilibrium point. The dynamic jellium model treats simultaneously the vibration modes of the ionic jellium background, the quantized electron motion and the interaction between the electronic and the ionic subsystems. In Ref. 23 the dynamical jellium model was applied for a consistent description of the physical phenomena arising from the oscillatory dynamics of ions.

The dynamic jellium model[23] allows calculation of the widths of the electron excitations in metal clusters caused by the dynamics of ions and their temperature dependence accounting for two mechanisms of the electron excitation line broadening, namely adiabatic and non-dynamic ones.

The adiabatic mechanism is connected with the averaging of the electron excitation spectrum over the temperature fluctuation of the ionic background in a cluster. This phenomenon has been also studied earlier in a number of papers.[23, 24, 41–45] The adiabatic linewidth is equal to

$$\Gamma = \sqrt{\frac{4\ln 2}{m\Omega} cth(\frac{\Omega}{2kT})} \left| \hat{V}_{nn} \right| . \tag{21}$$

Here m and Ω are the mass and frequency corresponding to the generalized oscillatory mode considered, T is the cluster temperature, k is the Bolzmann constant, $\left|\hat{V}_{nn}\right|$ is the matrix element of the electron phonon coupling, calculated for surface and volume cluster vibration modes in Ref. 24.

The mechanism of dynamic or non-adiabatic electron excitation line broadening has been considered for the first time in Refs. 23 and 24. This mechanism originates from the real multiphonon transitions between the excited electron energy levels. Therefore the dynamic linewidths characterize the real lifetimes of the electronic excitations in a cluster. The analytic expression for the non-adiabatic width obtained in Refs. 23 and 24 is rather cumbersome as compared to Eq. (21) and thus we do not present it here.

The adiabatic broadening mechanism explains the temperature dependence of the photoabsorption spectra in the vicinity of the plasmon resonance via the coupling of the dipole excitations in a cluster with the quadrupole deformation of the cluster surface.

The non-adiabatic linewidths characterize the real lifetimes of cluster electron excitations. Naturally, the non-adiabatic widths turn out to be much smaller than the adiabatic ones due to the slow motion of ions in the cluster. However, the adiabatic linewidths do not completely mask the non-adiabatic ones, because the two types of widths manifest themselves differently. The adiabatic broadening determines the pattern of the photoabsorption spectrum in the linear regime. The non-adiabatic linewidths are important for the processes, in which the real lifetime of electron excitations and the electron-ion energy transfer are essential. The non-adiabatic linewidths determined by the probability of multiphonon transitions are also essential for the treatment of the relaxation of electronic excitations in clusters and the energy transfer from the excited electrons to ions, which occurs after the impact- or photoexcitation of the cluster.

In Refs. 23 and 24, the role of volume and surface vibrations of the ionic cluster core in the formation of electron excitation linewidths was investigated. It was demonstrated that the volume and surface vibrations provide comparable contributions to the adiabatic linewidths, but the surface vibrations are much more essential for the non-adiabatic multiphonon transitions than the volume ones.

7. Concluding Remarks

We have considered a number of problems arising in fast and slow electron-cluster collisions. The choice of these particular problems was greatly influenced by the experimental efforts undertaken in the field. However, there are many more interesting problems in the field, which have been left aside in this paper, but deserve without doubt also careful theoretical and experimental consideration.

So far, most of the theoretical work on electron-cluster collisions was based on the jellium model as a very useful theoretical framework, allowing one to overcome the significant computational difficulties, but to take into account the essential features of the collision processes. However, this approximation has its limitations and it will be very interesting to study collision processes beyond the jellium model.

Acknowledgments

The author acknowledges support of this work by Russian Foundation for Basic Research (grant no. 03-02-16415-a), the Russian Academy of Sciences (grant no. 44) and the INTAS.

References

1. W.A. de Heer, *Rev. Mod. Phys.* **65**, 611 (1993).
2. C. Bréchignac, J.-P. Connerade, *J.Phys.B:At.Mol.Opt.Phys.* **27**, 3795 (1994).
3. H. Haberland (ed.), *Clusters of Atoms and Molecules, Theory, Experiment and Clusters of Atoms*, Springer Series in Chemical Physics, Berlin **52** (1994).
4. U. Näher, S. Bjørnholm, S. Frauendorf, F. Garcias and C. Guet *Physics Reports* **285**, 245 (1997).
5. W. Ekardt (ed.), *Metal Clusters*, Wiley, New York, (1999).
6. *Atomic Clusters and Nanoparticles*, NATO Advanced Study Institute, les Houches Session LXXIII, les Houches, 2000, edited by C. Guet, P. Hobza, F. Spiegelman and F. David (EDP Sciences and Springer Verlag, Berlin, 2001).
7. L.G. Gerchikov, J.-P. Connerade, A.V. Solov'yov and W. Greiner, *J.Phys.B:At.Mol.Opt.Phys.* **30**, 4133 (1997).
8. L.G. Gerchikov, A.N. Ipatov and A.V. Solov'yov, *J.Phys.B:At.Mol.Opt.Phys.* **30**, 5939 (1997).
9. L.G. Gerchikov, A.N. Ipatov, A.V. Solov'yov and W. Greiner, *J.Phys.B: At.Mol.Opt.Phys.* **31**, 3065 (1998).
10. L.G. Gerchikov, P.V. Efimov, V.M. Mikoushkin and A.V. Solov'yov, *Phys.Rev.Lett.* **81**, 2707 (1998).
11. L.G. Gerchikov, A.N. Ipatov, R.G. Polozkov and A.V. Solov'yov, *Phys.Rev.* A **62**, 043201 (2000).
12. J.-P. Connerade and A.V. Solov'yov, *J.Phys.B:At.Mol.Opt.Phys.* **29**, 365 (1996).
13. J.-P. Connerade and A.V. Solov'yov, *J.Phys.B:At.Mol.Opt.Phys.* **29**, 3529 (1996).
14. L.G. Gerchikov and A.V. Solov'yov, *Z.Phys.D:Atoms, Molecules, Clusters* **42**, 279 (1997).
15. A.V. Korol and A.V. Solov'yov, *Topical Review, J.Phys.B:At.Mol.Opt.Phys.* **30**, 1105 (1997).
16. J.-P. Connerade, L.G. Gerchikov, A.N. Ipatov and A.V. Solov'yov, *J.Phys.B:At.Mol.Opt.Phys.* **31**, L27 (1998).
17. L.G. Gerchikov, A.N. Ipatov and A.V. Solov'yov, *J.Phys.B:At.Mol.Opt.Phys.* **31**, 2331 (1998).

18. J.-P. Connerade, L.G. Gerchikov, A.N. Ipatov and A.V. Solov'yov, *J.Phys.B:At.Mol.Opt.Phys.* **32**, 877 (1999).
19. V. Kresin, A. Scheidemann and W.D. Knight, *Electron Collisions with Molecules, Clusters and Surfaces*, H. Eberhardt and L.A. Morgan (ed.) (New York: Plenum) 183 (1994).
20. V. Kasperovich, G. Tikhonov, K. Wong, P. Brockhaus and V.V. Kresin, *Phys. Rev.* A **60**, 3071 (1999).
21. V.V. Kresin and C. Guet, *Philosophical Magazine* **79**, 1401 (1999).
22. S. Sentürk, J.-P. Connerade, D.D. Burgess and N.J. Mason, *J.Phys.B: At.Mol.Opt.Phys.* **33**, 2763 (2000).
23. L.G. Gerchikov, A.V. Solov'yov and W. Greiner, *Int. Journal of Modern Physics* E **8**, 289 (1999).
24. L.G. Gerchikov, A.N. Ipatov, A.V. Solov'yov and W. Greiner, *J.Phys.B: At.Mol.Opt.Phys.* **33**, 4905 (2000).
25. A.G. Lyalin, S.K. Semenov, N.A. Cherepkov, A.V. Solov'yov and W. Greiner, *J.Phys.B:At.Mol.Opt.Phys.* **33**, 3653 (2000).
26. L.V. Chernysheva, G.F. Gribakin, V.K. Ivanov, M.Yu. Kuchiev, *J.Phys.B: At.Mol.Opt.Phys.* **21**, L419 (1988).
27. L.D. Landau and E.M. Lifshitz, *Quantum Mechanics*, Pergamon Press, London (1965).
28. C. Guet and W.R. Johnson, *Phys. Rev.* B **45**, 283 (1992).
29. J.W. Keller and M.A. Coplan, *Chem. Phys. Lett.* **193**, 89 (1992).
30. A.A. Lushnikov and A.J. Simonov, *Z.Physik* B **21**, 357 (1975).
31. A.A. Lushnikov and A.J. Simonov, *Z.Physik* **270**, 17 (1974).
32. C. Yannouleas and R.A. Broglia, *Ann.Phys.* **217**, 105 (1992).
33. C. Yannouleas, *Phys.Rev.* B **58**, 6748 (1998).
34. G.F. Bertsch, A. Bulcac, D. Tomanek, Y. Wang, *Phys.Rev.Lett.* **67**, 1991 (1992).
35. A.N. Ipatov, V.K. Ivanov, B.D. Agap'ev and W. Eckardt, *J.Phys.B: At.Mol.Opt.Phys.* **31**, 225 (1998).
36. P. Descourt, M. Farine, and C. Guet, *J.Phys.B: At.Mol.Opt.Phys.* **33**, 4565 (2000).
37. A. Bulgas and N. Ju, *Phys.Rev.* B **46**, 4297 (1992).
38. M. Lezius, P. Scheier and T.D. Mark, *Chem. Phys. Lett.* **203**, 232 (1993).
39. J. Huang, H.S. Carman and R.N. Compton, J. Phys. Chem. **99**, 1719 (1995).
40. Ch. Ellert, M. Schmidt, Ch. Schmitt, Th. Reiners and H. Haberland, *Phys. Rev. Lett.* **75**, 1731 (1995).
41. J.M. Pacheco and R.A. Broglia, *Phys. Rev. Lett.* **62**, 1400 (1989).
42. G.F. Bertsch and D. Tomanek, *Phys. Rev.* B **40**, 2749 (1989).
43. Z. Penzar, W. Ekardt and A. Rubio, *Phys. Rev.* B **42**, 5040 (1990).
44. B. Montag, T. Hirshmann, J. Mayer and P.-J. Reinhard, *Z. Phys.* D **32**, 124 (1994).
45. B. Montag and R.-J. Reinhard, *Phys. Rev.* B **51**, 14686 (1995).

PHOTOIONIZATION OF ALKALI NANOPARTICLES AND CLUSTERS

Kin Wong and Vitaly V. Kresin

Department of Physics and Astronomy, University of Southern California, Los Angeles, California 90089-0484, USA

We investigate the photoionization behavior of nanoscale alkali particles in a beam. Photoionization yield curves have been measured as a function of temperature. Near the ionization threshold, they can be fitted very well by finite-temperature Fowler plots, originally derived for bulk surfaces. The $T \rightarrow 0$ thresholds match the best literature work function values, and the temperature shifts of the work functions are in agreement with the predictions of recent theoretical models. The accuracy of the measurement demonstrates that the study of free nanoclusters offers an accurate, and economical, complement to traditional surface spectroscopy. At heights above the threshold, the yield functions begin to decline; the origin of this behavior is unclear at the moment, but is suggestive of inelastic scattering involving surface plasmon excitation. Furthermore, we demonstrate that the Fowler photoemission curves work very well even for smaller clusters (as tested on the available data for $K_{30\text{-}101}$). Both the ionization potentials and the internal cluster temperatures can be successfully extracted from such a fit. The results raise questions about the limits of applicability of bulk-derived models and highlights the need for the development of a comprehensive theory of metal cluster photoionization.

1. Introduction

The photoelectric effect provides one of the key routes to understanding the electronic structure of both bulk and nanoscale materials. Electron emission from metal surfaces has been studied for over a century,[1] from free metal clusters for about twenty-five years.[2-4] Whereas the basics of atomic and molecular photoionization and of solid-state photoemission are reasonably well understood, there does not appear to exist a general theory of the intermediate situation: cluster photoionization.

For example, many measurements of ionization thresholds for small and medium-sized metal clusters have been carried out. There is, however, no consensus on the most appropriate method of extracting ionization potentials (IP) from the data considering in particular, the omnipresent thermal smearing of the threshold region. For example, for alkali cluster spectra one generally finds fits to linear, exponential, error-function, or analogous shapes.[2-5] Theoretically, a few numerical publications are available,[6-9] but, as remarked above, no comprehensive theory.

In contrast, for bulk metallic surfaces the threshold behavior was understood back in the early 1930s. The so-called Fowler law[1,10] (see below) has been extensively used to determine the work functions of metallic compounds via photoeffect measurements. For surfaces, however, there exists a different sort of challenge: contamination. Since even a minute amount of impurity coverage can strongly shift the measured work function value, vacuum and sample preparation requirements are extremely demanding. For example, reliable work function measurements for the highly reactive alkali metals were not available until the 1970s, a remarkably late date.[11] This is a feature which can be alleviated by working with free nanoparticles: since the flight time in a molecular beam is very short (milliseconds) surface contamination can be reduced to a minimum.

Our recent work[5,12,13] has aimed at extracting mutually beneficial information from the intersection between surface and nanocluster photothreshold effects. Photoionization of free nanoscale alkali particles enabled us to obtain very accurate measurements of metal work functions, including their small temperature shifts, without the use of demanding ultra-high vacuum equipment. This is described in the next section. Conversely, we showed that the aforementioned Fowler equation, derived for bulk metal surfaces, turns out to provide a very satisfactory description of the ionization threshold spectra of much smaller alkali clusters. This is described in Section 3. These results testify once again that a lot of interesting physics, both experimental and theoretical, extrapolates productively from clusters to bulk materials and vice versa, over many orders of magnitude in size.

2. Photoionization of Free Alkali Nanoparticles

2.1 *Experiment*

Nanoparticles of lithium, sodium, or potassium are produced by evaporating metal from a crucible and quenching the vapor in a flow of cold helium gas, as shown in Fig. 1. Nanoclusters condense in the aggregation zone, exit the source through a nozzle, and form a collimated beam in free flight towards the detector. In order to control their temperature, a heated thermalization tube is attached to the nozzle exit. Similar arrangements have been demonstrated[14,15] to provide adequate control of the internal cluster temperature. For low-temperature measurements the heated tube is removed, and the clusters retain the temperature of the liquid nitrogen cooled nozzle.

At the end of the flight path the beam is ionized by focused monochromatic light: the output of a Hg-Xe arc lamp, passed through a monochromator adjusted to a bandwidth of 5 nm. The resulting ions are detected by an ion counter. The yield of positive ions is normalized to both the photon and nanocluster beam intensities. The latter was monitored by periodically illuminating the nanoclusters by a broadband 100 W halogen light bulb (in fact, even a pocket flashlight produced a countable signal rate!).

Since the particles are heavy, they are not size selected; the measurement is conducted on a relatively broad distribution with an average radius R~3–5 nm (~2×10^3–3×10^4 atoms), as determined by a time-of-flight (TOF) measurement and also by an independent electron attachment experiment.[16] The photoyield process is dominated by the larger clusters: first of all, their absorption cross sections are larger, secondly, there is evidence that much of the lighter population of the TOF spectrum comes not from the original beam constituents but from fragmentation induced by the TOF ionizing pulsed laser. The important point is that the shape of the photoionization threshold in this experiment derives from rather large particles, so that size-dependent shifts and corrections (typically $\propto R^{-1}$ [17]) are negligible.

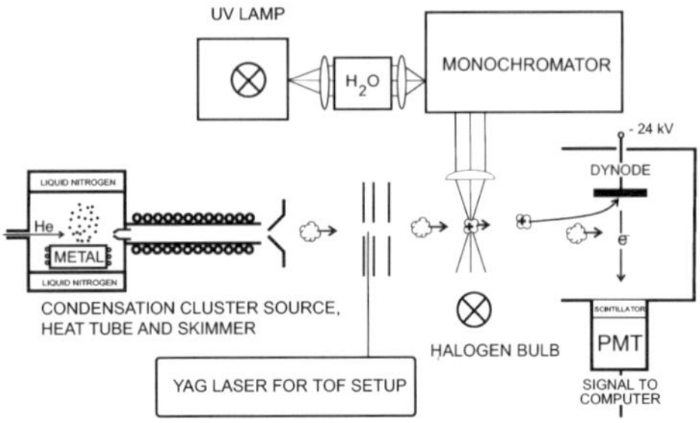

Fig. 1. Outline of the experimental setup (not to scale). The distance between the skimmer and the ionization region is ≈2 m. Nanoparticles are formed in a vapor condensation source and pass through a thermalization tube. Monochromatic light ionizes the particles, and the ion yield is measured by a positive-ion detector. The overall size distribution is monitored with a collinear time-of-flight setup which employs the tripled output of a pulsed Nd:YAG laser.

2.2 *The conundrum of inelastic scattering*

The data are shown in Fig. 2. Before concentrating on the threshold and its temperature dependence, we would like to remark that the origin of the decay in ionization efficiency at higher photon energies, seen in Fig. 2(a), is not understood at the moment. Such a drop has been observed in measurements on small alkali clusters (see, e.g., the references in Ref. 12) as well as on bulk surfaces.[18,19] The latter case has been interpreted within the framework of the "three-step model", whereby the photoemission process is viewed as consisting of three separate steps: (a) photoexcitation of an electron above the Fermi level, (b) transport of this electron to the surface, and (c) the electron dashing over the surface barrier into the vacuum region. The decay in yield at bulk surfaces has been analyzed quantitatively as associated with surface plasmon effects. Their role during step (a) can be to decrease light absorption by the metal, and thereby the electron excitation probability, and during step (b) to remove energy from the excited electron by inelastic scattering.

It is not clear how these considerations can be transferred to the case of clusters, whose dimensions are less than the wavelength of light, the bulk skin depth, and the mean free path. In such a situation it seems impossible to divide the ionization process into separate steps. This underscores the need for a fundamental analytical understanding of photoemission from finite metal particles and clusters.

a)
b)
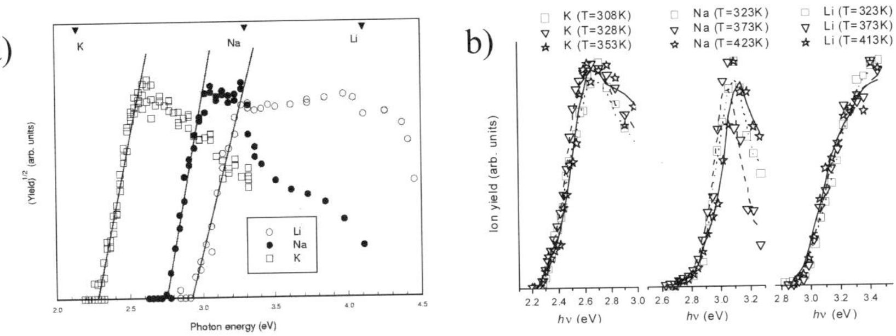

Fig. 2. (a) The square root of the nanoparticle photoionization yield for T=77K. The straight solid lines are least-square fits to the data near the threshold as is appropriate for the low-temperature limit of the Fowler law. The triangles at the top of the graph show the location of the surface plasmon resonance for each metal. (b) Threshold ion yield at selected temperatures.

2.3 *Ionization thresholds, work functions, and temperature shifts*

The aforementioned Fowler law for photoelectron yield from bulk metallic surfaces reads as follows:

$$\log\left(\frac{Y}{T^2}\right) = B + \log f\left(\frac{h\nu - \phi_0}{k_B T}\right), \tag{1}$$

where B is a constant and f is a known function. A plot of $\log(Y/T^2)$ vs. $(h\nu\text{-}\phi_0)/k_BT$ is known as a "Fowler plot", and by fitting the data to the universal curve $\log f(x)$ one extracts the work function ϕ_0. At T=0, the above equation reduces to $Y \propto (h\nu\text{-}\phi_0)^2$. The first use of the latter formula for nanoclusters involved coinage metal particles in aerosol gas suspensions, as reviewed in Ref. 20. We applied it to the variable-temperature alkali particle beam thresholds in Fig. 2(b).

The Fowler plots for some of the data are shown in Fig. 3(a), and the resultant work functions are shown in Fig. 3(b). Note, first of all, that the results extrapolate very well to the best literature polycrystalline bulk surface work functions at zero temperature.[11] Secondly, the weak temperature-induced work functions shifts are well resolved. This attests to the statement made in the Introduction: the study of free nanoclusters in beams can offer an accurate complement to traditional surface spectroscopy.

Many people are taken by surprise at the mention of a temperature-varying work function, because we are used to thinking of it as a single specified quantity. The variation is indeed not large, but obviously discernible. Its basic physical origin is easy to understand: thermal expansion of the metal results in a decrease of the electron density and a concomitant change of the Fermi energy and other electron gas parameters.

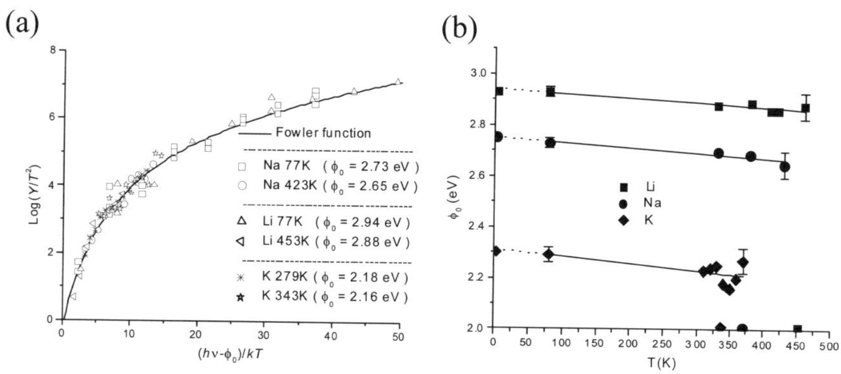

Fig. 3. (a) Fowler plots for several selected yield curves: the work function is determined by adjusting ϕ_0 to align the data with the Fowler function f. (b) The deduced temperature-dependent work functions. The solid straight lines are least-squares fits. The dotted lines are extrapolations to the polycrystalline work functions for bulk surfaces at zero temperature[11] (indicated by the dots on the y-axis). The marks on the x axis indicate the bulk melting point for each metal.

The $\Delta\phi_0/\Delta T$ values measured here are in excellent agreement with the recent theoretical calculations in Ref. [21]. This calculation is based on the image-charge approximation: the work function is represented as the amount of work, $\phi_0 = e^2/4d$, needed to remove an electron from a characteristic distance d outside the surface. Semiempirically,[22] d can be related to the nearest-neighbor distance in the crystal lattice, and therefore it is not surprising that by scaling ϕ_0 with the thermal expansion coefficient of the material, one can obtain a good first approximation to its temperature dependence.

3. Photoionization Thresholds of Smaller Clusters

Given the success of the Fowler formula with particles made up of thousands of atoms, it is tempting to inquire: might it also be useful for describing the photoionizaton yield of smaller metal clusters? Although such a leap would not be rigorous, the fact of the matter is that other fits currently in use (see the Introduction) are even more empirical. It also would offer the benefit of a coherent analytical framework for data analysis, encompassing both IP and temperature information. We decided to test the Fowler formalism on the spectra of $K_{30\text{-}101}$ clusters, obtained some time ago[23] by using a supersonic source and laser ionization. The near-threshold portions of the ion yield data from this work were digitized and fitted to Fowler plots in a manner analogous to that described above.

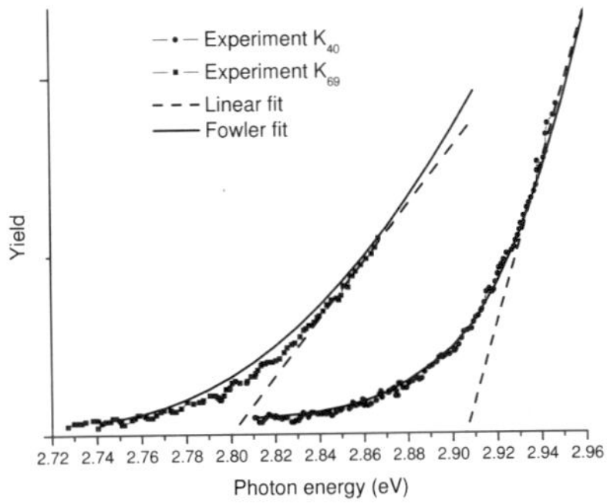

Fig. 4. Examples of ionization threshold curves. Dots: experimental data,[23] solid lines: finite-temperature Fowler functions, dashed lines: linear extrapolation employed in the original work.

The excellent quality of the resulting threshold curves is illustrated in Fig. 4 and compared with the linear fits from the original paper.[1]

The resultant IPs are shown in Fig. 5(a) where they are also compared with the values obtained in the original work by linear extrapolation. It is noteworthy that the Fowler-derived IPs for closed-shell spherical clusters extrapolate more accurately to the literature reference value for polycrystalline potassium surfaces, already encountered in Sec. 2.3 (see Fig. 5(b)).

---

[1] The fitting procedure used only a subset of the data above the ionization threshold. When the resulting values of *IP* and *T* were substituted into Eq. (1), they reproduced the experimental data over the entire photon energy range, as evidenced by Fig. 4. This provides additional indication that the success of the Fowler function for small clusters is more than a lucky coincidence.

It is also satisfying that the Fowler fits yield very reasonable internal cluster temperatures. According to evaporative ensemble theory,[24,25] the average internal temperature attained by clusters as they evaporate in flight is $k_B T \cong D/G$, where D is the cluster dissociation energy and G is the so-called Gspann parameter. Since dissociation energies of sequences of neutral alkali clusters have not been measured, we can estimate T by using the dissociation energies of cluster cations instead. For $K^{+}_{n\leq 25}$ D ranges from 0.6 eV to 0.75 eV,[26] as derived from unimolecular dissociation experiments and a rate coefficient expression which corresponds to G≅24–25 in the experiment analyzed here. Therefore the average temperature of the potassium clusters considered here is expected to be ~300-350 K. This agrees nicely with the temperature range derived from the Fowler photoionization plots: from 290 K for K_{30} to 335 K for K_{101}. Thus cluster temperature determination based on Eq. (1) provides a tractable, consistent, and more accurate alternative to exponential[23] and "pseudoexponential" fits,[27] yet another testimony to the efficacy of the approach.

One big question that remains, of course, is why does a theory developed for electron gas in a bulk solid appear to work so well for finite clusters with a curved surface and discrete energy levels? We hope that the presentation in this section will encourage a theoretical analysis of this and related issues in cluster ionization.

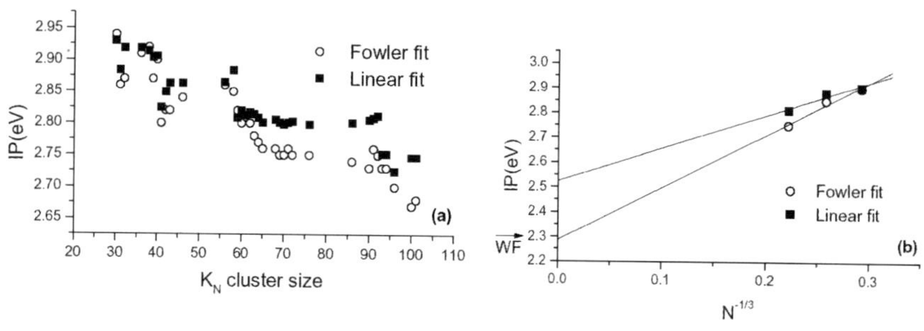

Fig. 5. (a) A comparison of cluster ionization potentials derived by linear extrapolation and by a Fowler fit. (b) Scaling of closed-shell cluster IPs to N→∞. The arrow denotes the bulk work function.

4. Summary

We have described two complementary investigations of the photoionization behavior of metal nanoclusters.

The first used a beam of nanoscale particles with variable internal temperature derived from a cluster condensation source. Their photoionization thresholds were analyzed according to the Fowler theory of surface photoemission, and accurate values of (temperature dependent) work functions were derived. Also, there are indications of surface plasmons influencing the photoionization process at frequencies above threshold; this issue awaits a further detailed investigation.

The second avenue involved applying the same Fowler formalism to a re-analysis of threshold photoionization data of small potassium clusters. Again, it turned out to be successful at yielding both the ionization potentials and the internal cluster temperatures for a range of particle sizes. The fit to the data is so good that it is unlikely to result from just a fortunate coincidence.

These results underscore the extent to which surface science and nanocluster research can benefit from each other, experimentally and theoretically. At the same time, they draw attention to the critical need for a comprehensive theoretical analysis of cluster photoionization behavior: to explore the limits of applicability of bulk-derived models, to account for inelastic process accompanying cluster ionization, to investigate the interplay between ionic structure and ionization potentials (including possible effects of melting), and so on.

Acknowledgements

We would like to thank G. Tikhonov and V. Kasperovich for their contributions to the experiment, and K. Bowen for advice on source construction. This work was supported by the Physics Division of the U.S. National Science Foundation.

References

1. M. Cardona and L. Ley, in *Photoemission in Solids*, ed. by M.Cardona and L.Ley, Springer, Berlin, (1978).
2. M.M. Kappes, *Chem. Rev.* **88**, 369, (1988).
3. A.W. Castleman and K.H. Bowen, *J. Phys. Chem.* **100**, 12911, (1996).
4. M.B. Knickelbein, *Philos. Mag.* B **79**, 1379, (1999).

5. K. Wong and V.V. Kresin, *J. Chem. Phys.* **118**, 7141, (2003).
6. W. Ekardt, *Phys. Rev.* B **65**, 6360, (1985).
7. B. Wästberg and A. Rosén, *Z. Phys.* D **18**, 267, (1991).
8. M. Koskinen and M. Manninen, *Phys. Rev.* B **54**, 14796, (1996).
9. M.E. Madjet and P.A. Hervieux, *Eur. Phys. J.* D **9**, 217, (1999).
10. R.H. Fowler, *Phys. Rev.* **38**, 45, (1931).
11. H.B. Michaelson, *J. Appl. Phys.* **48**, 4729, (1977).
12. K. Wong, V. Kasperovich, G. Tikhonov, and V.V. Kresin, *Appl. Phys.* B **73**, 407, (2001).
13. K. Wong, G. Tikhonov, and V.V. Kresin, *Phys. Rev.* B **66**, 125401, (2002).
14. C. Ellert, M. Schmidt, C. Schmitt, T. Reiners, and H. Haberland, *Phys. Rev. Lett.* **75**, 1731, (1995).
15. F. Chandezon, P.M. Hansen, C. Ristori, J. Pedersen, J. Westergaard, and S. Bjørnholm, *Chem. Phys. Lett.* **277**, 450, (1997).
16. V. Kasperovich, K. Wong, G. Tikhonov, and V.V. Kresin, *Phys. Rev. Lett.* **85**, 2729, (2000).
17. *Clusters of Atoms and Molecules*, ed. by H.Haberland (Springer, Berlin, 1994).
18. N.V.Smith and W.E.Spicer: Phys. Rev. **188**, 593 (1969).
19. R.J.Whitefield and J.J.Brady, Phys. Rev. Lett. **26**, 380 (1971).
20. H.Burtscher and H.C.Siegmann, in *Clusters of Atoms and Molecules*, Vol. II, ed. by H.Haberland, Springer, Berlin, (1994).
21. T. Durakiewicz, A.J. Arko, J.J. Joyce, D.P. Moore, and S. Halas, *Surf. Sci.* **478**, 72, (2001).
22. K. Wong, S. Vongehr, and V.V. Kresin, *Phys. Rev.* B **67**, 035406, (2003).
23. W.A. Saunders, K. Clemenger, W.A. de Heer and W.D. Knight, *Phys. Rev.* B **32**, 1366 (1985); W.A. Saunders, Ph.D. Thesis, University of California, Berkeley, 1985.
24. K. Hansen and U. Näher, *Phys. Rev.* A **60**, 1240, (1999).
25. P. Brockhaus, K. Wong, K. Hansen, V. Kasperovich, G. Tikhonov, and V.V. Kresin, *Phys. Rev.* A **59**, 495, (1999).
26. C. Bréchignac, Ph. Cahuzac, F. Carlier, M. de Frutos, and J. Leygnier, *J. Chem. Phys.* **93**, 7449, (1990).
27. U. Röthlisberger, M. Schär, and E. Schumacher, *Z. Phys.* D **13**, 171, (1988).

MAGNETIC EXCITATIONS INDUCED BY PROJECTILE IN FERROMAGNETIC CLUSTER

R-J. Tarento, P. Joyes

Laboratoire de Physique des Solides ,bât 510, Université de Paris-Sud, Orsay, 91405, France

R. Lahreche, D.E. Mekki

Faculté des Sciences, Département de Physique, Université d'Annaba B.P12, Annaba, Algérie

We investigate the diffusion of a spin polarized projectile on a ferromagnetic linear cluster. The interaction between the projectile and the target is described with a Heisenberg Hamiltonian which excludes the charge degree of freedom during the process. Our calculation is based on a real time description of the spins of the two interacting systems. The spin excitations induced by the spin of a projectile-atom are investigated versus the cluster size, the trajectory parameters (projectile velocity, impact parameter). Nonadiabatic behaviour during the collision has been characterized by the spin temperature at the end of the collision process. The effects of the phonons on the spin excitations will be discussed.

1. Introduction

Many experimental[1] and theoretical[2] studies have been devoted to interaction between a fast atomic particle and a metallic cluster showing various kinds of non-atomic effects like electronic excitations, ionization and capture.

Other experiments have been performed on the spin diffusion on bulk magnetic targets like the spin depolarization of low energy polarized electrons through ferromagnetic films.[3] Though, up to now, only bulk targets are employed, similar experiments could be made in the near future on finite media.

We will use a time dependent Heisenberg formalism which will be presented in Sec 2. We discuss our results on the spin diffusion on clusters, on the nonadiabatic effects and on the phonon magnon coupling effect in Sec. 3.

2. The Heisenberg Hamiltonian

The most simple Hamiltonian describing the time-dependent behavior of spin excitations on a spin-polarized nanostructure is the Heisenberg Hamiltonian:

$$H^{(Heis)}=-\sum_{i,j} J\,(r_{ij})\,\mathrm{S_i}\,\mathrm{S_j} \qquad (1)$$

where S_i refers to the spin of particle i. In the present study one of the sites, say site 1, is moving along a given trajectory, while the others are assumed fixed. Between two close target atoms the J_{ij} interaction term remains constant and is J_o. The dependence of J_{1i} on the r_{1i} distance between the target atom i and the incident atom 1 can be taken as

$$J_{1i} = J_{i1} = J(r_{1i}) = J_o^{'} \exp\left(-\frac{r_{1i}}{r_o}\right). \qquad (2)$$

$J_o^{'}$ has been chosen in our calculation so that r_{1i} is equal to the nearest target distance, and J_{1i} is equal to J_o. r_o ($r_o \cong 1$Å), a typical interaction length. The distance r_{1i} between the incident and the target particles depends on time t. It is assumed that in the studied speed range the motion of the particle 1 does not depend on spin evolution during interaction. Only linear trajectories with constant V velocities will be considered.

$H^{(Heis)}$ has been widely used in various solid state problems, for instance, in the aggregate case to investigate π electron magnetic properties in polyenes[4]. But the use of $H^{(Heis)}$ implies no electron transfer: i.e no charge exchange between the incident particle and the target during the collision and no electron population changes in the target during collision either.

3. Results for a Small Linear Cluster

We consider the collision of an up incident spin on a ferromagnetic down spin target both initially polarized along the Z-axis. Therefore we take a J_o=0.1eV positive value so that the initial target is the ground state. The target, called C_n, is a linear chain of n atoms which are located along the Y-axis (the atoms are located at sites 2, 4 … , 2 n (Å)).

The wave function $|\Psi\rangle$ is written as $|\Psi\rangle=\sum_i \alpha_i|\phi_i\rangle$,

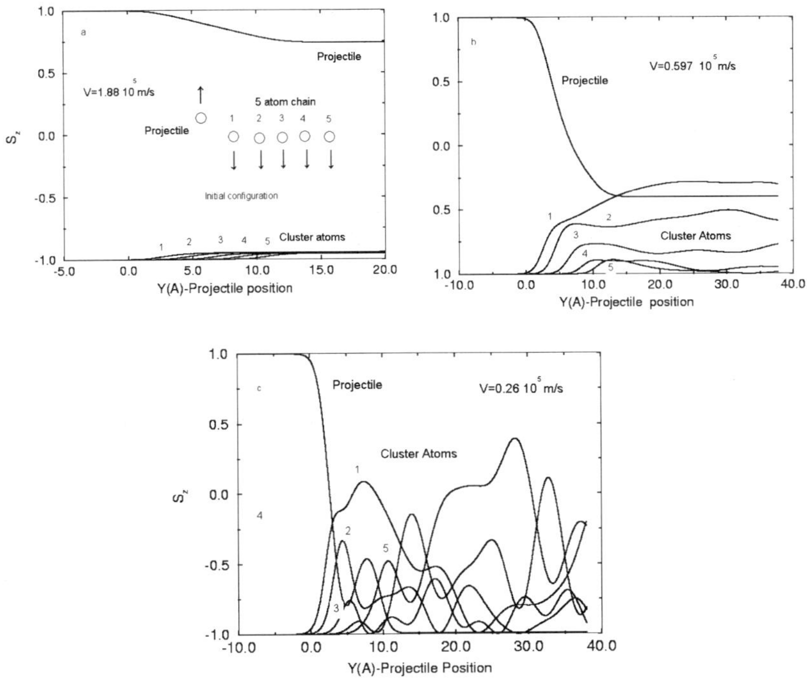

Fig. 1. Evolution of the z spin component S_z for the projectile and 5 atom chains atoms versus the projectile velocity V. Y is the projectile position for an arbitrary trajectory parallel to the y-axis with an impact parameter b=2Å. The initial spin configuration is displayed in Fig.1a.

$\{\phi_i\rangle\}$ are the states $|\uparrow;\downarrow,...,\downarrow\rangle, |\downarrow;\uparrow,...,\downarrow\rangle$, with only one up spin for the projectile and cluster system (the first spin is referred to the projectile one). The time dependence α_i are obtained by solving numerically the Schrödinger equation with $|\uparrow;\downarrow,...,\downarrow\rangle$ for the initial configuration. At every time of the dynamics the conservation of the total spin S_z and S^2 of the system has been checked.

In Fig. 1 we report three different behaviours of S_z^i for the projectile and atom cluster spin versus the projectile velocity V corresponding to an energy for a silver atom of 20Kev, 2KeV and 400eV. During a first phase T1, the projectile spin is varying like the cluster ones. In a second phase T2 only the excited cluster has varying spins. Notice that for this velocity range, during T1 the cluster atom positions are frozen. The projectile trajectory is parallel to the cluster axis and the impact parameter b=2Å. The evolution depends on the two characteristic time $\tau=\hbar/J_0$, which describes the intra spin in the chain and $\tau'(V)$ the travelling time along one atomic distance. For $\tau'<\tau$ (Fig. 1a), the target spins vary significantly when the projectile is above them. Each site behaves as if it were isolated. For $\tau'\approx\tau$ (Fig. 1b), the spin amplitudes of variation are larger due to a larger projectile-site interaction. For $\tau'>\tau$ (Fig. 1c), the incident spin flip almost completely. The target spin has a complex evolution. The projectile is interacting with a fully excited target.

To get some insight into the dynamics let us report the 6 adiabatic surfaces derived by diagonalizing H^{Heis} (Fig. 2) for the same condition of Fig. 1. The adiabatic surfaces are V-independent. The degeneracy of some of them is removed during the collision. Before the collision our system is in the ground state, during the diffusion process the surface 1 is described by the vector $|1\rangle=\sum(\frac{1}{6})^{1/2}|\varphi_i\rangle$. With such a choice, the initial state of the system $|\uparrow;\downarrow,\ldots,\downarrow\rangle$ has components on the |1> and |2> (Fig. 3). The wave function |Ψ> can be expanded on the adiabatic surface states. The weight W_i of different adiabatic states (Fig. 3) displays a quite different evolution versus V. For large V, all adiabatic surfaces are involved in process dynamics and the W_i display symmetric variations with final values nearly equal to the initial ones. For the low values the symmetry disappears. Notice that the W_i begin to vary or have changed significantly when two adiabatic surfaces are crossing.

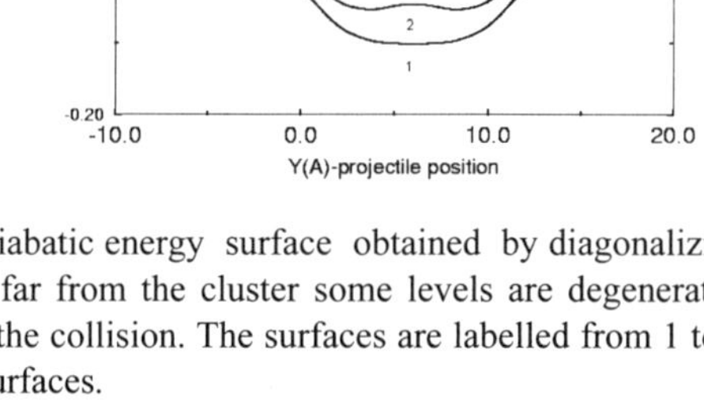

Fig. 2. The 6 adiabatic energy surface obtained by diagonalizing H^{Heis}. Note that when the projectile is far from the cluster some levels are degenerate, but the degeneracy is removed during the collision. The surfaces are labelled from 1 to 6 i.e from the lowest to highest energy surfaces.

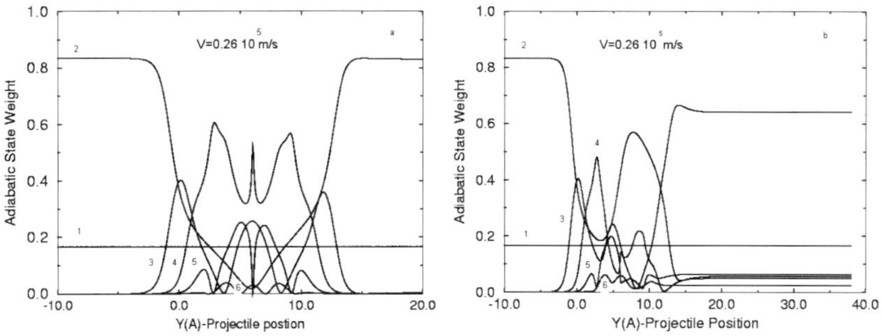

Fig. 3. The evolution of the weights of the projected wavefunction $|\psi\rangle$ on the adiabatic states for different projectile velocity V. The initial spin configuration is displayed in Fig. 1.

The excitation energy Ex is defined as E-E_{ad} where the system energy E is given by $\langle\Psi|\,H^{Heis}|\,\Psi\rangle$ and E_{ad} is the energy of the adiabatic state associated to the initial state (i.e :$(|1\rangle+3^{1/2}\,|2\rangle)$). In Fig. 4. Ex is displayed during the phase T1 and the beginning of the phase T2. At the beginning of the collision, Ex is increasing with V. We can see that the energy transfer at the end of the collision (i.e T1 phase) is small at low and high velocity and displays a maximum for V=0.6 10^4 m/s. The time evolution of Ex is quite complex. Ex is not only increasing with time. This effect is linked to the nonadiabatic process involved during the collision. Phase T2 is discussed below.

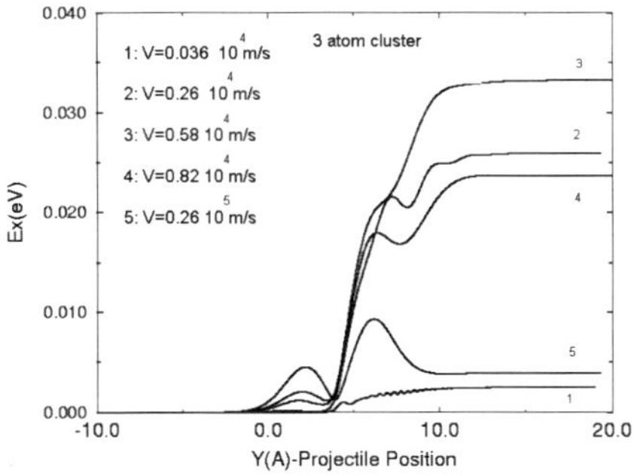

Fig. 4. Excitation energy evolution during collision between a 3 atom chains cluster and a projectile with velocity V. The initial spin configuration is displayed in Fig. 1.

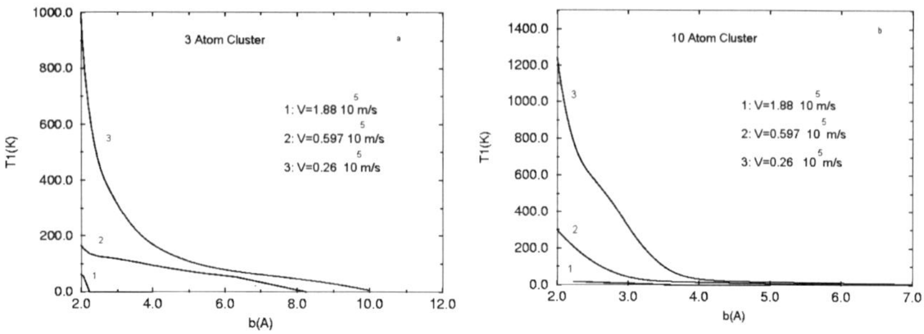

Fig. 5. Spin temperature in the cluster at the end of the collision in the 3 or 10 atom chains atoms versus the impact parameter b versus the projectile velocity V. The initial spin configuration is displayed in Fig. 1a.

The spin temperature in the cluster at the end of the phase T1 or the beginning of T2 is another way of characterizing the excitation in the system. In Fig. 5 the spin temperature is displayed versus the impact parameter. The spin temperature is increasing when b is decreasing. At high velocity, as the energy transfer is small, the spin temperature is also small. In the case of the 10 atom chain the temperature is significant for b<4.5Å but for the 3 atom chain it is for b<10Å. For small b the temperature is large for intermediate velocity. For larger time we can see below that the cluster could excite other degrees of freedom.

At the end of phase T_1, the cluster spin is only excited if the projectile speed is not too small (the atom nucleus are still nearly frozen). The set of excitations produced in the cluster just after the collision is reported in Fig. 6 versus the cluster size. Increasing n, the excitation spectra forms a band varying between 0 et $8\omega_o$ ($\omega_o = \frac{J}{4\hbar}$) which is associated to the classical chain spin wave dispersion ε_k:

$$\varepsilon_k = 4JS\,(1-\cos(ka)). \tag{3}$$

The lowest frequency ω_{min} in Fig. 6 goes to zero with n. ω_{min} could be derived from Eq.3. by introducing an infrared cut-off $k_c=2\pi/L$ (L is the cluster size).

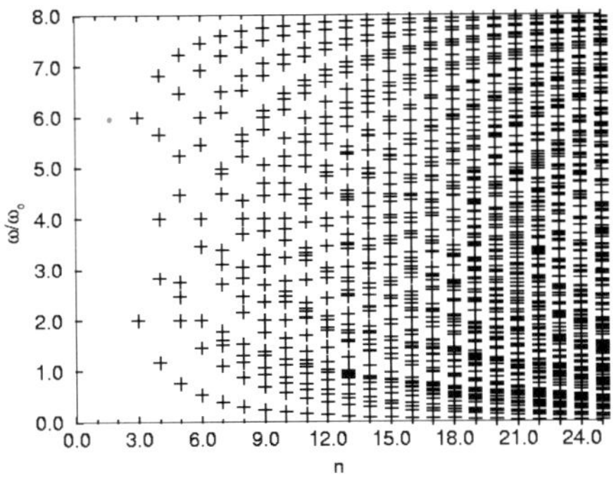

Fig. 6. Set of the excitation frequencies in the cluster just after the collision (i.e. at the beginning of the phase T2) versus the size of the cluster $\omega_0 = \frac{J}{4\hbar}$.

In fact Fig. 6 demonstrates for a given size all the possible excitation frequencies. But only the collision conditions control if a frequency is really excited. This fact is illustrated on the 3 atom cluster. In this case two frequencies are possible: $\omega_2=2\omega_o$ and $\omega_6=6\omega_o$ for the spin excitations of the two atoms located at the beginning and end of the chain and only one frequency (ω_6) for the atom at the chain centre. In Table 1 the ratio of the excitation frequencies ω_6/ω_2 for the atom located at the chain end is reported versus the collision impact parameter b and the projectile velocity V (the trajectory is along the chain axis). For large V the ratio is zero, meaning that only ω_2 is involved. If we decrease V, ω_6 is excited and the ratio ω_6/ω_2 increases when b decreases.

Let us discuss what is happening in the cluster during the phase T2. During the phase T1 the cluster is spin-excited. Its excitation has been previously characterized by Ex, or the spin temperature. During the phase T2 the excess of excitation energy for the spins is distributed to the vibration degrees of freedom i.e. the nuclei are no more frozen. This phenomena generate, a coupling between the magnon and the phonon. In our calculation the nuclei have been treated classically like in the Carr-Parrinello procedure. A part of the spin excitation is transformed into vibration excitation. Table 2 shows the spin temperature $T2^{in}$ ($T2^{f}$) at the beginning (at the end) of the phase T2 which follows the collision. $T2^{in}$ and $T2^{f}$ increase when b decreases. But the difference $T2^{in}$-$T2^{f}$ decreases with b, meaning that the vibration amplitude is smaller when decreasing.

Table 1. Ratio of the excitation frequencies ω_6/ω_2 appearing just after the collision in the 3 atom chain cluster versus the impact parameter b and the projectile velocity V (in x 10^5m/s unit). The initial collision trajectory conditions are reported in Fig. 1.

ω_6/ω_2	b=2Å	b=3Å	b=4Å	b=5Å	b=6Å
V=0.26	0.52	0.57	0.23	0.15	0.11
V=0.597	0.35	0.52	0.30	0.25	0.23
V=1.88	0	0	0	0	0

Table 2. Phonon pulsation ω_{ph} (x10^{12} s^{-1}) and spin temperatures $T2^{in}$($T2^f$) at the beginning (at the end) of phase T2 following the collision of a projectile and the 3 atom chain cluster versus the impact parameter b with the projectile velocity V=0.44 10^5m/s. The initial collision trajectory conditions are reported in Fig. 1.

	b=2Å	b=3Å	b=4Å	b=5Å	b=6Å
$T2^{in}$(K)	293	172	121	87.5	65.5
$T2^f$(K)	268.5	166	118.5	86	64
ω_{ph}	1.39	01.46	01.49	1.495	1.5

For a projectile trajectory parallel to the chain axis only the stretching vibration mode is excited. The effect of the phonon-magnon coupling decreases the stretching vibration mode ω_{ph} (Table 2) from its free value when b is decreasing.

3. Conclusions

The calculation has shown the importance of the spin excitation during the collision phase, which induces vibration excitations coupled to them after the collision.

Acknowledgements

The authors wish to thank the Egide for their support (Procope:03196UB).

References

1. C. Bréchignac, Ph. Cahuzac, B. Concina, J. Leygnier and I. Tignières, *Eur. Phys.* D **12,** 185, (2000).
2. F. Martin, M.F. Politis, B. Zarour and P.A. Hervieux, *Phys.Rev.* A **60**, 4701, (1999).
3. D. Oberli, R. Burgermeister, S. Riesen, W. Weber and P.A. Siegmann, *Phys.Rev Lett.* **81,** 4228, (1998).
4. D. Maynaud, Ph. Durand, J.P. Daudey and J.P. Malrieu, *Phys.Rev.* A **28**, 3193, (1983).

Clusters in Laser Fields

COLLISION AND LASER INDUCED DYNAMICS OF MOLECULES AND CLUSTERS

Rüdiger Schmidt, Thomas Kunert and Mathias Uhlmann
Institut für Theoretische Physik
Technische Universität Dresden, 01062 Dresden, Germany
E-mail: schmidt@physik.tu-dresden.de

Collision and laser induced dynamics in atomic many-body systems are investigated within a common microscopic framework, the so-called Nonadiabatic Quantum Molecular Dynamics (NA-QMD). It is shown that the collision induced electronic and vibrational excitation mechanism depends crucially on the projectile mass in ion-fullerene collisions. In addition, its velocity-dependence is qualitatively different from that in metal-cluster collisions. In first fully three-dimensional calculations of molecules in a laser field, dramatic differences in the alignment of fragmenting H_2^+ and H_2 are obtained.

1. Introduction

The theoretical and experimental investigation of nonadiabatic many electron problems coupled to nuclear degrees of freedom represents one of the most challenging problems of actual research. Experimentally, large progress has been made during the last years in the study, in particular, of collisions with atomic clusters, as well as the interaction of clusters and molecules with intense laser fields. Theoretically, much less is known about the underlying mechanisms in both cases due to the generally very crude approximations or model assumptions made in order to handle a nonadiabtic many-electron system coupled to various nuclear degrees of freedom. Thus, there is a strong need to develop fully microscopic, universal approaches which will apply for both situations.

In this contribution, we present theoretical investigations of ion-cluster collisions and laser-molecule interactions obtained within a common microscopic framework, the so-called Nonadiabatic Quantum Molecular Dynamics (NA-QMD) developed[1] and extended[2,3] recently. This approach is based on a mixed classical-quantum description of the coupled nuclear and electronic degrees of freedom, respectively. The NA-QMD formalism is briefly sketched in Sec. 2. It will be applied to different ion-fullerene collisions (H^+ + C_{60}, Ar^+ + C_{60}) as well as metal cluster-ion collisions (Na + Na_9^+) in Sec. 3. It will be demonstrated that the excitation mechanisms are basically different in all three cases. The results are also compared with available, recent experimental data. First, fully three dimensional calculations of the excitation, ionization and fragmentation of H_2^+ and H_2 molecules in intense laser fields are presented in Sec. 4. Dramatic differences in the dynamics (in particular in the alignment of the resulting fragments) between both molecules are found. The results for initially aligned H_2^+ molecules are also compared to that of available fully quantum-mechanical calculations.

2. NA-QMD Formalism

The NA-QMD method represents a rather universal approach to studying the nonadiabatic dynamics in atomic many-body systems. In this theory, electronic and vibrational degrees of freedom are treated simultaneously and self-consistently by combining classical molecular dynamics (MD) for the nuclei with time-dependent density functional theory (TD-DFT) for the electrons using a *finite basis set expansion* for the TD-Kohn-Sham orbitals. Originally, the formalism[1] has been worked out for conservative systems, in particular to investigate adiabatic and nonadiabatic collisions involving molecules and atomic clusters.[4,5] Recently, it has been extended[2] to describe also the interaction of finite atomic many-body systems with external laser fields. In this case, and in contrast to collisions, one has to treat simultaneously bound (core and valence) electrons as well as practically free electrons in the continuum. To this end, we have developed a special method[3] which combines a linear combination of atomic orbitals (LCAO) for the bound electrons with grid-like Gauss-functions arranged on a chain along the laser polarization axis for the electrons in the continuum (see also Sec. 4.1). In this paper we present also an alternative expansion method[6] consisting of an ordinary LCAO but including distances up to 70 a.u. away from the nuclei (see Sec. 4.2).

3. Cluster Collisions

In Fig. 1, the calculated total kinetic energy loss ΔE (TKEL) as well as its electronic contribution (resulting from excitation, direct ionization and charge transfer processes) as a function of the impact velocity for H^+ +

C_{60} and $Ar^+ + C_{60}$ collisions is shown.

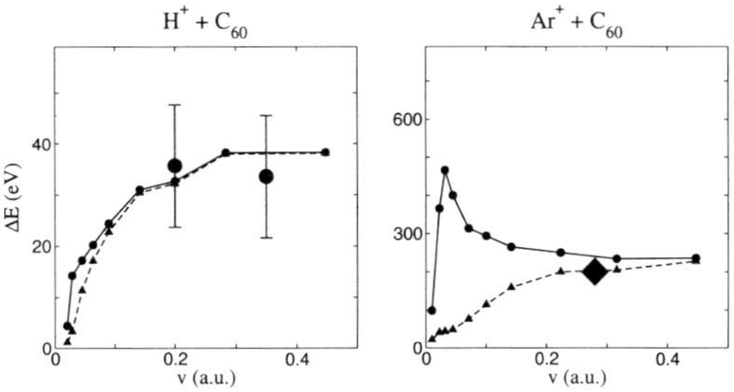

Fig. 1. Calculated total kinetic energy loss ΔE (circles) and its electronic contribution (triangles) as function of impact velocity v for different ions (H^+, Ar^+) on C_{60} at fixed collision geometry with b=2 a.u. . Recent experimental data for H^++C_{60} (filled circles with error bars) and very recent data for Ar^++C_{60} (filled diamond) are also shown.

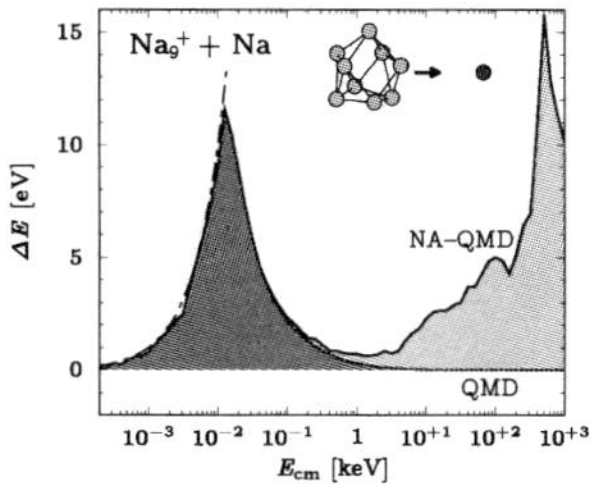

Fig. 2. Calculated total kinetic energy loss ΔE (full line) as well as its electronic (light grey) and vibrational (dark grey) contributions as function of impact energy for Na_9^+ + Na collisions at fixed collision geometry with b=0 a.u. .

Note, first of all, the quantitative differences in the absolute values of ΔE for the ions, which must result in qualitatively different fragmentation patterns. Second, the transferred energy ΔE increases smoothly with v in the low velocity range (v$\lesssim$0.3 a.u.) only for the proton collisions whereas a distinct maximum of ΔE is observed for the heavier projectile, resulting from first dominating and then decreasing vibrational energy transfer. Third, and most surprisingly, the microscopic calculations provide velocity-independent excitation energies with dominating electronic nature in the high velocity range of v $\approx$0.25...0.5 a.u. for both ions ("saturation effect").

Obviously, the calculated absolute values of ΔE are in excellent agreement with available experimental data in Refs. 7 to 9 respectively. Moreover, the present calculations explain nicely other experimental observations:

i) the very different fragmentation patterns obtained for light[7,10] and heavy projectiles,[11,12] respectively;
ii) the dominating electronic excitation mechanism observed with light projectiles;[7,10]
iii) the velocity-independent fragmentation patterns measured for protons and Ar^+ collisions[7,11] in the range of $0.2 \lesssim v \lesssim 0.4$.

In Fig. 2, the calculated TKEL, as well as its electronic and vibrational contributions as function of the c.m. impact energy $E_{c.m.}$, is shown, but for metallic Na_9^+ + Na collisions. The results are in striking contrast to that obtained for fullerene collisions (Fig. 1):

i) There is a wide range of $E_{c.m.}$ between 10^{-3} keV and 10^{-1} keV (corresponding to velocities of $\approx 0.04 \ldots 0.4$ a.u.) where pure vibrational excitations occur (adiabatic regime).
ii) Electronic excitations start to occur at larger impact energies and dominate for $E_{c.m.} \gtrsim 1$ keV.
iii) In the intermediate range, where both excitation mechanisms compete, a deep minimum of ΔE is observed ("transparency effect").

Unfortunately, for this kind of collisional system, experimental data for comparison is still not available.

4. Laser Interaction with Molecules

4.1. *Excitation, dissociation and ionization of H_2^+*

We will consider here the interaction of a laser with the H_2^+ molecule, which is *aligned* along the laser polarization axis $\mathbf{e}_z$. For this simplest possible case full quantummechanical solutions do exist.[13]

For the basis expansion we use a LCAO consisting of the 1s and 2s eigenstates of the H-atom, centered at each nucleus. These functions are expanded in s-type Gaussians $g(\mathbf{r};\sigma) = \left(2/\sigma^2\pi\right)^{\frac{3}{4}} \exp\left(-r^2/\sigma^2\right)$. Additionally we introduce chain-like arranged Gaussian functions with three different widths ($\sigma = 1.5; 3.0; 5.0$ a.u.) spread from $z = -30$ a.u. to 30 a.u.[3] This allows us to calculate all matrix elements analytically. Finally, in order to compare the results with that obtained from full quantum mechanical calculations,[13] the nuclei are assumed to be in their sixth vibronically excited state (see below) and the laser field is chosen to be identical that of,[13] i.e. $V_L(\mathbf{r},t) = E_0(t)\, z\, \sin(\omega t)$ with the frequency $\omega = 5.7$ eV, a trapezoidal envelope and a rise/fall time t_0=1 fs.

Figure 3 shows the electronic density along the z-axis for two trajectories calculated with a LCAO basis *only* and our extended (LCAO plus additional grid functions) basis. To make the initial state of these calculations visible we have plotted it from $t = -10 \ldots 0$ fs. In both pictures the electron is initially tightly bound to the nuclei. However, when the laser is switched on, the dynamics dramatically differ:

i) the electron is always bound to the nuclei for the LCAO only (left)

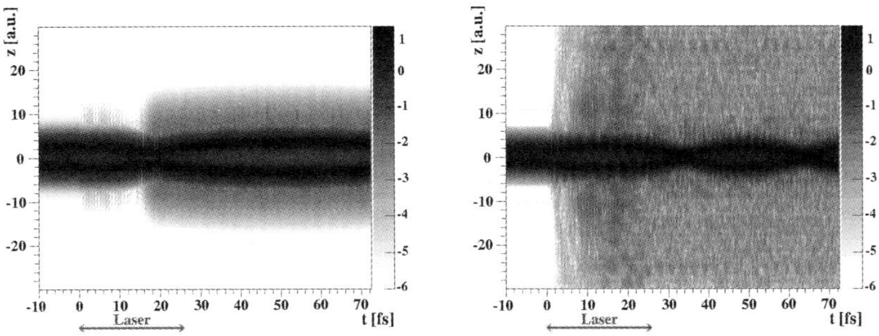

Fig. 3. Electronic density of H_2^+ along the z-direction (i.e. laser polarization direction) for sample trajectories calculated with the LCAO basis only (left) and the extended (LCAO + additional grid of Gaussians) basis (right) exposed to a 26fs, 212nm laser pulse with an intensity of $3.5 \cdot 10^{13} \frac{\text{W}}{\text{cm}^2}$.

but is spread over the whole grid for the larger basis (right);

ii) the quiver motion of the electron is observed only for the larger basis;

iii) the necessity of introducing absorbing boundary conditions can be clearly seen as the electron is spread up to the ends of the grid (the absorbing scheme is described elsewhere[3]).

Figure 4 shows the calculated curves for the dissociation and ionization probabilities and a comparison with the results of Chelkowski *et al.*[13] With the larger basis the full quantum-mechanical results[13] are very well reproduced. Even the kink in the dissociation probability around 20 fs is given correctly. This is in sharp contrast to the LCAO results where neither the ionization probability (it is zero all the time) nor the dissocation probability (it is overestimated by more than a factor of 2) agree with the full quantum mechanical results.

4.2. *Alignment dynamics in H_2 and H_2^+*

In this section we present first full three dimensional calculations of H_2^+ and H_2 molecules, initially *not aligned* with respect to the laser polarization axis. The H_2 calculations were performed using a Hartree-Fock x-potential. The basis used in this section consists of s- and p-type Gaussians with widths up to ≈ 70 a.u. centred at the nuclei with a total basis size of 24 functions per nucleus.[6]

Figure 5 shows the behaviour of an H_2 molecule in a 500 fs $\sin^2$, 266 nm

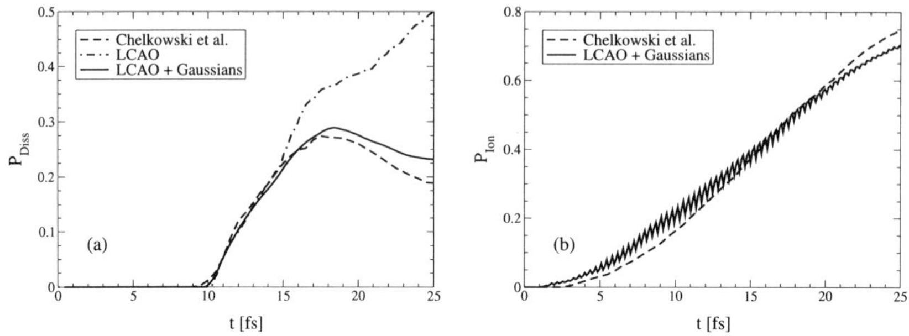

Fig. 4. Dissociation (left) and ionization (right) probability for H_2^+ exposed to a 26fs, 212nm laser pulse with an intensity of $1.0 \cdot 10^{14} \frac{W}{cm^2}$ calculated with the LCAO + grid basis (full line) and the LCAO basis only (dash-dotted line). The full quantum mechanical results of Chelkowski et al. are shown as dashed lines for comparison.

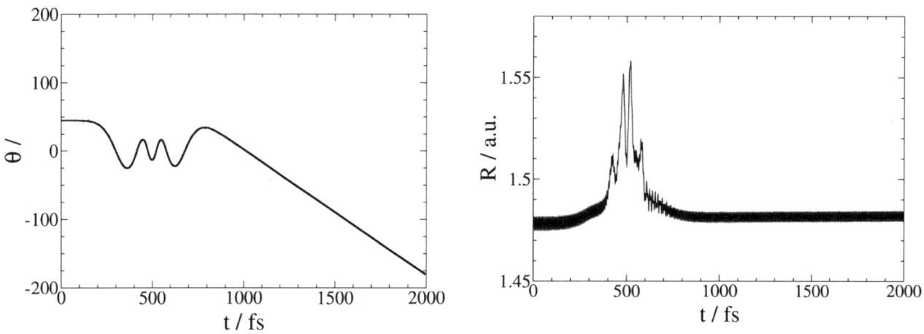

Fig. 5. Angle (left) and internuclear distance (right) as function of time for H_2. An intensity of $6 \cdot 10^{13}$ $\frac{W}{cm^2}$, a wavelength of 266 nm and a $\sin^2$ pulse with a duration of 500 fs is chosen.

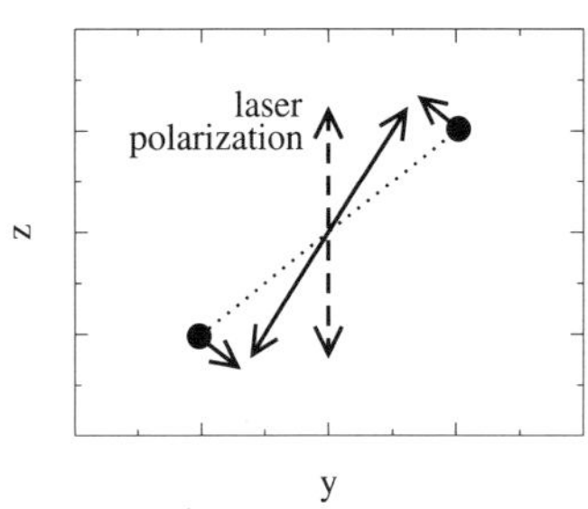

Fig. 6. Direction of the dipole moment (solid line) with fixed nuclei (filled circles), an angle of 45° between molecular (dotted line) and laser axis (dashed line) at a wavelength of 266 nm in the case of H_2. The arrows denote the effective torque.

laser pulse with an intensity of $6 \cdot 10^{13}$ $\frac{W}{cm^2}$. The initial angle between molecular and laser polarization axis (i.e. the direction of the $\vec{E}$-field) is 45°. The laser induces a dipole moment and thus generates an effective torque that turns the molecule towards the laser axis (see Fig. 6).[14] Since this torque is always directed towards the laser axis the molecule performs a pendular motion. However, the amplitude of the oscillation is much smaller than the initial angle. After the laser is switched off (at $t = 1000$ fs) the molecule starts to rotate freely. This calculation shows an example of dynamical alignment (the molecules are turned towards the laser axis) with an adiabatic pulse turn-on/off. During the duration of the laser pulse the molecule is stretched (right part of Fig. 5). This stretch is maximal at the maximum of the laser field (at $t = 500$ fs).

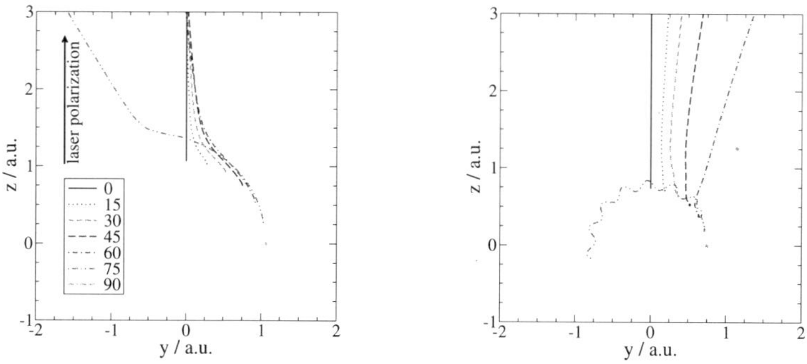

Fig. 7. For different initial angles ($\Theta = 0 \ldots 90°$) and H_2^+ (left) and H_2 (right) the trajectories of one proton in the yz-plane are shown. The molecules are exposed to a 50 fs $\sin^2$ pulse with a wavelength of 266 nm and an intensity of $1.5 \cdot 10^{14}\ \frac{W}{cm^2}$.

Fragmenting trajectories for H_2^+ and H_2 in the yz-plane (see Fig. 6) are shown in Fig. 7. In both cases the parallel oriented molecules ($\Theta = 0°$) fragment directly and the perpendicular oriented ones ($\Theta = 90°$) do not, and are in addition, completely unaffected by the laser pulse. This is the so-called geometric alignment effect which results from the fact that the molecule is much easier to excite when parallel to the field than when perpendicular.[15] All trajectories (except that with $\Theta = 0°$) are turned towards the laser polarization axis. Furthermore, all other trajectories show fragmentation in the case of H_2^+. For H_2, we observe one further stable trajectory. The striking difference between the two molecules is the final angle under which the fragments are observed. For H_2^+ all trajectories except the $\Theta = 75°$ show fragments that come out under nearly $0°$. This is in sharp contrast to H_2 where the fragments are observed within a wide range of final angles.

5. Summary

We have investigated the nonadiabatic coupled electron-nuclear dynamics in systems as small as H_2^+ (with one electron) up to systems as large as C_{60} (with at least 240 electrons) within one and the same microscopic and universal framework. The predictive power of this NA-QMD theory has been demonstrated by confronting the calculations with experimental data for various C_{60} collisions, as well as by comparing the NA-QMD results with that of full quantum mechanical calculations for H_2^+ in intense laser fields.

A further elaboration of the method to treat also larger systems in intense laser fields (heavier diatomic molecules, organic molecules or C_{60}) is still in progress.

We gratefully acknowledge the allocation of computer resources from the Zentrum für Hochleistungsrechnen of the Technical University Dresden. We thank the Deutsche Forschungsgemeinschaft (DFG) for support.

References

1. U. Saalmann and R. Schmidt, *Z. Phys. D* **38**, 153 (1996).
2. T. Kunert and R. Schmidt, *Eur. Phys. J. D* **25**, 15 (2003).
3. M. Uhlmann, T. Kunert, F. Grossmann and R. Schmidt, *Phys. Rev. A* **67**, 013413 (2003).
4. U. Saalmann and R. Schmidt, *Phys. Rev. Lett.* **80**, 3213 (1998).
5. T. Kunert and R. Schmidt, *Phys. Rev. Lett.* **86**, 5258 (2001).
6. M. Uhlmann, T. Kunert and R. Schmidt, to be published.
7. J. Opitz et al., *Phys. Rev. A* **62**, 022705 (2000).
8. We are grateful to Bernd Huber for providing us with the temperature-corrected experimental data.
9. S. Martin, private communication (2003).
10. T. Schlathölter, O. Hadjar, R. Hoekstra, and R. Morgenstern, *Phys. Rev. Lett.* **82**, 73 (1999).
11. A. Reinköster, U. Werner, and H. O. Lutz, *Eur. Phys. Lett.* **43**, 653 (1998).
12. H. Cederquist et al., *Phys. Rev. A* **61**, 022712 (2000).
13. S. Chelkowski, Tao Zuo, O. Atabek and A. D. Bandrauk, *Phys. Rev. A* **52**, 2977 (1995).
14. Bretislav Friedrich and Dudley Herschbach, *Phys. Rev. Lett.* **74**, 4623 (1995).
15. K. Codling, L. J. Frasinski and P. A. Hatherly, *J. Phys. B* **22**, L321 (1989).

PROBING THE DYNAMICS OF IONIZATION PROCESSES IN CLUSTERS

A. W. Castleman, Jr. and T. E. Dermota

Departments of Chemistry and Physics, The Pennsylvania State University, University Park, Pennsylvania 16802 USA; E-mail: awc@psu.edu

Phenomena associated with the electronic excitation, relaxation and ionization of clusters can span a wide range of time scales. In the case of weakly bound systems with high ionization potentials, processes in the femtosecond time scale dominate, while in the case of strongly bound clusters with low ionization potentials, delayed ionization extending to microseconds and beyond can be operative. Additionally, electron excitation in clusters arising from short laser pulses can contribute to the formation of highly charged species. Examples of each of these potentially important processes will be discussed, with attention focused on quantifying the cluster properties and laser excitation responsible for their dominance.

In particular, three classes of systems will be discussed. Firstly the influence of laser fluence, wavelength, and pulse duration will be presented for the case of van der Waals clusters, showing the effects on the formation of high charge states. The possibility of using ensuing coulomb explosion as a way of arresting intermediates in fast reactions will be discussed. Secondly the formation of ion-pairs and concomitant rearrangement of solvent molecules around the ions will be presented for the case of acid solvation phenomena and thirdly Met-Cars, unique early transition metal-carbon clusters of composition M_8C_{12}, displaying both fast excitiation and relaxation dynamics and operative competitive delayed ionization that depends on the laser excitation characteristics.

1. Introduction

Ascertaining the influence that solvation has on the dynamics of chemical reactions[1-8] is one of the scientifically challenging problems in the field of chemical physics. Studies of clusters at selectively increased degrees of aggregation offer the opportunity to explore the changes in molecular properties that are brought on by solvation. Moreover, the study of certain systems of finite size provides the opportunity to determine the effects that can arise from quantum restriction.

The most widely used techniques to quantify the size-dependent properties of large clusters involve ionization accompanied by subsequent mass spectrometric detection, making an understanding of ionization mechanisms crucial. In addition to providing details on the energetics of interactions, studies of cluster ions and ionization mechanisms yield information on basic mechanisms of ion reactions within a cluster.

Herein we provide an overview of three different regimes of cluster ionization processes. The first involves ionization in Met-Cars, a unique semiconductor early transition metal-carbon cluster system comprised of 8 metal atoms bound to 12 carbons, in a cage-like structure. The system displays two rather different ionization mechanisms, one involving delayed ionization extending into the microsecond time domain, while the other occurs in the femtosecond time period in competition with electron-electron and electron-scattering processes. The second phenomena discussed in this overview pertains to ion-pair formation processes acquired through the dissolution process of an acid, namely HBr in the present example. Finally, we consider the phenomenon of coulomb explosion in cluster systems subjected to intense femtosecond optical fields. Application to arresting reactions in fast reactions is presented.

2. Electronic Excitation, Relaxation, and Ionization of Met-Cars

Several years ago, during the course of undertaking detailed studies of dehydrogenation reactions of hydrocarbons induced by titanium ions, atoms, and clusters, we discovered the formation of an unusually abundant and stable cationic species having a molecular weight of 528 amu, which thereafter was established to contain eight titanium atoms and twelve carbons.[9-12] Subsequent work revealed the existence of the neutral molecular cluster, its anion analogue, the stability of other transition metal-carbon complexes of identical stoichiometry, and thereafter a general class of caged molecular clusters comprised of early transition metals bound to carbon atoms in the same stoichiometric ratio,[9-55] M_8C_{12} (where M denotes an early transition metal). These have been termed metallocarbohedrens, or Met-Cars for short. Recent experiments and theory tend to support a T_d D_{2d} or C_{3V} structure.[55,56] rather than the one originally proposed one of T_h symmetry.[9]

2.1. *Delayed ionization*

One of the interesting aspects of Met-Cars is the large degree of delayed ionization which they undergo when subjected to pulsed lasers of nanosecond time duration. The phenomenon of delayed ionization in clusters has been reported for several systems including the Met-Cars,[47-49] pure transition metals,[57-59] metal oxides,[60-62] metal carbides,[15] and the fullerenes.[63-66] Currently there is considerable interest in the origin of the various operative phenomena, prompting extensive theoretical and experimental investigations for various systems.[49,53,67] In order for a cluster to display delayed ionization in a manner analogous to bulk phase thermionic emission, the ionization potential of the cluster must be less than its dissociation energy and all of phase space must be accessed by the system. The first requirement ensures that ionization, as an energy dissipation mechanism, will be more favorable than dissociation. Theoretical calculations and experimental results show that Met-Car clusters meet this requirement. The second requirement enables the system to temporarily store energy in excess of the cluster's ionization potential through statistically sampling a large number of accessible vibrational and electronic states.

Studies conducted in our laboratory employing nanosecond lasers revealed that some degree of ionization occurred on a time scale orders of magnitude longer than that which is characteristic of normal photoionization that obeying the photoelectric effect (see Fig. 1). In order for the delayed ionization to be observed for Met-Cars, the clusters must have a way to accommodate the energy necessary for the ionization to occur, while at the same time not undergo dissociation into smaller cluster fragments. Metallocarbohedrenes are thus ideal systems to exhibit this behavior, because when comparing the experimentally measured ionization potential (IP) and the theoretically predicted value for the dissociation energy (E_{diss}), a favorable relationship ($IP/E_{diss} < 1$) exists for this family of cluster molecules. This favorable relationship and the large density of electronic states for these transition metals-carbon species, may allow for the clusters to "store" the energy gained during the excitation and delay ionization on a long time scale characteristic of the experiment.

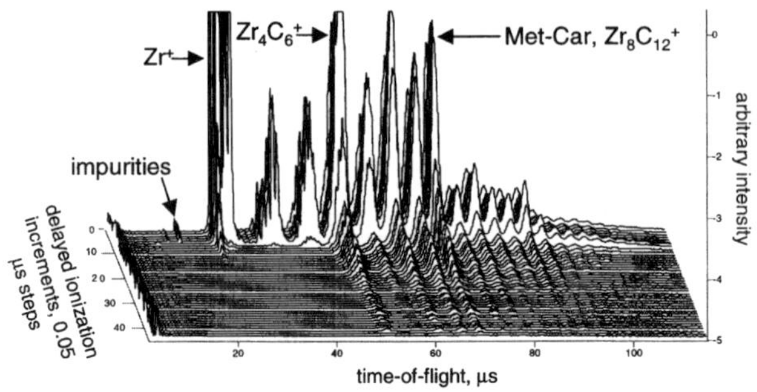

Fig. 1. 3-D data plot of the mass spectra obtained in a delayed ionization experiment on the zirconium-carbon system. The delay between spectra is 0.05 ms. Note the long time scale of the delayed ionization in the larger metal-carbon species.

In order for the delayed ionization to be well described by the thermionic emission model, adequate time must be available for the energy associated with the electronic excitation to be distributed throughout all accessible states; namely, that there is sufficient time for all of phase space to be sampled. In the case of excitation via femtosecond laser pulses, the time between successive absorptions may be insufficient for this requirement to be attained.

2.2. *Ultrafast dynamics*

To shed further light on the phenomena involved, femtosecond pump-probe experiments were performed on vanadium Met-Cars.[50-51] A comparison between the lifetimes of V_mC_n clusters and the vanadium Met-Car was performed. In addition, the influence of pump wavelength on the excited state lifetime has been explored.

The femtosecond dynamics of vanadium-carbon clusters generated from a laser vaporization source were investigated employing a 400 nm pump and 620 nm probe laser pulses and detection by mass spectrometry.[50] Fitted transients for a series of vanadium-carbon clusters, including V_8C_{12}, are presented in Fig. 2. The Met-Car response with a fitted width of 225 fs (FWHM) is noticeably longer than the autocorrelation width of the laser pulses with a cross correlation measured to be 86 fs. This measurable difference indicates that a state (or band of states) with an appreciable lifetime was being accessed. The pump-probe response was thus attributed to the temporal dependence of the electronic relaxation behavior of this system.

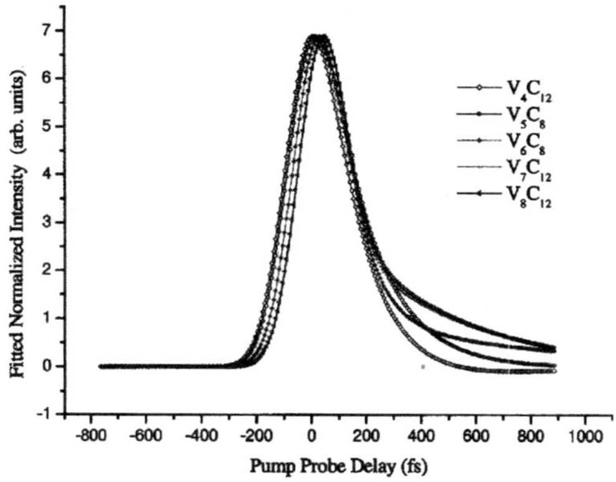

Fig. 2. The fitted pump-probe transients of the vanadium-carbon clusters: V_4C_6, V_5C_8, V_6C_8, V_7C_{12}, and V_8C_{12}. Pump: 400 nm, 50 fs, 24 mJ; probe: 620 nm, 50 fs, 20 mJ. Adapted from Ref. 50.

Similar relaxation behavior was also observed at other wavelengths, as well. For example, a typical pump-probe transient of the vanadium Met-Car pumped at moderate laser fluence by a range of wavelengths and ionized by an 800 nm probe is shown in Fig. 3.[51] These findings begin to reveal the large density of excited electronic states involved, and the concomitant effect on the relaxation dynamics. Observations indicate that small changes in the pump wavelength dramatically affect the behavior of the system.

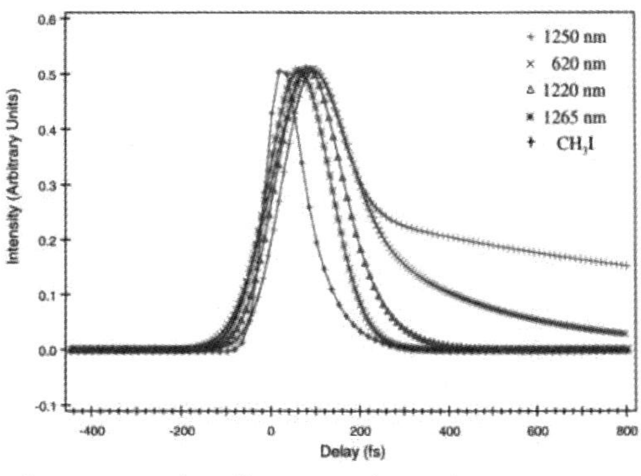

Fig. 3. Normalized pump–probe fit comparison for V_8C_{12} as the implemented wavelengths are tuned closer in resonance with an excited electronic state or band of coupled states. The fitted response for CH_3I is reproduced for comparison. Adapted from Ref. 51.

The dynamics studies provide clear evidence of electronic energy coupling to lattice modes being operative, with findings that at long delay times the latter could contribute to the thermionic emission that has been observed. Importantly, these new findings show that the delayed ionization process, typically seen only under excitation via nanosecond pulses, might be revealed in experiments conducted with laser fluences that enable relaxation into electronic bands during a femtosecond excitation process. This work is providing a new understanding of the process generally termed "thermionic" ionization/emission, in terms of the microscopic quantum effects of the influence of the dynamics of relaxation into electronic bands that govern the temporal characteristics of the phenomena. With the vast number of electrons dictating the behavior observed in these clusters, relaxation from presumably closely spaced excited states would likely occur on a short timescale (subpicosecond) given the vast number of electrons involved and the subsequent interaction and energy redistribution processes.

Met-Cars display unexpectedly low ionization potentials (V: 5.5eV; Ti: 4.40eV; Zr: 3.95eV), especially considering the comparatively high ionization potentials of the constituent metal (V: 6.74eV; Ti: 6.82eV; Zr: 6.84eV) and carbon (11.26eV). These species contain relatively increased delocalized electron character, in comparison to species of smaller size, and hence the findings are in accord with expectations that the relaxation processes likely include efficient electron-phonon coupling and electron-electron scattering.[57,68]

3. Dynamics of Acid Dissolution

The study of acid dissolution is one of the most fundamental areas of solution phase chemistry. Although thoroughly studied and understood in terms of bulk properties, the interaction between acids and their accompanying solvent on the molecular scale continues to be an area of intense scientific study driven by both scientific curiosity and practical importance.

A recent investigation of the dynamics of HBr dissolution in mixed clusters with water[69,70] is of particular interest. Clusters were formed by expansion of HBr molecules diluted in Ar backing gas. The resulting HBrn clusters then interacted with a continuous flow of water vapor, which resulted in the replacement of HBr molecules from the cluster by the concomitant incorporation of water molecules. This technique ensured that the HBr molecules remaining were within the solvent cluster rather than attached to the outside. The clusters were then pumped into the bound Rydberg C state of molecular HBr by a 271 nm laser pulse. This excitation left the HBr molecule in a bound excited state with insufficient energy to undergo the transition to the V valence state, where the ion-pair is formed (Fig. 4), which has been found to occur with excitation wavelengths in the area of 256 nm.[71] However, when the HBr chromophore is in a cluster, the energy of the V state is believed to be lowered allowing the transition from the C state to the V state thus enabling ion-pair formation.

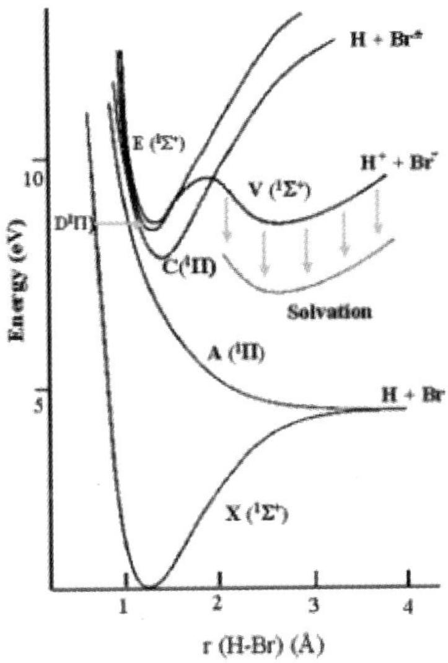

Fig. 4. Potential Energy Surface (PES) of HBr with downward arrows qualitatively indicating the solvation of the $V(^1S^+)$ valence state. Adapted from Ref. 69.

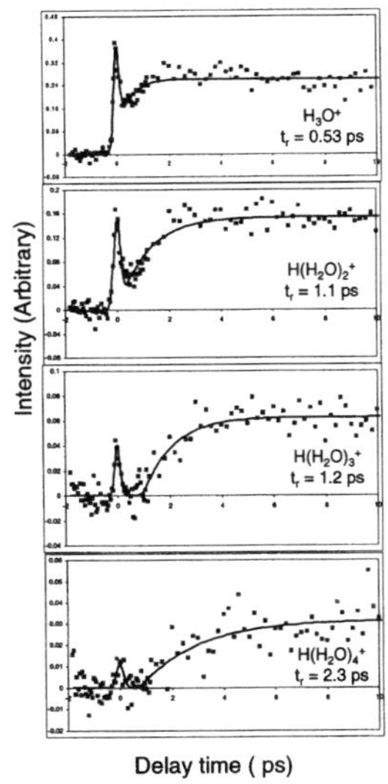

Fig. 5. Pump-probe temporal response of protonated water clusters that originate from $HBrH_2O_n$ clusters that were pumped with two photons of 271 nm light. Adapted from Ref. 70.

Probing the excited state population by multiphoton ionization, as a function of time, produces pump-probe transients that contain temporal information about the ion-pair formation process (see Fig. 5). After an initial Gaussian-shaped response at zero delay time, the signal resulting from ionization of the $HBr_mH_2O_n$ clusters rose with increasing delay time. This rise in the signal was attributed to a reorganization of the solvent around the newly formed H^+Br^- ion-pair. In the case of the HBr solvated by one to four water molecules, the time constant of the rise increased from 0.5 to 2.3 ps, respectively. This increase in rise time with cluster size indicated that a significant amount of time was needed for each additional solvent molecule to reorganize around the newly formed ionic species. In addition, since detection of cluster ions required the resonant excitation of an HBr molecule, no ion signal would be detected for cluster species where the ion-pair has already formed in the ground state. Thus, the absence of any ion signal in the mass-spectrum that results from an HBr molecule with more than four solvent molecules, indicates that five solvent molecules are sufficient to form the H^+Br^- ion-pair in the ground state, independent of whether they were HBr or water molecules, as long as one water molecule was present.[69]

4. The Phenomenon of Coulomb Explosion

Coulomb explosion (CE) is a class of reaction that has been observed when clusters become multiply charged.[72-79] Ionization involving multiple electron loss is not new, having been first observed in various systems ionized via electron impact techniques. Findings concerning the phenomenon of coulomb explosion come from a plethora of investigations of the interaction involving intense laser fields with atoms,[80-88] molecules,[89-92] and clusters.[93-104]

High flux femtosecond laser pulses provide short bursts of light, leading to the delivery of many photons to the cluster and almost instantaneously the loss of many electrons. These initially ionized free electrons can further ionize the inner-shell electrons of the atomic constituents of the cluster via intracluster inelastic electron-atom collisions. During the photon absorption and subsequent ionization processes, there is little time for nuclear motion. The highly charged atomic ions of the cluster are formed in very close proximity and hence the system thereafter rapidly explodes due to the Coulomb repulsion from like charges, releasing atomic ions with high kinetic energies,[93-95] Early studies of CE induced by femtosecond laser pulses, showing the role of clustering, were reported by our group[93-95] for high charge state species generated in ammonia clusters and in xenon clusters, where high charge states in xenon were reported by Rhodes and coworkers[101-104] at about the same time. An example of the high charge states that can be acquired as well as the degree of coulomb explosion that can be generated as evidenced by the peak splitting is seen in Fig. 6. At the present time, the exact mechanisms leading to the phenomenon are a subject of intense current interest among theoreticians as well as experimentalists, and are not fully resolved.

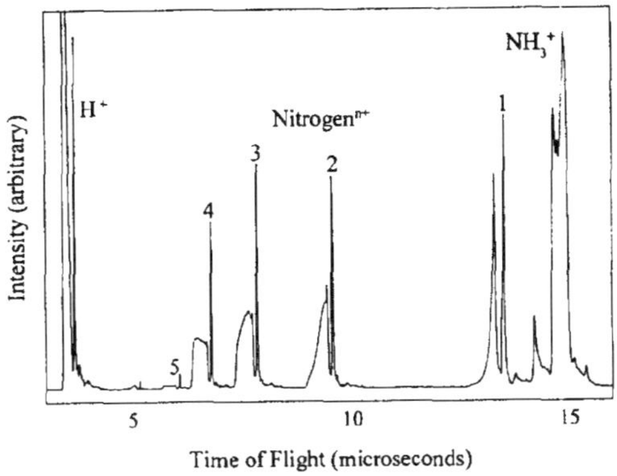

Fig. 6. Characteristic peak splitting and high charge states observed in the time-of-flight mass spectrum resulting from coulomb explosion, in this case of ammonia clusters. Adapted from Ref. 93.

4.1. Theoretical Models

Several theoretical approaches have been introduced to model the CE process in molecules and clusters. Among them are the ionization ignition model (IIM), the coherent electron motion model (CEMM), and the charge resonance enhanced ionization (CREI) and the related dynamical charge resonance enhanced ionization model. In all these models, barrier suppression plays an important role in the first stage of cluster ionization. For a laser beam with an intensity of ~ 1015 W/cm2, the field strength is on the order of 10 V/Å, which is strong enough to field ionize at least one of the valence electrons of all the nuclei in the cluster.

In the IIM picture,[101,102] at the very beginning of the ionization events, the nuclei in the cluster are considered frozen while the ionized electrons are quickly removed from the cluster by the laser field. In the CEMM model,[103-106] interaction between the laser field and clusters can enter a regime of strong electromagnetic coupling, which arises from the coherent motion of the field ionized electrons (Z) induced by the external laser field. These coherently coupled electrons behave like quasi-particles each with a charge Ze and a mass Zme. Subsequent ionization can be envisioned as the electron impact ionization by these coherently energized electrons. The CREI model[101,102,107-113] predicts that the rate of ionization is a highly irregular function of the internuclear distance. In the CREI model for diatomic molecules,[107-112] an inner potential barrier existing between the two nuclei rises with the internuclear distances. A dynamic CREI model was later proposed by Jortner and coworkers,[113,114] in which the electron energy was shown to increase roughly at the same rate as the level of the barrier height. The dynamic model predicts higher ionization efficiency than the static CREI model.

4.2. Experimental studies

Several different types of experiments have provided definitive evidence for the role of clusters in effecting the facile formation of highly charged species at modest laser fluences. The first conclusive evidence was obtained from experiments employing a covariance analysis approach.[82,116] Probing various cluster distributions in a pulsed molecular beam, and correlating the findings of high charge state formation with cluster size through covariance methods, has served to further elucidate the mechanisms responsible for the generation of multiply ionized constituents.

Other direct evidence for the role of clusters has been acquired in a series of experiments conducted in a manner which focused on the composition of the molecular beam which produced the most intense highly charged species at a given laser fluence.[99,100] One experiment involved fixing the focal position and changing the delay of the pulse nozzle relative to the laser pulse. This experiment enables scanning of the entire neutral cluster and unclustered molecule packet. Several different factors were examined: charge state distribution, kinetic energy release, and integrated signal intensity as a function of the delay between the laser and pulsed valve. All three exhibited the same trend, namely showing that the high charged species were only present, and with high kinetic energy release, when clusters dominated the beam. Essentially no highly charged species were observed when monomer alone was present, even at very high concentrations and laser intensities.

Another aspect of the CE process has been learned from the study of acetone clusters[95] where highly charged carbon and oxygen atoms were found to be produced. A unique feature of the work performed on this system was the utilization of pump-probe techniques in the ionization process as a test of theoretical predictions of the possible mechanisms involved. In this case, the pump laser at 624 nm has a power density of ~ 3 x 10^{14} W/cm^2, and the probe laser, also at 624 nm, was slightly (~10 %) weaker. The pump-probe transients of the oxygen fragments (O^{n+}, $1 \leq n \leq 5$) are shown in Fig. 7. There are several significant features that should be noted. A large dip was observed in the O_+ transient at zero delay; and a peak begins to grow in for each subsequently higher charged species. For the higher charged species (O^{n+}, $2 \leq n \leq 5$), the ion signal drops to a local minimum, then returns to a local maximum at some pump-probe delay. At longer delay times, the O^{n+}, $3 \leq n \leq 5$, show clear signs of an oscillatory pattern. The unequal intensities of the maxima at positive and negative delay times can be attributed to the unequal pump and probe intensities, but significantly, the maxima in O^+ and O^{5+} occur at positive delay times, whereas the maxima in O^{2+}, O^{3+}, and O^{4+} occur at negative delay times. C^{n+} ($1 \leq$ n ≤ 4) observed in the experiment was found to behave in a similar fashion. These findings point to the acquisition of some quantum control in the area of high intensity laser irradiation of clusters.

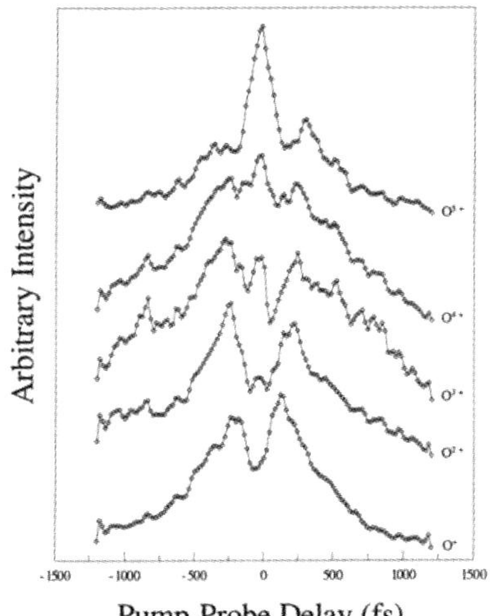

Fig. 7. Pump-probe transients of oxygen atoms with multiple charges produced by the coulomb explosion of acetone and acetone clusters. Adapted from Ref. 94.

4.3. Coulomb explosion imaging

The ability to selectively dissociate molecules via CE has been developed into a technique termed CE imaging, which takes advantage of this phenomenon to gain information on the time evolution of the reaction intermediates.[117-118] In the CE imaging technique, a first femtosecond laser pulse was employed to launch the system to an excited state of interest. A second intense femtosecond laser pulse was used to initiate the CE process in the reaction intermediates, and the Coulomb exploded fragments were detected at defined delay times. Calculations[119] have placed the approximate separation times of the multiple charged species in the neighborhood of 25 fs, so that little change occurs after the initial CE event. Therefore, information regarding the reaction intermediates can be extrapolated from the time-resolved detection of fragments. This technique was successfully applied to the study of the tautomerization reaction in the 7-azaindole dimmer,[117-118] as indicated by the data in Fig. 8 where the mass 119 fragment is an indication of the first proton transfer leading to the increase of the mass 119/118 ratio with the fist proton transfer process.

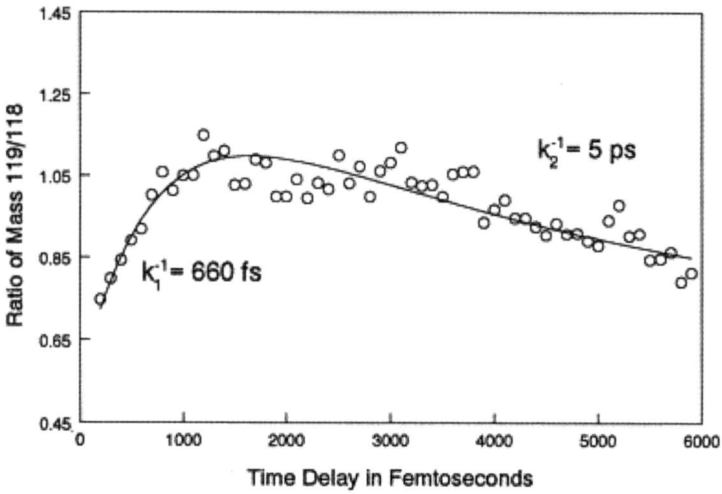

Fig. 8. Ratio of mass 119 to mass 118 throughout a typical pump–probe experiment. The open circles represent experimental data. The solid line represents a fit of the data as described in the text. Time constants of around 660 fs and 5 ps were obtained for the first and second proton transfer, respectively. Adapted from Ref. 116.

5. Future Prospects

The field of gas-phase cluster dynamics is a diverse one that reaches into many aspects of fundamental and application driven science ranging from solvation phenomena to nanoscale technology. Of particular importance are studies involving the investigation of the properties of electronic excited states that serve to provide an ever increasing understanding of quantum systems of restricted size. Understanding the role of ionization dynamics plays an important role in elucidating much of the behavior of interest and we can expect this field to continue to develop with the many prospects offered by the range of developments in femtosecond pump-probe spectroscopy.

Acknowledgements

Financial support by: the U. S. Air Force Office of Scientific Research Grant No. F49620-01-1-0122, the Atmospheric Sciences and Experimental Physical Chemistry Divisions of the U.S. National Science Foundation, Grant No. NSF-0089233, and the Department of Energy Grant No. DE-FG02-97ER14258 is gratefully acknowledged.

References

1. A. W. Castleman, Jr., R. G. Keesee, *Chem. Rev.* **86**, 589 (1986).
2. A. W. Castleman, Jr., R. G. Keesee, *Science* **241**, 36 (1988).
3. H. Haberland (Ed.), *Clusters of Atoms and Molecules*, Springer-Verlag: Berlin, Heidelberg, (1994).
4. R. E. Miller, *Acc. Chem. Res.* **23**, 10 (1990).
5. J. Jortner, R. D. Levine, B. Pullman, (Eds.), *"Reaction Dynamics in Clusters and Condensed Phases"*, Kluwer Academic: Dordrecht, Holland, (1994).
6. E. R. Bernstein, *Annu. Rev. Phys. Chem.* **46**, 197 (1995).
7. A. W. Castleman, Jr., K. H. Bowen, Jr., *J. Phys. Chem.* **100**, 12911 (1996).
8. A. W. Castleman, Jr., in *"Advances in Gas-Phase Ion Chemistry"*, (Eds.) L. M. Babcock, N. G, Adams, JAI Press: Stamford, CT: **3**, 185 (1998).
9. B. C. Guo, K. P. Kerns, A. W. Castleman, Jr., *Science* **255**, 1411 (1992).
10. S. Wei, B. C. Guo, J. Purnell, S. A. Buzza, A. W. Castleman, Jr., *J. Phys. Chem.* **98**, 9682 (1994).
11. S. Wei, B. C. Guo, J. Purnell, S. A. Buzza, A. W. Castleman, Jr., *J. Phys. Chem.* **96**, 4166 (1992).
12. S. Wei, B. C. Guo, J. Purnell, S. A. Buzza, A. W. Castleman, Jr., *Science* **256**, 818 (1992).

13. J. S. Pilgrim, M. A. Duncan, *J. Am. Chem. Soc.* **115**, 6958 (1993).
14. B. C. Guo, A. W. Castleman, Jr., in *"Advances in Metal and Semiconductor Clusters"*, (Ed.) M. Duncan, JAI Press: **2**, 137 (1994).
15. S. F. Cartier, B. D. May, A. W. Castleman, Jr., *J. Chem. Phys.* **100**, 5384 (1994).
16. S. F. Cartier, B. D. May, A. W. Castleman, Jr., *J. Am. Chem. Soc.* **116**, 5295 (1994).
17. H. T. Deng, B. C. Guo, K. P. Kerns, A. W. Castleman, Jr., *Int. J. Mass Spectrom. Ion Processes* 138, 275 (1994).
18. B. C. Guo, K. P. Kerns, A. W. Castleman, Jr., *J. Am. Chem. Soc.* **115**, 7415 (1993).
19. S. Wei, B. C. Guo, J. Purnell, S. A. Buzza, A. W. Castleman, Jr., *J. Phys. Chem.* **97**, 9559 (1993).
20. S. F. Cartier, B. D. May, B. J. Toleno, J. Purnell, S. Wei, A. W. Castleman, Jr., *Chem. Phys. Lett.* **98**, 9682 (1994).
21. S. Wei, B. C. Guo, H. T. Deng, K. Kerns, J. Purnell, S. A. Buzza, A. W. Castleman, Jr., *J. Am. Chem. Soc.* **116**, 4475 (1994).
22. S. Wei, A. W. Castleman, Jr., *Chem. Phys. Lett.* **227**, 305 (1994).
23. J. Purnell, S. Wei, A. W. Castleman, Jr., *Chem. Phys. Lett.* **229**, 105 (1994).
24. K. P. Kerns, B. C. Guo, H. T. Deng, A. W. Castleman, Jr., *J. Am. Chem. Soc.* **117**, 4026 (1995).
25. H. T. Deng, B. C. Guo, K. P. Kerns, A. W. Castleman, Jr., *J. Phys. Chem.* **98**, 13373 (1994).
26. K. P. Kerns, B. C. Guo, H. T. Deng, A. W. Castleman, Jr., *J. Chem. Phys.* **101**, 8529 (1994).
27. H. T. Deng, K. P. Kerns, A. W. Castleman, Jr., *J. Am. Chem. Soc.* **118**, 446 (1996).
28. S. F. Cartier, B. D. May, A. W. Castleman, Jr., *J. Phys. Chem.* **100**, 8175 (1996).
29. S. Lee, N. G. Gotts, G. von Helden, M. T. Bowers, *Science* **267**, 999 (1995).
30. C. S. Yeh, S. Afzaal, S. A. Lee, Y. G. Byun, B. S. Freiser, *J. Am. Chem. Soc.* **116**, 8806 (1994).
31. R. W. Grimes, J. D. Gale, *J. Chem. Soc. Chem. Commun.* **17**, 1222 (1992).
32. B. V. Reddy, S. N. Khanna, P. Jena, *Science* **258**, 1640 (1992).
33. T. Rantala, D. A. Jelski, J. R. Bowser, X. Xia, T. F. George, *Z. Phys.* D **26**, 255 (1993).
34. (a) M. Methfessel, M. van Schilfgaarde, M. Scheffler, *Phys. Rev. Lett.* **70**, 20 (1993); (b) *Phys. Rev. Lett.* **71**, 209 (1993).
35. R. W. Grimes, J. D. Gale, *J. Phys. Chem.* **97**, 4616 (1993).
36. A. Ceulemans, P. W. Fowler, *J. Chem. Soc., Faraday Trans.* **88**, 2797 (1992).
37. M. Rohmer, P. De Vaal, M. Bénard, *J. Am. Chem. Soc.* **114**, 9696 (1992).
38. Z. Lin, M. B. Hall, *J. Am. Chem. Soc.* **114**, 10054 (1992).
39. I. Dance, *J. Chem. Soc., Chem. Commun.* 1779 (1992).
40. H. Chen, M. Feyereisen, X. P. Long, G. Fitzgerald, *Phys. Rev. Lett.* **71**, 1732 (1993).
41. M.-M. Rohmer, M. Bénard, C. Henrite, C. Bo, J.-M. Poblet, *J. Chem. Soc., Chem. Commun.* **15**, 1182 (1993).
42. L. Pauling, *Proc. Natl. acad. Sci. U.S.A.* **89**, 8175 (1992).
43. I. B. Dance, *Aust. J. Chem.* **46**, 727 (1993).
44. M.-M. Rohmer, M. Bénard, C. Bo, J.-M. Poblet, *J. Am. Chem. Soc.* **117**, 508 (1995).
45. A. Khan, *J. Phys. Chem.* **99**, 4923 (1995).

46. S. F. Cartier, Z. Y. Chen, G. J. Walder, C. R. Sleppy, A. W. Castleman, Jr., *Science* **260**, 95 (1993).
47. B. D. May, S. F. Cartier, A. W. Castleman, Jr., *Chem. Phys. Lett.* **242**, 265 (1995).
48. S. F. Cartier, B. D. May, A. W. Castleman, Jr., *J. Chem. Phys.* **104**, 3423 (1996).
49. S. E. Kooi and A. W. Castleman, Jr., *J. Chem. Phys.* **108**, 8864 (1998).
50. S. E. Kooi, B. D. Leskiw and A. W. Castleman, Jr., *Nano Letters* **1**, 113 (2001).
51. B. D. Leskiw, K. L. Knappenberger, A. W. Castleman, Jr., *J. Chem. Phys.* **117**, 8321 (2002).
52. E. S. Wisniewski, B. D. Leskiw, S. M. Hurley, T. E. Dermota, D. P. Hydutsky, K. L. Knappenberger, Jr., M. Hershberger, A. W. Castleman, Jr., *World Scientific* (2001).
53. H. Sakurai, S. E. Kooi and A. W. Castleman, Jr., *J. Cluster Science* **10**, 493 (1999).
54. H. Sakurai and A. W. Castleman, Jr., *J. Phys. Chem.* **101**, 7695 (1997).
55. H. Sakurai and A. W. Castleman, Jr., *J. Phys. Chem.* A **102**, 10486 (1998).
56. M.-M. Rohmer, M. Bénard, J.-M. Poblet, *Chem. Rev.* **100**, 495 (2000).
57. T. Leisner, K. Athanassenas, O. Echt, D. Kreisle, E. Rechnagel, *J. Chem. Phys.* **99**, 9670 (1993).
58. T. Leisner, K. Athanassenas, O. Echt, O. Kandler, D. Kreisle, E. Rechnagel, *Z. Phys.* D **20**, 127 (1991).
59. B. A. Collings, A. H. Amrein, D. M. Rayner, P. A. Hackett, *J. Chem. Phys.* **99**, 4174 (1993).
60. B.C. Nieman, E. K. Parks, S. C. Richtsmeier, K. Liu, L.G. Pobo, S.L. Riley, *High Temp. Sci.* **22**, 115 (1986).
61. K. Athanassenas, T. Leisner, U. Frenzel, D. Kreisle, *Ber. Bunsenges. Phys. Chem.* **96**, 1192 (1992).
62. A. Amrein, R. Simpson, P. Hackett, *J. Chem. Phys.* **95**, 1781 (1991).
63. E. E. B. Campbell, G. Ulmer, I.V. Hertel, *Phys. Rev. Lett.* **67**, 1986 (1991).
64. D. Ding, J. Huang, R.N. Compton, C.E. Klots, R.E. Haufler, *Phys. Rev. Lett.* **73**, 1084 (1994).
65. E. E. B. Campbell, K. Hoffmann, I. V. Hertel, *Eur. Phys. J.* D **16**, 345 (2001).
66. H. Lin, K.-L. Han, Y. Bao, E. B. Gallogly, W. M. Jackson, *J. Phys. Chem.* **8**, 12495 (1994).
67. C. E. Klots, *Chem. Phys. Lett.* **186**, 73 (1991).
68. C. Voisin, N. Del Fatti, D. Christofilos, F. Vallée, *J. Phys. Chem.* B **105**, 2264 (2001).
69. S. M. Hurley, T. E. Dermota, D. P. Hydutsky, A. W. Castleman, Jr., *J. Chem. Phys.* **118**, 9273 (2003).
70. S. M. Hurley, T. E. Dermota, D. P. Hydutsky, A. W. Castleman, Jr., *Science* **298**, 202 (2002).
71. S. M. Hurley, Q. Zhong, A. W. Castleman, Jr., *J. Chem. Phys.* **112**, 4644 (2000).
72. D. Kreisle, O. Echt, M. Knapp, E. Recknagel, K. Leiter, T. D. Mark, J. J. Saenz, J. M. Soler, *Phys. Rev. Lett.* **56**, 1551 (1986).
73. C. Brechignac, P. Cahuzae, F. Carlier, M. de Frutos, *Phys. Rev. Lett.* **72**, 1636 (1994).
74. A. K. Shukla, C. Moore, A. Stace, J. Chem. Phys. Lett. **109**, 324 1984; A. Stace, *J. Phys. Rev. Lett.* **61**, 306 (1988).

75. D. Kreisle, K. Leiter, O. Echt, T. D. Mark, *Z. Phys.* D **3**, 319 (1986).
76. P. Scheier, T. D. Mark, *J. Chem. Phys.* **86**, 3056 (1987).
77. M. Lezius, T. D. Mark, *Chem. Phys. Lett.* **155**, 496 (1989).
78. P. Scheier, A. Stamatovic, T. D. Mark, *J. Chem. Phys.* **88**, 4289 (1988).
79. P. Cheier, B. Dunser, T. D. Mark, *Phys. Rev. Lett.* **74**, 3368 (1995).
80. K. Codling, L. J. Frasinski, P. Hatherly, J. R. M. Barr, *J. Phys.* B **20**, L525 (1987).
81. L. J. Frasinski, K. Codling, P. Hatherly, J. Barr, I. N. Ross, W. T. Toner, *Phys. Rev. Lett.* **58**, 2424 (1987).
82. L. J. Frasinski, K. Codling, P. A. Hatherly, *Science* **246**, 1029 (1989).
83. L. A. Lompre, G. Mainfray, in *"Multiphoton Processes"*, (Eds.) P. Lambropoulos, S. J. Smith, Springer-Verlag: Berlin, 23 (1984).
84. U. Johann, T. S. Luk, I. A. McIntyre, A. McPherson, A. P. Schwarzenbach, K. Boyer, C. K. Rhodes, *"Multiphoton Ionization in Intense Ultraviolet Laser Fields", Proceedings of the Topical Meeting on Short Wavelength Coherent Generation*, (Eds.) D. T. Attwood, J. Bokor, AIP Conference Proceedings, AIP: New York **147**, 157 (1986).
85. A. l'Huillier, L. A. Lompre, G. Mainfray, C. Manus, *Phys. Rev.* A **27**, 2503 (1983).
86. T. S. Luk, U. Johann, H. Egger, H. Pummer, C. K. Rhodes, *Phys. Rev.* A **2**, 214 (1985).
87. K. Boyer, T. S. Luk, J. C. Solem, C. K. Rhodes, *Phys. Rev.* A **39**, 1186 (1989).
88. C. Cornaggia, J. Lavancier, D. Normand, J. Morellec, H. X. Liu, *Phys. Rev.* A **42**, 5464 (1990).
89. K. Codling, L. J. Frasinski, P. A. Hatherly, *J. Phys.* B **21**, L433 (1988).
90. P. A. Hatherly, L. J. Frasinski, K. Codling, A. J. Langley, W. Shaikh, *J. Phys.* B **23**, L291 (1990).
91. J. Lavancier, D. Normand, C. Cornaggia, J. Morellec, H. X. Liu, *Phys. Rev.* A **43**, 1461 (1991).
92. D. Normand, C. Cornaggia, J. Lavancier, J. Morellec, H. X. Liu, *Phys. Rev.* A **44**, 475 (1991).
93. E. M. Snyder, S. Wei, J. Purnell, S. A. Buzza, A. W. Castleman, Jr., *Chem. Phys. Lett.* **248**, 1 (1996).
94. J. Purnell, E. M. Snyder, S. Wei, A. W. Castleman, Jr., *Chem. Phys. Lett.* **229**, 333 (1994).
95. E. M. Snyder, S. A. Buzza, A. W. Castleman, Jr., *Phys. Rev. Lett.* **77**, 3347 (1996).
96. T. Ditmire, J. Zweiback, V. P. Yanovsky, T. E. Cowan, G. Hays, K. B. Wharton, *Nature* **398**, 489 (1999).
97. T. Ditmire, E. Springate, J. W. G. Tisch, Y. L. Shao, M. B. Mason, N. Hay, J. P. Marangos, M. H. R. Hutchinson, *Phys. Rev.* A **57**, 369 (1998).
98. T. Ditmire, J. W. G. Tisch, E. Springate, M. B. Mason, N. Hay, R. A. Smith, J. P. Marangos, M. H. R. Hutchinson, *Nature* **386**, 54 (1997).
99. J. V. Ford, Q. Zhong, L. Poth, A. W. Castleman, Jr., *J. Chem. Phys.* **110**, 6257 (1999).
100. J. V. Ford, L. Poth, Q. Zhong, A. W. Castleman, Jr., *Int. J. Mass Spectrom. Ion Processes* **192**, 327 (1999).

101. C. Rose-Petruck, K. J. Schafer, C. P. J. Barty, in *"Application of Laser Plasma Radiation II"*, (Eds.) M. C. Richardson, G. A. Kyrala, SPIE, Bellingham **2523**, 272 (1995).
102. C. Rose-Petruck, K. J. Schafer, K. R. Wilson, C. P. J. Barty, *Phys. Rev.* A **55**, 1182 (1997).
103. A. McPherson, T. S. Luk, B. D. Thompson, K. Boyer, C. K. Rhodes, *Appl. Phys.* B **57**, 337 (1993).
104. A. McPherson, T. S. Luk, B. D. Thompson, A. B. Borisov, O. B. Shiryaev, X. Chen, K. Boyer, C. K. Rhodes, *Phys. Rev. Lett.* **72**, 1810 (1994).
105. A. McPherson, B. D. Thompson, A. B. Borisov, K. Boyer, C. K. Rhodes, *Nature* **370**, 631 (1994).
106. K. Boyer, B. D. Thompson, A. McPherson, C. K. Rhodes, *J. Phys. B: At. Mol. Opt. Phys.* **27**, 4373 (1994).
107. T. Seideman, M. Y. Ivanov, P. B. Corkum, *Phys. Rev. Lett* **75**, 2819 (1995).
108. W. Constant, H. Stapelfeldt, P. B. Corkum, *Phys. Rev. Lett.* **76**, 4140 (1996).
109. S. Chelkowski, A. Conjusteau, T. Zuo, A. D. Bandrauk, *Phys. Rev.* A **54**, 3235 (1996).
110. S. Chelkowski, A. D. Bandrauk, *J. Phys. B: At. Mol. Opt. Phys.* **28**, L723 (1995).
111. H. Yu, A. D. Bandrauk, *Phys. Rev.* A **56**, 685 (1997).
112. H. Yu, A. D. Bandrauk, *J. Phys. B: At. Mol. Opt. Phys.* **31**, 1533 (1998).
113. T. Zuo, A. D. Bandrauk, *Phys. Rev.* A **52**, R2511 (1995).
114. I. Last, I. Schek, J. Jortner, *J. Chem. Phys.* **107**, 6685 (1997).
115. I. Last, J. Jortner, *Phys. Rev.* A **58**, 3826 (1998).
116. D. A. Card, D. E. Folmer, S. Sato, S. A. Buzza, A. W. Castleman, Jr., *J. Phys. Chem.* **101**, 3417 (1997).
117. D. E. Folmer, L. Poth, E. S. Wisniewski, A. W. Castleman, Jr., *Chem. Phys. Lett.* **287**, 1 (1998).
118. D. E. Folmer, E. S. Wisniewski, A. W. Castleman, Jr., *Chem. Phys. Lett.* **318**, 637 (2000).
119. L. Poth, A. W. Castleman, Jr., *J. Phys. Chem* A **102**, 4075 (1998).

CLUSTERS IN INTENSE LASER FIELDS

Ch. Siedschlag, U. Saalmann and J. M. Rost
Max-Planck-Institute for the Physics of Complex Systems
Nöthnitzer Str. 38, 01187 Dresden, Germany
Email: us@mpipks-dresden.mpg.de

We have investigated the interaction of short, intense laser pulses with rare gas clusters in two different frequency regimes. At optical frequencies, we find for small clusters an enhancement of energy absorption which can be explained by the enhanced ionization mechanism for small clusters. For larger clusters the dynamics of quasi-free electrons, held back by the clusters space charge, determines the cluster ionization. At X-ray frequencies, however, the energy absorption turns out to be much less efficient than in the purely atomic case, due to the space charge of the cluster in combination with a small quiver amplitude and delocalization of electrons in the cluster.

1. Rare Gas Clusters in Optical Fields

Since the first experiments on clusters in intense femtosecond pulses at optical frequencies a wealth of theoretical papers have been published in order to explain the physical mechanism behind the observation of an unexpectedly large absorption of energy.[1–3] Soon it emerged that there were two possible candidates: *ionization ignition*, where, in an avalanche process, the first ionized atoms of the clusters lead to a rapid increase of the total field strength inside the cluster, thereby assisting other atoms to become ionized too; and the *plasma resonance* picture, according to which the initially only weakly ionized cluster starts to expand due to the Coulomb repulsion of the ions, until the plasma frequency of the electron cloud gets into resonance with the applied laser frequency, leading to a rapid heating and subsequent ionization of these electrons.

Both of these mechanisms can play a role in laser-cluster interaction, depending on the size of the clusters. We show that for small clusters the avalanche picture can be understood as a generalization of the *enhanced ionization* mechanism originally developed for small molecules,[4] rendering the interatomic distance in the cluster to be the most important variable (see Sec. 1.1). As to large clusters (of more than 10^5 atoms), it has been proposed that strong laser pulses create a nanoplasma inside the cluster and resonantly absorb energy when the (density-dependent) plasma frequency matches the laser frequency due to cluster expansion.[1] However, already at medium-sized clusters (> 100 atoms) the ionization dynamics may fundamentally differ from the mechanism mentioned above, because in these clusters the space charge built up early in the pulse is large and prevents direct ionization (see Sec. 1.2).

1.1. *Small clusters and enhanced ionization*

We describe the interaction of small rare gas clusters with intense laser light using a mixed quantum-classical approach.[5] Briefly, the cluster initially consists of neutral atoms whose positions are determined by a relaxation procedure assuming pairwise Lennard-Jones interactions. Once the laser pulse sets in, electrons are allowed to tunnel sequentially out of the atomic levels (*inner ionization*) and are then treated classically. The tunneling in-

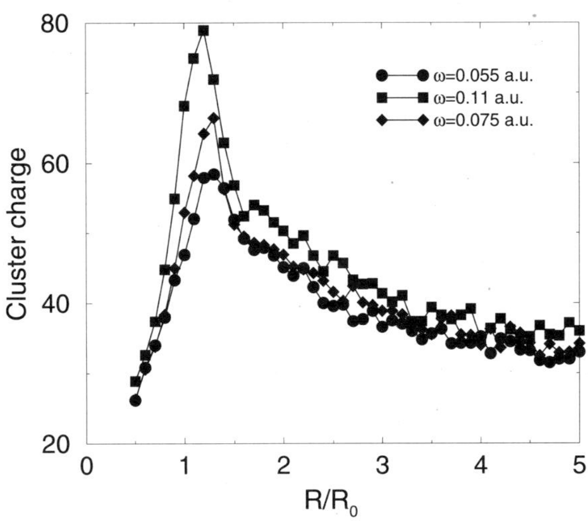

Fig. 1. Ionization yield of Ar_{16} as a function of the scaled internuclear distance for three different laser frequencies. Pulse parameters: $f(t) = F\sin^2 \pi/Tt \sin\omega t, F = 0.16a.u., T = 2285\ a.u.$

tegral is evaluated semiclassically using the full potential landscape, and the atomic energy levels are shifted by the laser field and the surrounding charges. Once a few electrons have left the cluster (*outer ionization*), the ions start to repel each other, leading to a slow increase of the internuclear distance $\bar{R}$.

To study the influence of $\bar{R}$ on the ionization process, we first performed calculations using scaled ionic positions $\vec{R}_{\text{ion}} = \lambda \vec{R}^0_{\text{ion}}$, but keeping the ions fixed during the calculation. The ionization yield as a function of the scaled internuclear distance for three different laser frequencies is shown in Fig. 1. One clearly observes a maximum in the ionization yield, whose position turns out to be almost independent of the applied frequency, which rules out any kind of resonance mechanism for this small cluster size. The yield itself, however, increases with the applied frequency: due to the lower ponderomotive energy at higher frequencies more inner-ionized electrons stay in the cluster environment and help to field-ionize other electrons before outer ionization finally takes place. This optimal cluster radius with static nuclei has important consequences on the ionization mechanism when the nuclei are allowed to move. To investigate the dependence of the ionization yield on the nuclear expansion, we chose different laser pulse lengths while keeping the total pulse energy fixed, i.e. $\int f^2(t)\ dt$=const. The results of such a calculation is shown in Fig. 2.

For very short pulses, the ionization yield first decreases when increasing the pulse length (i.e. decreasing the field strength). At a pulse length of about 2000 a.u., this trend is, however, reversed, and the yield *increases* with increasing pulse length until a maximum at $T \approx 4200$ a.u. is reached. For even longer pulse lengths, the yield decreases again. This behavior can easily be explained with the existence of the optimal cluster radius R_{crit} from Fig. 1: the cluster expands during the pulse because of the nuclear repulsion once the first electrons have left the cluster. When the applied pulse is very short, the nuclei have almost no time to move during the pulse, so that the behavior for short pulses is more like in the single atom case, where one also observes a decrease in the ionization yield with increasing pulse length. With longer pulses, however, the expansion of the cluster has progressed more and more at the time of the pulse maximum, until for a certain pulse length the maximum of the pulse and the optimal cluster radius coincide, which leads to optimal absorption. For even longer pulses, R_{crit} is always reached during the pulse, but with less and less laser energy available due to the energy normalization, so that the ionization yield decreases again.

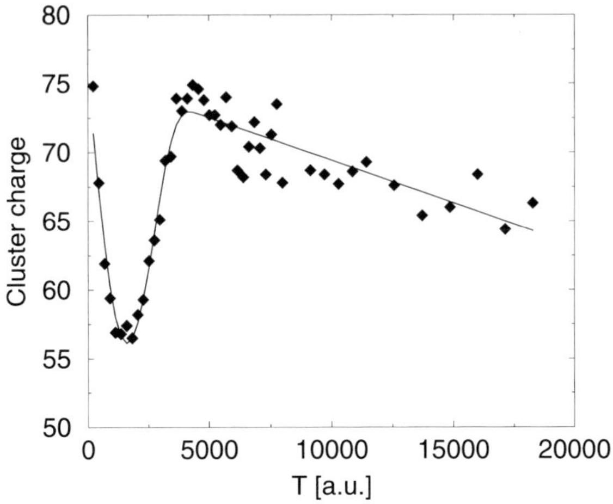

Fig. 2. Ionization yield of Ar_{16} as a function of the energy normalized pulse length. The reference pulse had a field strength of $F = 0.16$ a.u. at a pulse length of $T = 2284$ a.u.; the frequency was $\omega = 0.055$ a.u.

1.2. *Space charge effects and quasi-free electrons*

To elucidate the effect of the high attractive space charge in medium-sized clusters (~ 1000 atoms) we study the quasi-free electrons. These electrons are already excited from their atomic bound states (inner ionization), but have not yet escaped from the cluster as a whole to the continuum (outer ionization). The lower panel of Fig. 3 shows the amount of quasi-free electrons for three different cluster sizes in a 100-fs-pulse with $I = 9 \cdot 10^{14}\,\mathrm{W/cm^2}$. Whereas in the case of Xe_{13} almost all electrons are ionized directly (and no quasi-free electrons are created), the larger clusters Xe_{147} and in particular Xe_{1415} can hold an appreciable number of quasi-free electrons. This is reflected by a lower ionization rate for these cluster at $t \approx -25 \ldots -15\,\mathrm{fs}$, (see upper panel of Fig. 3). Only later in the pulse, in a second ionization step, these quasi-free electrons are ionized through a driven collective motion inside the cluster.[6] For Xe_{147} this happens around $t \approx 5\,\mathrm{fs}$ as can be seen in the middle panel of Fig. 3 at the second peak of the ionization rate and the drop down of the number of quasi-free electrons. (For Xe_{1415} the pulse was too short.) Optimal conditions for this ionization step depend on cluster charge and size, seen by the quasi-free electrons, and consequently are connected with the cluster expansion.[6]

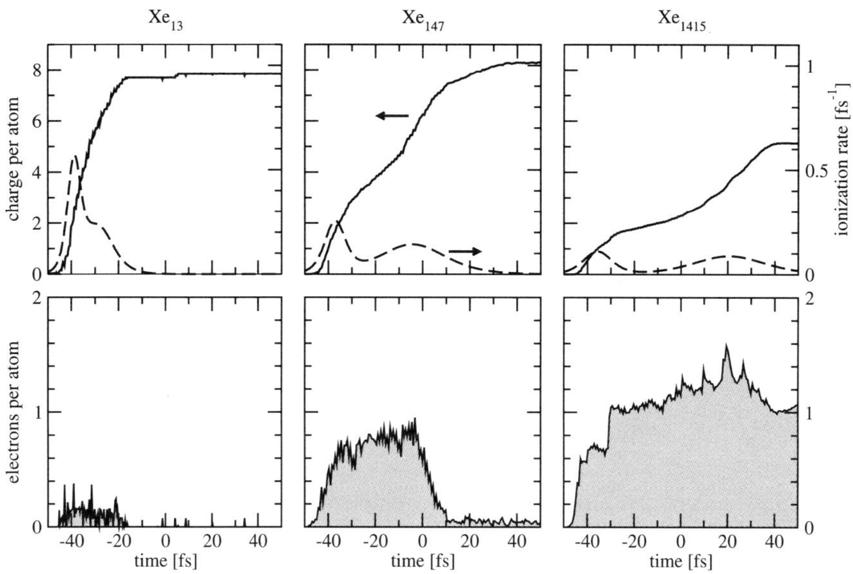

Fig. 3. Electron dynamics for three different cluster sizes (Xe_{13}, Xe_{147}, and Xe_{1415}) in a strong laser pulse ($\lambda = 780\,\text{nm}$, $I = 9 \cdot 10^{14}\,\text{W/cm}^2$, rise and fall time 20 fs, plateau for $t = -30 \ldots + 30\,\text{fs}$). It is shown as a function of time t in the *upper panel*: average charge per atom (*solid line, left axis*) and corresponding rate (*dashed line, right axis*), and in the *lower panel*: quasi-free electrons per atom (*gray filled line*).

2. Clusters in Intense X-ray Laser Pulses

In the short-wavelength regime[7] the interaction is notably different from long-wavelength pulses. This is already evident from the ponderomotive energy $E_{\text{pond}} \sim I/\omega^2$ which represents the average kinetic energy of a free electron in a laser field. For the laser parameters discussed in this section one finds $E_{\text{pond}} \approx 0.4\,\text{meV} \ldots 4\,\text{eV}$, vastly different from $E_{\text{pond}} \approx 20\,\text{eV} \ldots 200\,\text{keV}$ for a 780-nm-laser at the same intensities. For such small ponderomotive energies the laser-atom interaction is, despite the high intensities, of non-relativistic and perturbative nature. Nevertheless, one may expect exhaustive ionization leading subsequently to an enormous Coulomb explosion of the irradiated argon clusters due to the following scenario: firstly, inner-shell electrons are photoionized as their cross sections at X-ray wavelengths are considerably higher than those of the valence electrons. Even multiple (single-photon) ionization is possible since the inverse rates are much smaller than the pulse length. Furthermore, the inner shells are rapidly refilled by Auger-like decay processes. Thus the atoms can be efficiently "pumped dry", which occurs "inside-out" and is the exact opposite to the ionization mechanism in the visible wavelengths regime (see Sec. 1).

In the following we discuss first results[10] of X-ray ($\lambda \approx 3.5\,\text{nm}$) laser interaction with small argon clusters at intensities of $I \approx 10^{-2} \ldots 10^{+2} I_0$ with $I_0 = 3.51 \cdot 10^{16}\,\text{W/cm}^2$. The laser wavelength λ was chosen to allow for single-photon ($\hbar\omega = 350\,\text{eV}$) ionization of the L shell of Argon ($E_{\text{bind}} = 326,\ 249\,\text{eV}$). We use a mixed quantum/classical approach similar to those successfully applied in studying rare gas clusters in strong pulses of visible light[5] and already presented in Sec. 1. It differs, however, in two important aspects: on one hand, the influence of the laser field on the free particle dynamics is negligible due to the small quiver amplitude $\Delta x \sim \sqrt{I}/\omega^2$ of the electrons (here $\Delta x \approx 0.0003 \ldots 0.03\,\text{Å}$). On the other hand, one has to incorporate inner-shell electrons and intra-atomic processes rendering the ionization/excitation dynamics much more complex. This has been accomplished[10] via parametrized cross sections for photoionization and branching ratios and decay rates from Hartree-Fock calculations for decay cascades. Interatomic coulombic decay, theoretically proposed[11] and experimentally found in Neon clusters,[12] is expected to be of minor importance in these highly charged clusters and therefore neglected.

It is particularly interesting to see how the cluster environment affects the laser interaction since the coupling to the laser is an individual atom-photon interaction. Therefore; Fig. 4 compares final charge and absorbed energy per atom for two different clusters, Ar_{13} and Ar_{55}, and single Argon atoms in a 100-fs-pulse for different field strengths f. From the left panel of Fig. 4 one can deduce that at lower field strengths ($f = 0.1\,\text{au}$) only a single photoionization event with one subsequent Auger decay per atom occurs independently of the cluster size. After a quite steep rise the final charge starts to saturate in the atomic case for stronger fields ($f > 0.3\,\text{au}$) due to the fact that single photoionization becomes impossible beyond a certain charge state of the ion at the given photon frequency. The rise of the final charge per atom in the cluster is notably weaker. Therefore — in striking contrast to optical laser-cluster interaction — clusters are less effectively ionized at high fields than atoms. The reduction is even more pronounced for the larger cluster (see the left panel of Fig. 4). One reason for the reduced ionization in the cluster compared to the atom is the much larger space charge produced in a cluster. Such a space charge suppresses ionization, because the absorption of one single photon transfers only a fixed amount of energy to the electron. In our model, these post-collisional interaction effects are taken into account by the propagation of

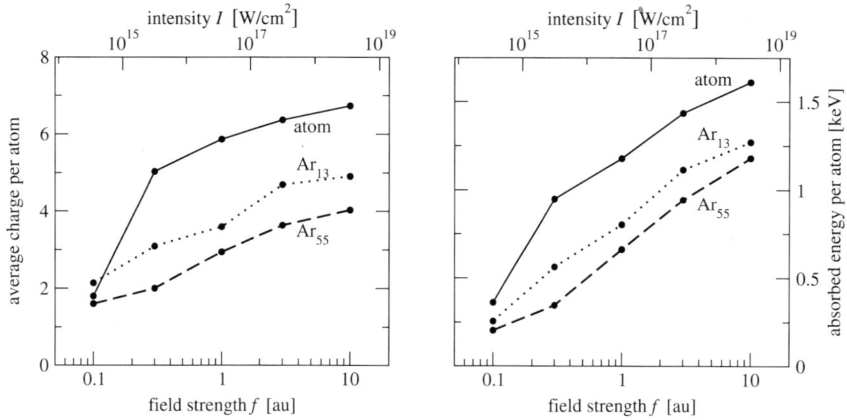

Fig. 4. Average charge (*left panel*) and absorbed energy (*right panel*) per atom for two cluster sizes Ar_{13} and Ar_{55} produced by an XFEL pulse ($\hbar\omega = 350\,\text{eV}$, rise and fall time: 20 fs, plateau time: 60 fs) as a function of the field strength f compared to atomic argon as target. The points are averages from 10 simulations.

the photoionized electrons. For larger cluster and higher fields, i. e. higher space charges, one finds that an increasing number of electrons are bound at the end of the pulse to one of the fragment ions.

The right panel of Fig. 4 shows the energy absorbed from the laser, which provides direct insight into the photoionization since all the other processes (intra-atomic decay, intra-cluster screening) are not influenced by the laser due to its high frequency. Surprisingly, for all field strengths considered, the absorption of photons itself is reduced in the environment provided by the cluster. This was unexpected because of the fact that predominantly deep-lying inner-shell electrons are affected and possible effects on the photoionization rates of these strongly localized electrons from neighbouring ions are unlikely and in fact not contained in the model. The time evolution of the cluster dynamics reveals that a delocalization of the valence electrons is indirectly responsible. It has mainly two effects: firstly, the photoionization cross sections become very small since the electrons are far away from the nucleus. Secondly, the Auger decay rates are also reduced because the overlap with the core holes becomes smaller. We only mention here that these effects can be quantified by analyzing the electron dynamics inside the clusters.[10]

References

1. T. Ditmire, T. Donnelly, A.M. Rubenchik, R. W. Falcone, and M.D. Perry, *Phys. Rev.* A **53**, 3379 (1996).
2. J. Daligault and C. Guet, *Phys. Rev.* A **64**, 043203 (2001).
3. K. Ishikawa and T. Blenski, *Phys. Rev.* A **62**, 063204 (2000).
4. T. Seideman and M.Y. Yvanov and P.B. Corkum, *Phys. Rev. Lett.* **75**, 2819 (1995).
5. Ch. Siedschlag and J.M. Rost, *Phys. Rev. Lett.* **89**, 173401 (2002); *Phys. Rev.* A **67**, 013404 (2003).
6. U. Saalmann and J.M. Rost, to be published (2003).
7. The new X-ray free electron laser (XFEL) sources, under construction at DESY in Hamburg[8] and at the LCLS in Stanford,,[9] will deliver intense laser pulses at high frequencies (from VUV to hard X-ray).
8. The X-ray free electron laser, Vol. V in *TESLA Technical Design Report*, edited by G. Materlik and T. Tschentscher (DESY, Hamburg, 2001).
9. J. Arthur, *Rev. Sci. Instrum.* **73**, 1393 (2002).
10. U. Saalmann and J.M. Rost, *Phys. Rev. Lett.* **89**, 143401 (2002).
11. R. Santra, J. Zobeley, L.S. Cederbaum, and N. Moiseyev, *Phys. Rev. Lett.* **85**, 4490 (2000).
12. S. Marburger, O. Kugeler, U. Hergenhahn, and T. Möller, *Phys. Rev. Lett.* **90**, 203401 (2003).

LEARNING ABOUT CLUSTERS BY TEACHING LASERS TO CONTROL THEM

A. Lindinger, A. Bartelt, C. Lupulescu, M. Plewicki, and L. Wöste

Institut für Experimentalphysik, Freie Universität Berlin, Arnimallee 14, D-14195 Berlin, Germany

E-mail: *lindin@physik.fu-berlin.de*

We have performed multi-photon ionization experiments on small alkali clusters in order to probe their wave packet dynamics. The observed processes were highly dependent on the irradiated pulse parameters such as wavelength range or its phase and amplitude, emphasing the importance of employing a feedback control system for generating the optimum pulse shapes. Their spectral and temporal behavior reflects interesting properties about the investigated system and the irradiated photo-chemical process.

The controllability of three-photon ionization pathways is investigated on the model-like system NaK. A closed learning loop for adaptive feedback control is applied to find the optimal fs pulse shape. We examined the frequency dependent closed loop optimization of fs pulse shapes for the NaK dimer. The obtained ion yield depending on the employed central laser frequency shows a maximum in the range of 770-780 nm. By investigating the modifications of the optimal pulse shapes for different wavelengths we could gain information about the associated ionization path during the control process. Characteristic motions of the involved wave packets are proposed as an attempt to explain the optimized dynamical processes. Moreover, sinusoidal parameterizations of the spectral phase modulation are applied with regard to the obtained optimal field. By reducing the number of parameters and thereby the complexity of the phase modulation, optimal pulse shapes can be generated that carry fingerprints of the molecule's dynamical properties. This enables us to find "understandable" optimal pulse forms and offers the possibility to gain insight into the photo-induced control process.

1. Introduction

Research on metal clusters has been strongly motivated by the goal to bridge the gap between the understanding of small molecules and the bulk.[1] This development was favoured by the advent of tunable femtosecond laser sources, which have opened fascinating perspectives allowing an insight into elementary intramolecular processes like bond shaking, breaking and making.[2] Such processes can now be monitored on a real-time basis by means of pump-probe spectroscopy.[3] Moreover, the photo-induced dynamical sequences can even be directly influenced by using adequately shaped fs-laser pulses.[4] Alkali metal clusters are excellent model systems for time-resolved investigations in this regard: they exhibit pronounced absorption bands,[5] which are well located within the tuning range of the available fs-laser sources; their detachment and photoionization energies[6] are also well within their reach, and the vibrational frequencies can well be resolved with the available fs-pulse lengths. Based on a suggestion of Rabitz *et. al.*[7] we have, therefore, carried out active control experiments, in which we excited differently branching fragmentation and ionization pathways of photo-excited Na_2K, its corresponding fragment NaK[8] and various organo metallic compounds.[9] By employing an evolutionary algorithm for optimizing the phase and amplitude of the applied laser field, the yield of the resulting parent and fragment ions could significantly be influenced and interesting features about the investigated system and the irradiated photo-chemical process could be drawn from the obtained pulse shapes.

In most studies, the aim is to increase the efficiency of photo-chemical reactions, but a new major goal of closed loop experiments is to gain information about the photo-induced control process itself.[10] The observable suitable for this inversion is the acquired optimal pulse shape, which in principle carries all information about the propagation processes of the created wave packets on the involved energy surfaces. Due to the fact that the obtained pulse shapes are usually complicated, one should firstly investigate simple model systems, where the number of possible pathways is limited. This will simplify the interpretation of the optimized process. Slight changes of the laser carrier frequency should allow one to get information about the involved energy surfaces, e.g. the wave packet chooses different ionization paths. The optimization process itself will reveal the best ionization path coded within the pulse shape. This path may be very sensitive to the tuning of the laser carrier frequency, since other transition states could also be reached within the laser bandwidth. The wave packet dynamics of some of NaK has already been investigated in our group by means of fs pump-probe experiments. It exhibits oscillations periods of T_{osc} (NaK) = 440 fs on the $A(2)^1\Sigma^+$ state in a three-photon process.[11,12] Indications of half and integer values of this period within the optimized pulse shapes would give hints about the employed optimized ionization path.

In many cases free optimizations lead to optimal pulse shapes, which are very complicated and impossible to interpret. Some components of the pulse might not contribute to the effect and are therefore not considered by the optimization.[13] Hence, a very important step towards understanding the acquired pulses is to reduce the unimportant components of the light field. In this contribution we use a sinusoidal spectral phase parameterization and reduce the range of a particular parameter value.

The experimental set-up is presented in the next section. The results of the frequency dependent optimizations are shown and discussed in the third section. In the fourth part the parametric sinusoidal optimization method is described for the ionization process of NaK.

2. Experimental Setup

In the experimental setup (shown in Fig. 1) the laser beam passes a pulse shaper system, which allows simultaneous modulation of the phase and amplitude of the laser pulses by applying voltages to a double mask liquid crystal spatial light modulator (SLM) consisting of 2 x 128 pixels. The SLM is placed in the Fourier plane of a zero dispersion compressor, which is a linear setup of two gratings (1200 l/mm) and two plano-convex lenses (f = 200 mm) in a 4f-arrangement.

We apply an algorithm based on evolutionary strategies[14] to find the pulse shape, which yields to the highest feedback signal and performs exclusively phase optimizations in order to keep the pulse intensity constant. The produced optimal pulse shapes are analyzed by intensity cross-correlation and XFROG technique which provide intuitive and direct information about the time ordering of the pulse elements. The reference pulse directly taken from the Ti:Sapphire laser is characterized by SHG-FROG.

The NaK clusters are produced in an adiabatic co-expansion of mixed alkali vapor and argon carrier gas through a nozzle of 70 μm diameter into the vacuum. The stagnation conditions of the molecular beam determine the cluster size distribution. The oven temperature was set to 650 °C and the argon pressure to 2 bars. For further details of the alkaline cluster production see Ref. 10. The fs laser beam is focused onto the cluster beam in order to excite and ionize the neutral particles. The resulting photo-ions are extracted into a quadrupole mass spectrometer and detected by a secondary electron multiplier. A Ti:sapphire oscillator (Tsunami; Spectra Physics) generates pulses of 100 fs duration (FWHM) and an energy of 10 nJ per pulse with a repetition rate of 80 MHz. This high repetition rate of the MHz system assures a good duty cycle. The intensity of the laser pulses within the focus at the interaction region with the molecular beam is estimated to be of the order of 1 GW cm^{-2}. Thus, one could state that the experiments are carried out in the weak field regime. The experiments are carried out by employing central wavelengths in the range of λ_0=760-790 nm. The spectral width of the laser is about $\Delta\lambda$ 8 ± 0.3 nm at FWHM.

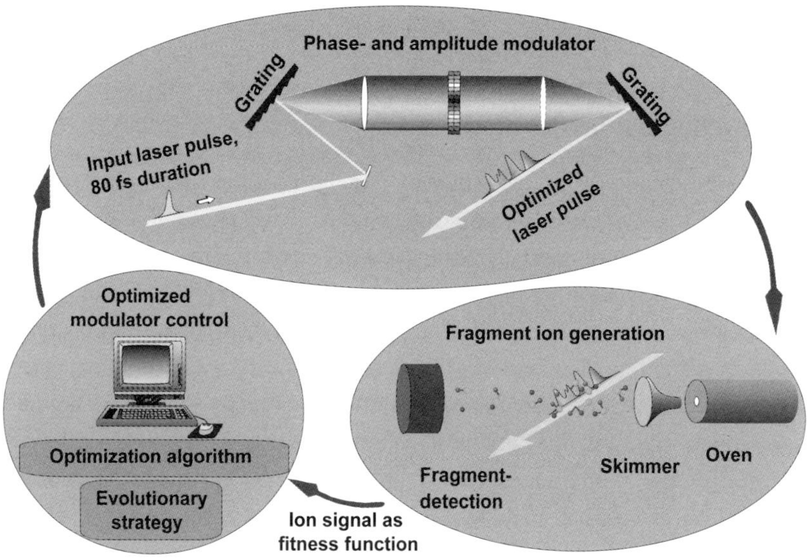

Fig. 1: Schematic view of the feedback loop for optimizing the ion yields of particular reactive channels by employing a self-learning algorithm.

3. Results and Discussion

3.1. Frequency dependent optimization factors for transient three-photon ionization

We present the results of the frequency dependent free optimization factors (ratio of optimized and unshaped pulse ion yields I_{opt}/I_{us}) for the photo-ionization of NaK. The experimental conditions for the preparation of the molecular beam were set for exclusive production of NaK dimers; hence, fragmentation from Na_2K and other larger clusters into NaK^+ does not occur. By recording the ion yield for each generation, a learning curve can be obtained. A typical learning curve of the optimization measurement is shown in Fig. 2a for the case of λ_0 = 775 nm. The three points for each generation show the best, the mean and the worst of all values in one generation. In the beginning, the ion signal is very low because the phase values of the first generation are randomly chosen. This leads to complex pulse trains spread in time over several ps, which yield to low ion signals. With successive iterations, a pronounced rise in the NaK^+ signal is visible. The algorithm converges after about 120 generations, when the optimal pulse shape is found.

The mass spectrum (Fig. 2b) shows the distribution of cluster ions produced by an unshaped pulse. The NaK^+ intensity is dominant, while the signals of the trimer ions Na_2K^+, K_2Na^+ and K_3^+ are not significant. Larger clusters are not detected in the beam. The ion distribution after optimization is represented in Fig. 2c. From the comparison of both mass spectra, a rise of about 60 % in the NaK^+ ion yield can be obtained. The intensities of K_2^+ and K^+ are receded after optimization, because the optimal pulse for NaK^+ does not effectively ionize these species.

Figure 3a shows the obtained optimization factors in the wavelength range between 760 and 790 nm. Only the maximal optimization factor recorded for each wavelength is plotted. The reason is that the respective values have been obtained for the most stable cluster beam conditions and therefore the least disturbed optimization process. The error bars show the cluster beam fluctuations in determining the optimization factor.

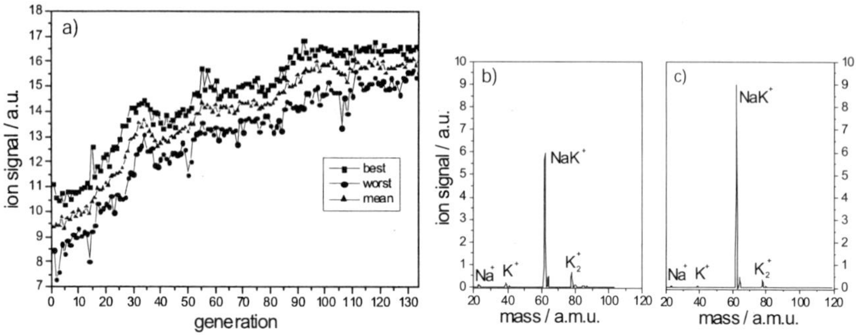

Fig. 2 a) Progression of the NaK^+ ion signal during optimization. At the beginning the ion yield is small, since the initial pulses are randomly formed. After approximately 120 generations the algorithm converges. b) The mass spectrum measured with an unshaped pulse exhibits no trimers and larger clusters. c) The mass spectrum recorded with an optimized pulse reveals a NaK^+ yield of $I_{opt}/I_{us} = 1.6$.

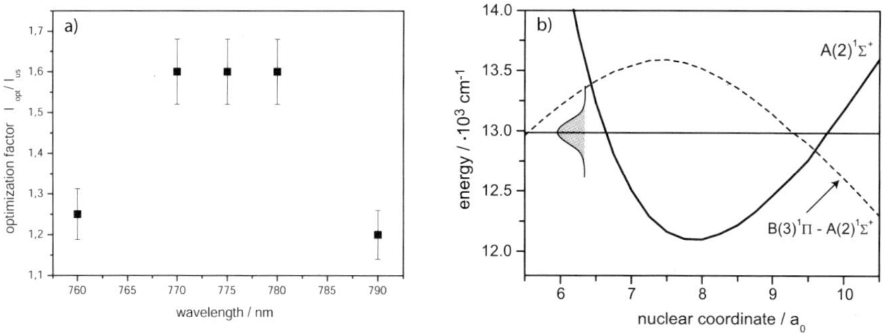

Fig. 3 a) Frequency dependent optimization factor. The factor is calculated by comparing the ion yield produced by an optimal pulse with the ion signal produced by an unshaped pulse within the experimental error. b) Scheme of the involved potential energy curves of NaK.

The viewgraph shows a flat maximum between 770 nm and 780 nm with an optimization factor I_{opt}/I_{us} of 1.6. The optimization factor decreases to 1.25 at 760 nm and 1.2 at 790 nm, respectively. Qualitatively, this behavior can be understood by taking the potential energy diagram shown in Fig. 3b into account. The most relevant part of the NaK potential energy diagram is presented here. It shows the first excited $A(2)^1\Sigma^+$ state and the difference potential $B(3)^1\Pi$-$A(2)^1\Sigma^+$ together with the spectral distribution of the laser pulse at the central wavelength of λ_0 = 770 nm.[15,16] Both curves cross at an energy of 12800 cm^{-1} ($\lambda \sim$ 780 nm), which yields the best resonance condition at the outer turning point for the wave packet propagation in the excited state. The optimized ionization path, therefore, uses this Franck-Condon window to achieve the maximal ion signal. At lower photon energies, the outer turning point of the wave packet is located more inward and the $B(3)^1\Pi$ state can not be reached resonantly, which results in lower optimization factors. For higher photon energies the resonance condition to the $B(3)^1\Pi$ state is located at smaller nuclear distances, where the Franck-Condon factors are lower. This and the fact that the laser frequency is closer to the resonance condition at the inner turning point, which is more favorable for an unshaped pulse, is due to a diminished optimization factor at shorter wavelengths.

The time dependent intensity of the pulse shape is measured by intensity cross-correlation. While the maximized ion yield stays almost the same for repeated optimization runs, the optimized pulse shapes can slightly differ. Here, we show the pulse forms under the given experimental conditions that we have most frequently obtained in the optimization runs for the best individual, respectively.

Figure 5 displays the cross-correlation signals of the optimal pulses for 790 nm (a), 775 nm (b), and 760 nm (c), respectively. Their joint features are the occurrence of triple pulse structures with predominantly a most intense central pulse and separations of multiple Δ t = 250 fs between them. These separations can be assigned to multiples of half oscillation periods in the excited $A^1\Sigma^+$ state, respectively.[11]

We propose that the first pulse creates a wave packet in the $A(2)^1\Sigma^+$ state. The second, most intense pulse, arriving approximately $1.5 \cdot T_{osc}^{NaK}$ later, ionizes at the outer turning point and additionally transfers supplementary population from the ground to the first excited state. The last pulse, which arrives $1.5 \cdot T_{osc}^{NaK}$ later, leads to ionization of the residual population from the $A(2)^1\Sigma^+$ state via a resonant transition state. This process is most pronounced for the highest optimization factor at 775 nm and is disturbed at going to higher and lower wavelengths, visible in the intensity cross-correlations. In order to verify our proposed ionization mechanism, theoretical simulations which will monitor the evolution of the wave packet on the potential energy surfaces induced by the obtained optimal pulses are of an undoubted necessity.

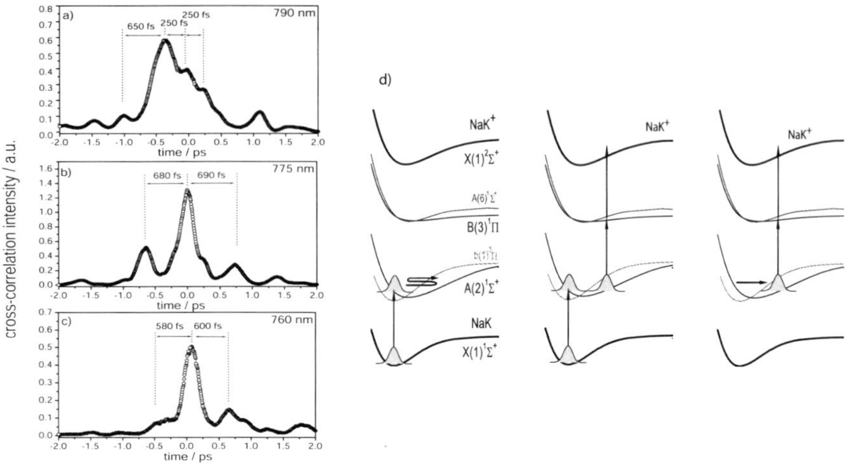

Fig. 4 Intensity cross-correlations of optimal pulse forms obtained for the ionization of NaK are displayed on the left. The figures a), b), and c) show the cross-correlation signals for 790 nm, 775 nm, and 760 nm, respectively. d) Optimal photo-induced transition from the ground neutral state of NaK to its ground ionic state.

3.2. Sinusoidal spectral phase modulation

The spectral field distribution $E(\omega)$ in the Fourier plane is modulated with a sinusoidal phase having the form $\psi(\omega)=a\cdot sin((\omega-\omega_0)\tau+c)$. The shaped pulse form is given as the Fourier transform of the modulated spectral field $E_{mod}(\omega)=E_{in}(\omega)\cdot e^{i\psi(\omega)}$. Due to the periodicity of the modulation a series of pulses are created. The time separation of the pulses within the pulse train increases linearly with the frequency of the modulation. This frequency is governed by the parameter τ and leads to pulse separations equal to τ.

In contrast to the free optimization, where 128 discrete phase components of the spectral field are optimized, here only the three parameters of the sinusoidal function serve as optimization parameters. Thus, searching space is dramatically reduced and convergence is reached in 5-8 generations. The results of the optimization for λ_0 = 770 nm is shown in Fig. 5 (1a+b). A double pulse is obtained with a time separation of Δt=290 fs with the second being more intense than the first one. The XFROG trace exhibits a clear frequency separation between the two pulses. The first pulse is red shifted, while the second pulse is slightly blue shifted. The optimization factor κ=1.36 depicted on the right side is reduced compared to that of the free optimization mentioned above.

In order to further reduce complexity of the pulse form, restrictions are introduced to the parameter τ governing the pulse separation of the pulses within the pulse form. The optimal pulse forms in the second and third row are obtained for parameter values τ restricted to numbers larger than a threshold value $\tau > \tau_0$. While $\tau_0=0$ resulted in a double pulse of Δt=290 fs separation (1a+b), the optimization for τ >185 fs results in a pulse train consisting of three evenly spaced pulses. The time separation of the pulses is exactly Δt=220 fs (2b in Fig. 5). The efficiency of the pulse form compared to the result for $\tau_0=0$ is slightly reduced with an optimization factor of κ=1.34. The pulses are unchirped (see XFROG trace in 2a of Fig. 5). The central pulse is more intense than the other two pulses. Here, an optimal pulse form was found by the algorithm, which clearly describes and exploits the dynamical properties of the NaK molecule as measured in the wave packet oscillation period of T_{osc}(NaK)=440 fs.[11] An explanation for the optimal pulse form can be given in the view of time-dependent Franck-Condon factors: the first pulse excites a wave packet in the excited $A^1\Sigma_u^+$ state of NaK. The second pulse arrives after $\Delta\tau = \frac{1}{2}T_{osc}$(NaK)=220 fs, which is identical to half the wave packet oscillation period in the excited state. Meanwhile, the same pulse can also excite new molecules and create new wave packets, which can be transferred into the ion state by the third pulse arriving again $\Delta\tau$ =220 fs later.

If $\Delta\tau$ =220 fs is not allowed, the pulse separation increases strongly. In the pictures 3a and 3b of Fig. 5 the optimal pulse form for the restriction of τ >275 fs are shown: the pulse separation of the pulse train is now $\Delta\tau$ = 560 fs. The ion yield is further reduced to an optimization factor of κ = 1.29. The separation between the first and the third pulse is $\Delta\tau$ = 1100 fs. Since 2.5· T_{osc}(NaK) gives also t = 1100 fs, the pulse form could be explained by a delayed pump-probe cycle between the first and the third pulse. A wave packet excited by the first pulse, could effectively be transferred into the ion state by the third pulse. The contribution of the second pulse is unclear in this explanation.

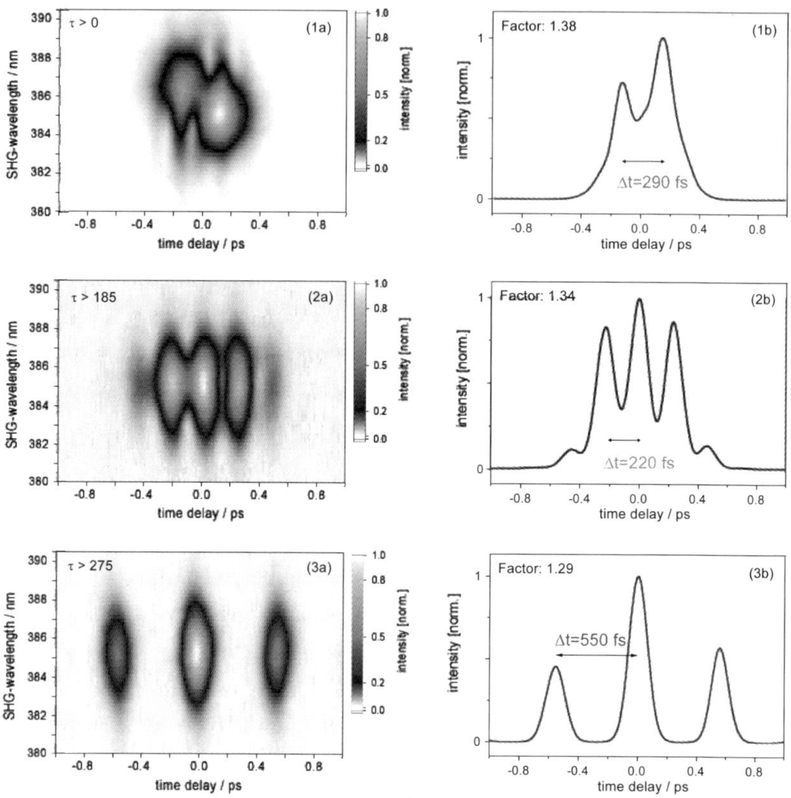

Fig. 5. Optimized pulse forms for maximal NaK^+. Restrictions are introduced to the pulse separation τ. For $\Delta\tau > 185$ fs the three-parameter-optimized pulse form consists of three pulses with a time separation of 220 fs each. For $\tau > 275$ fs the pulses within the optimized pulse train are separated by 550 fs each.

4. Conclusion

Adaptive feedback control of three-photon ionization processes in the model system of NaK has been investigated by means of photon energy dependence. A closed learning loop was applied in order to control the wave packet dynamics. Shaped fs pulse forms were generated by spectral phase modulation and optimization was achieved by maximizing the NaK^+ ion yield.

The recorded optimization factor, dependent on the employed central laser frequency, showed a maximum in the range of 770-780 nm. Information about wave packet propagation at the ionization path, oscillation periods and the involved energy curves could be gained from the pulse shapes. For all employed wavelengths three subpulses were obtained, whereby the most intense subpulse is predominantly located at the central position. Schematic ionization pathway models were proposed in order to explain the observed features within the pulse shapes. Thereby, only the dynamics in the first excited state $A(2)^1\Sigma^+$ was taken into account, which may oversimplify the optimization process. Certainly, a detailed theoretical verification is needed in order to understand this process. The frequency dependent optimization can be regarded as an automated molecular monitor to explore the involved Hamiltonians, although with our present experimental data this goal cannot be achieved. If enough experimental data would be available, the potential energy surfaces could be inferred by solving an inversion problem.[17]

Further, we investigated the adaptive feedback control results of parametrically phase shaped fs pulses. The spectral phase pattern of the pulse shapes were expressed by sinusoidal functions. The obtained optimal pulse shapes differed significantly, depending on the range of values of the restricted parameter. It was possible to eventually extract a triple pulse train with pulse separations of half of the oscillation period $\frac{1}{2}T_{osc}$ of the respective molecular wave packet propagation. Parametric optimizations serve as appropriate methods of learning about the involved molecular system and are promising methods for further coherent control studies.

Acknowledgements

We thank Dr. Š. Vajda for his participation in the earlier experiments. In addition we thank Prof. H. Rabitz and Prof. V. Bonačić-Koutecký for stimulating discussions. The generous support of this research project by the Deutsche Forschungsgemeinschaft in the frame of the SFB 450 research project is gratefully acknowledged. A. Lindinger thanks the DFG for financial support.

References

1. H. Haberland (ed.), *Clusters of Atoms and Molecules, vol. I-II,* Springer Verlag, (1994).
2. J. Manz, L. Wöste (eds.), *Femtosecond Chemistry, vol. I-II,* VCH Publishers, Inc., New York (1994).

3. E.D. Potter, J.L. Herek, S. Pedersen, Q. Liu, A.H. Zewail, *Nature* **355,** 66 (1992).
4. N. Dudovich, D. Barak, T. Gallagher, S.M. Faeder, Y. Silberberg, *Phys. Rev. Lett.* **86,** 47 (2001).
5. S. Vajda, T. Leisner, S. Wolf, L. Wöste, *Philos. Mag.* B **79**, 1353 (1999).
6. A. Herrmann, S. Leutwyler, E. Schumacher, L. Wöste, *Chem. Phys. Lett.* **52**, 413 (1977).
7. R.S. Judson, H. Rabitz, *Phys. Rev. Lett.* **68**, 1500 (1992)
8. S. Vajda, A. Bartelt, E.-C. Kaposta, T. Leisner, C. Lupulescu, S. Minemoto, P. Rosendo-Francisco, L. Wöste, *Chem. Phys.*, **267**, 231 (2001).
9. C. Daniel, J. Full, l. Gonzalez, C. Kaposta, M. Krenz, C. Lupulescu, J. Manz, S. Minemoto, M. Oppel, S. Vajda, L. Wöste, *Chem. Phys.* **267**, 247 (2001).
10. Š. Vajda, A. Bartelt, C. Kaposta, T. Leisner, C. Lupulescu, S. Minemoto, P. Rosendo-Francisco, and L. Wöste, *Chem. Phys.* **267**, 231 (2001).
11. J. Heufelder,H. Ruppe, S. Rutz, E. Schreiber, and L. Wöste, *Chem. Phys. Lett.* **269**, 1 (1997).
12. L.-E. Berg, M. Beutter, and T. Hansson, *Chem. Phys. Lett.* **253**, 327 (1996).
13. J. M. Geremia, W. Zhu, and H. Rabitz, *J. Chem. Phys.* **113**, 10841 (2000).
14. I. Rechenberg, *Evolutionstrategie*, Frommann-Hozboog Stuttgart (1994).
15. S. Magnier, M. Aubert-Frecon, Ph. Millie, *J. Molec. Spec.* **200**, 96 (2000).
16. R.S. Mulliken, *J. Chem. Phys.* **55,** 309 (1971).
17. H. Rabitz and W. Zhu, *Acc. Chem. Res.* **33,** 572 (2000).

Ion-Cluster Collisions

COLLISION OF METAL CLUSTERS WITH SIMPLE MOLECULES: ADSORPTION AND REACTION

M. Ichihashi and T. Kondow

Cluster Research Laboratory of Toyota Technological Institute,
East Tokyo Laboratory, Genesis Research Institute Inc,
717-86 Futamata, Ichikawa, Chiba 272-0001, Japan
E-mail: kondow@clusterlab.jp

Collisional reactions of isolated metal cluster ions (Cu and Cr) were studied at collision energies less than 2 eV by use of a tandem-type mass spectrometer. In the reaction of Cu_n^+ with a methanol molecule, dominant reactions were methanol chemisorption, demethanation, and H(OH) formation on the cluster ions. The absolute cross sections of each reaction were measured, and found to change drastically with the cluster size; the demethanation proceeds efficiently on Cu_{6-8}^+, H(OH) formation on $Cu_{4,5}^+$, and the chemisorption on Cu_n^+ ($n \geq 9$). Structures of the reaction intermediates were calculated by use of the density functional method, and the origin of the size-specific reactivity was examined. In the reaction of Cr_n^+ and Cr_nO^+ ($n \geq 2$) with an ethylene molecule, the abstraction of one Cr atom from the cluster by the ethylene molecule was found to proceed dominantly through ethylene chemisorption on Cr_n^+. On the other hand, it was found that one ethylene molecule chemisorbs on CrO^+ and $CrOH^+$ under a single collision condition, and one more ethylene molecule chemisorbs further under a multiple collision condition. The density functional calculation and the measured threshold energy for the elimination of the chemisorbed ethylene molecules indicate that the chemisorbed ethylene molecules polymerize on $CrOH^+$ and CrO^+

1. Introduction

In a collision of a molecule with a metal cluster ion, the molecule is weakly trapped by the metal cluster ion due to an electrostatic interaction, and then is chemisorbed and reacts. Obviously, the cross sections of these processes change critically with the size of the cluster ion, because the reaction potential is expected to change with the cluster size. Although it is not practically possible to obtain reaction potential by any means, we may utilize several key parameters, such as the electronic structure (such as HOMO-LUMO gap) and the geometrical structure of the cluster in comparison with the size-dependence of the reaction cross sections. We studied low-energy collisions of Cu_n^+ with methanol and of Cr_n^+, Cr_nO^+ and $CrOH^+$ with ethylene in comparison with the electronic and geometrical structures of reaction intermediates calculated.

2. Experiment

Copper and chromium cluster ions were produced by 15-keV Xe^+ sputtering on copper and chromium targets and were allowed to pass through in a cooling cell containing He gas of $\approx 10^{-3}$ Torr. After being mass-selected in the first quadrupole mass filter (QMS), cluster ions were allowed to react with methanol or ethylene in a reaction cell. The product ions were mass-analyzed by the second QMS. Reaction cross sections were obtained from the dependence of the intensity of an ion concerned on the pressure of a reactant gas.

3. Results and Discussion

3.1. Copper cluster ion with methanol

The mass spectra of the product ions measured under single collision conditions showed that the following reactions occur:

$Cu_n^+ + CH_3OH \rightarrow Cu_n^+(CH_3OH)$ (chemisorption),

$Cu_n^+ + CH_3OH \rightarrow Cu_nO^+ + CH_4$ (demethanation),

$Cu_n^+ + CH_3OH \rightarrow Cu_{n-1}^+(H)(OH) + CuCH_2$ (H(OH) formation).

The absolute cross sections for these reactions were found to depend critically on the parent cluster size, *n*. As shown in Fig. 1, chemisorption cross section increases with the cluster size in the size range of $n \geq 8$, the demethanation occurs efficiently at the sizes of 6–8, and H(OH) formation proceeds only at the sizes of 4 and 5.

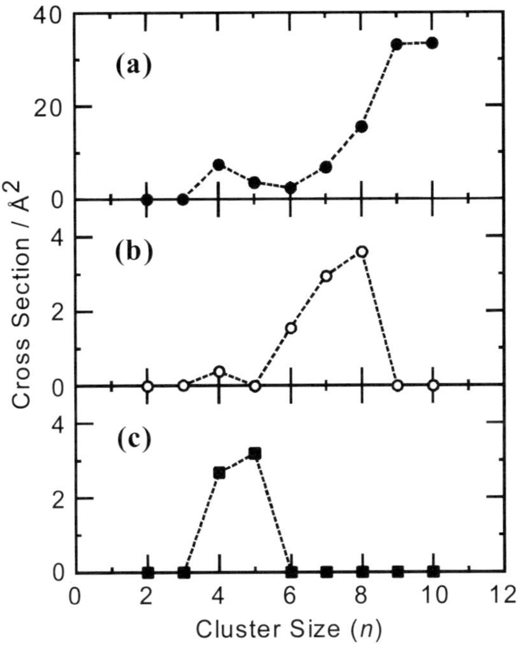

Fig. 1. Cross sections for the production of $Cu_n^+(CH_3OH)$ [panel (a)], Cu_nO^+ [panel (b)] and $Cu_{n-1}^+(H)(OH)$ [panel (c)] as a function of the cluster size. The collision energy is 0.2 eV.

In order to explain such size-dependent characteristics, the structures and the energies of reaction intermediates were calculated by the density functional method, and a schematic reaction potential was obtained. For instance, Fig. 2 is a chemisorption reaction potential obtained for the Cu_4^+ + CH_3OH obtained. As shown in Fig. 2, several reaction intermediates are obtained; $Cu_4^+(CH_3OH)$, $Cu_4^+(OCH_3)(H)$, $Cu_4^+(OH)(CH_3)$, and $Cu_4^+(H)(OH)(CH_2)$ in addition to the final product, $Cu_3^+(H)(OH)$ + $CuCH_2$. Firstly, the copper tetramer ion having a rhombic structure chemisorbs a methanol molecule on an on-top site of Cu_4^+.

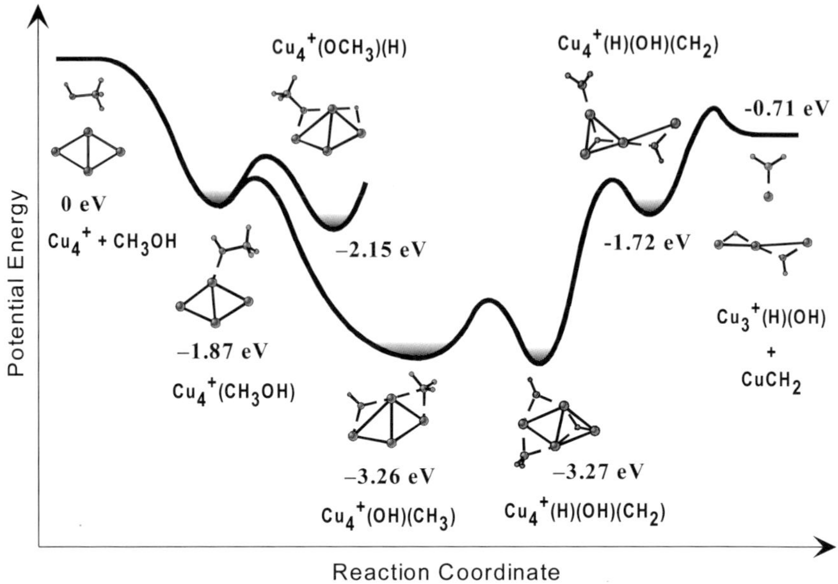

Fig. 2. Schematic potential energy surface along the reaction coordinate of Cu_4^+ with methanol. Minimum-energy structures were calculated by the density functional method. The energies are given with respect to that of the initial state.

The methanol dissociates into $-OCH_3$ and –H, or to –OH and $-CH_3$. The hydroxy intermediate, $Cu_4^+(OH)(CH_3)$, has a lower energy than the methoxy intermediate, $Cu_4^+(OCH_3)(H)$, has. The hydroxy intermediate is supposed to be dominant in the reaction process. On this hydroxy intermediate, the methyl group dissociates further into –H and $-CH_2$. Secondly, the frame of Cu_4^+ begins to dissociate into $Cu_3^+(H)(OH)$. This H(OH) formation proceeds exothermically. The exothermicity is confirmed by the dependence of the reaction cross section on the collision energy. The cross sections for the chemisorption and for the H(OH) formation decrease with the collision energy monotonically.

In the collision of Cu_6^+ with a methanol molecule, chemisorption and demethanation were found to proceed, but H(OH) formation was not. These findings are consistent with the energetics calculated; the chemisorption and the demethanation are exothermic. Note that Cu_6^+ has a planar structure which is the most stable.

The present calculation predicts that the demethanation reaction on Cu_9^+ is exothermic, but the experiment shows that the reaction does not proceed. This contradiction suggests that the energy barrier between the chemisorption and the demethanation is high. The high energy barrier may originate from the fact that Cu_9^+ has a three-dimensional rigid structure in its ground state and hence is not flexible enough to change into a precursor from which a methane molecule is released.

3.2. *Chromium cluster with ethylene*

Figure 3 shows typical mass spectra of the product ions in the reaction of a chromium cluster ion, Cr_n^+, with an ethylene molecule, C_2H_4, at the collision of 0.15 eV. As shown in Fig. 3, Cr_n^+ reacts with C_2H_4 into Cr_{n-1}^+ ($2 \leq n \leq 4$) and $Cr_{n-1}^+(C_2H_4)$ (n=2). The following reaction is likely to proceed, since the collision energy is too low to dissociate Cr_n^+ into $Cr_{n-1}^+ + Cr$ (see Ref. 4):

$$Cr_n^+ + C_2H_4 \rightarrow Cr_n^+(C_2H_4) \rightarrow Cr_{n-1}^+ + Cr(C_2H_4).$$

Note that no chromium cluster ions chemisorbed with C_2H_4. This was not observed even under multiple collision conditions.

The reaction is strongly influenced by introduction of O and OH to Cr_n^+. In the collision of CrO^+ with C_2H_4, $Cr_{n-1}O^+$ is observed similarly to the collision of Cr_n^+ with C_2H_4, except for CrO^+ and $CrOH^+$; CrO^+ and $CrOH^+$ react with C_2H_4 into $CrO^+(C_2H_4)$ and $CrOH^+(C_2H_4)$ respectively, under a single collision condition, and even into $CrO^+(C_2H_4)_2$ and $CrOH^+(C_2H_4)_2$ respectively, under a multiple collision condition.

The energy for the elimination of $(C_2H_4)_2$ from $CrOH^+(C_2H_4)_2$ was found to be 2.5 eV by using a method of collision-induced elimination (see Fig. 4 for the elimination of $(C_2H_4)_2$ from $CrOH^+(C_2H_4)_2$). The present calculation predicts that the energy for the elimination of two C_2H_4 molecules from $CrOH^+(C_2H_4)_2$ is 3.83 eV and that for the elimination of one butene molecule, C_4H_8, from $CrOH^+(C_4H_8)$ is 3.19 eV. The comparison of the energies obtained experimentally with the theoretical values leads us to conclude that the hydrocarbons on $CrOH^+(C_2H_4)_2$ is a butene molecule resulting from polymerization of the two C_2H_4 molecules on $CrOH^+$. A similar polymerization reaction proceeds in $CrO^+(C_2H_4)_2$.

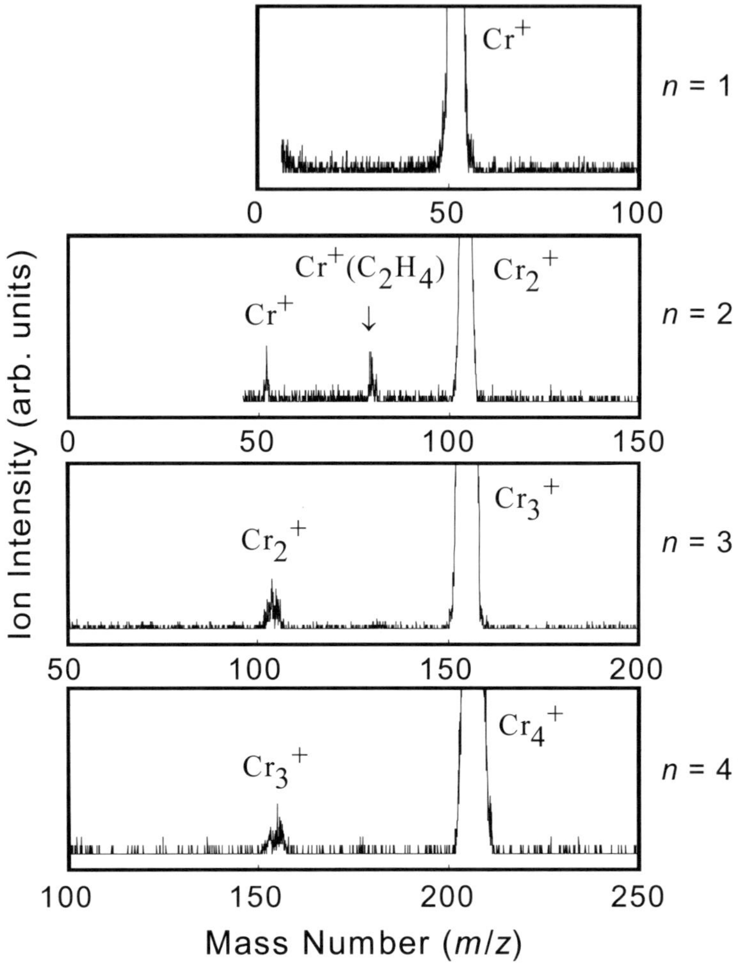

Fig. 3. Mass spectra of ions produced in the collision of Cr_n^+ (n=1–4) with C_2H_4 at a collision energy of 0.15 eV.

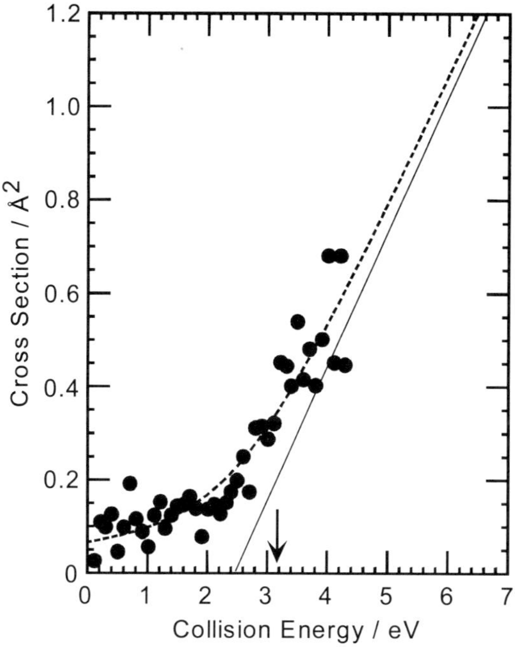

Fig. 4. Cross sections (filled circle) for the production of $CrOH^+$ from $CrOH^+(C_2H_4)_2$ by Xe impact as a function of the collision energy. A hyperbola (dashed line) including the effect of the internal energy of $CrOH^+(C_2H_4)_2$ was fitted to the obtained cross sections. The intersection of its asymptote (solid line) and the abscissa gives the threshold energy of 2.5 eV. The arrow shows the energy of the reaction of $CrOH^+(C_4H_8) \rightarrow CrOH^+ + C_4H_8$ calculated by the density functional method.

Acknowledgements

The authors thank Dr. C. A. Corbett and Prof. J. M. Lisy in the experiment of copper cluster ions, Mr. T. Hanmura in the experiment of chromium cluster ions, and Mr. T. Monoi and Dr. K. Matsuura in the calculation of chromium cluster ions. A part of the calculation for copper clusters was performed on NEC HPC and SX7 computers of the Research Center for Computational Science, Okazaki National Research Institute. Special thanks are due to Prof. L. H. Wöste and his group co-workers for improving the present experimental apparatus.

References

1. R. L. Whetten, M. R. Zakin, D. M. Cox, D. J. Trevor and A. Kaldor, *J. Chem. Phys.* **85**, 1697 (1986).
2. J. Conceição, R. T. Laaksonen, L.-S. Wang, T. Guo, P. Nordlander and R. E. Smalley, *Phys. Rev.* B **51**, 4668 (1995).
3. H. Kietzmann, J. Morenzin, P.S. Bechthold, G. Ganteför and W. Eberhardt, *J. Chem. Phys.* **109**, 2275 (1998).
4. C.-X. Su and P.B. Armentrout, *J. Chem. Phys.* **99**, 6506 (1993).

STABILITY AND FRAGMENTATION OF HIGHLY CHARGED FULLERENE CLUSTERS

B. Manil, L. Maunoury and B.A. Huber

Centre Interdisciplinaire de Recherche Ions Lasers – CEA-CNRS-ENSICAEN Rue Claude Bloch, BP 5133, F-14070 Caen Cedex 05, France

J. Jensen, H.T. Schmidt, H. Zettergren and H. Cederquist

Physics Department, Stockholm University, SCFAB, S-106 91 Stockholm, Sweden

S. Tomita and P. Hvelplund

Department for Physics and Astronomy, University of Aarhus, DK-8000 Aarhus C, Denmark

The stability and the fragmentation of highly charged fullerene clusters has been studied with high-resolution mass spectrometry. Fullerene clusters which are produced in a cluster aggregation source are multiply charged in slow collisions with highly charged ions ($Xe^{20+,\,30+}$). The observed appearance sizes are as small as 5, 10, 21 and 33 for charge states ranging from 2 to 5, respectively. The correlation measured between different fragment ions indicates that the charge is mobile within the cluster, i.e. the van der Waals cluster of fullerenes becomes conducting as soon as it is multiply charged. This is in contrast to recent results for Ar clusters, where charge localization effects have been observed.

1. Introduction

Van de Waals clusters have been studied for many years, often concentrating on rare gas systems because of the ease with which they can be generated. At a first approximation the binding energies can be described as a sum of the pair-wise interactions between the neighbouring constituents. When charged, the excess charge is concentrated on a core surrounded by neutral atoms, in contrast with metal clusters, where the charge is mobile and equally distributed on the cluster surface. In a recent study of highly charged Ar clusters,[1] charge localization effects showed up in the fragmentation spectra. The fragmentation of highly charged Ar clusters may lead among other reactions to very asymmetric charges of the atomic fragments. Thus, Ar^{7+} fragments have been measured in correlation with singly charged Ar ions, indicating that an electric communication between the neighbouring Ar atoms does not exist and that van der Waals clusters are typically insulators.

Another van der Waals-type cluster with larger constituents and which may have other properties can be formed from fullerenes like C_{60} or C_{70}. First studies have been performed with singly and doubly charged $(C_{60})_n^{q+}$ clusters[2,3] and with singly charged pure and mixed $(C_{60})_n(C_{70})_m$ clusters.[4] These studies concentrated on the occurrence of shell effects and the analysis of the geometrical structure of the fullerene clusters. It was demonstrated that the closed shell clusters with n=13 and n=55 do have an icosahedral structure, whereas incomplete geometrical shells undergo structural changes similar to those observed for rare gas clusters.[2] A higher cluster temperature may favour also decahedral or closed-packed structures, which do not depend on the charge state of the cluster ion (q=1 and 2). The experimentally observed structures have been confirmed by global-minima calculations for $(C_{60})_n$ clusters.[5] Furthermore, the experiments have shown a very similar behaviour patterns for pure and mixed clusters, as regards the cluster formation, the structure and the binding energies.[4]

In the present contribution we will describe experiments with fullerene clusters in higher charge states, which are found to be (meta)stable in charge states up to q=5 for sizes as small as 33. The main points of discussion concern the appearance sizes and the analysis of the correlated fragmentation spectra, which contain information on the charge mobility within the fullerene clusters.

2. Experimental Method

Neutral fullerene clusters are produced in a cluster aggregation source.[6] A powder of C_{60} and C_{70} is heated in an oven to a temperature between 500 and 570 °C. The C_{70} content amounts to about 5%. When leaving the oven, the fullerenes enter a region of cold He-gas (p~10 mbar, T~77 K) where condensation and cluster formation occurs. The formed clusters leave the source region through a 2 mm-aperture and pass several differential pump stages before interacting with a beam of highly charged ions. The neutral cluster distribution contains monomers, dimers and larger clusters. Clusters containing up to 40 fullerene molecules have been observed for optimum temperature and He-pressure.

Highly charged ions are produced in an ECR ion source. Typically the ion beam is accelerated to a voltage of 20 kV before it intercepts the cluster beam at the object point of a linear time-of-flight mass spectrometer. The ion beam is pulsed with a repetition rate of several kHz and a pulse width of 1 to 10 μs.

The ionized clusters and the charged fragments formed during the ion-cluster collisions are extracted perpendicular to the two beams with a voltage of ~7 kV over a distance of 14 cm and their mass/charge ratio is determined with a 1m-linear Wiley-McLaren type spectrometer. The extraction voltage is pulsed (pulse length ~ 60 μs) and synchronized with the ion beam pulse. After passing the field free drift region the ions are post-accelerated towards a conversion plate (-25 kV) in order to increase the detection efficiency. The secondary electrons emitted from the conversion plate are focused on a channelplate detector, the pulses of which are treated in a multi-hit TDC (time-to-digital converter) on an event-by-event basis. Thus, spectra for a given number of fragments (stops or multiplicity) can be analyzed separately allowing for classification of the type of collision (distant and close).

3. Spectra of Multiply Charged $(C_{60})_n{}^{q+}$ and $(C_{60})_n(C_{70})_m{}^{q+}$ Clusters

The measured mass spectra consist of two different parts. One is showing intact singly and multiply charged clusters of fullerenes. This part we will discuss first by analyzing the appearance sizes and the relative intensities. The second part, which shows contributions due to fragmentation processes, localized at mass/charge ratios equal or below the singly charged monomer, will be analyzed later, with respect to the correlation between different fragment ions.

Figure 1 shows a typical cluster distribution obtained in collisions with Xe^{20+} projectiles at a collision energy of 400 keV corresponding to a relative velocity of ~ 0.3 a.u. The dominant peaks correspond to the position of singly charged fullerene clusters. However; as a wide primary distribution is used, the peaks also contain contributions from multiply charged clusters.

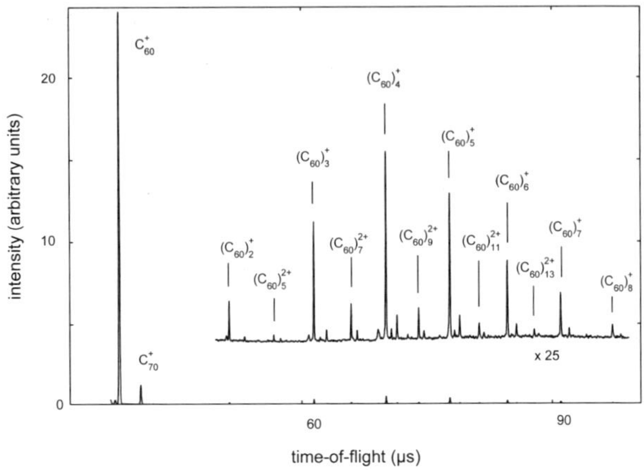

Fig. 1. Time-of-flight mass spectrum showing fullerene clusters singly and doubly charged in collisions with Xe^{20+} ions at 400 keV. (T_{oven} = 525 °C).

The dominance of the singly charged monomer is due to the abundance in the primary cluster distribution and to essential contributions from fragmentation processes. The spectrum shows singly as well as doubly charged clusters $(C_{60})_n^{q+}$ for n<16, and in addition clusters which contain one C_{70} molecule are present. The limited size range is due to the low oven temperature (T=525 °C). When increasing T to 560 °C, the measured spectrum becomes more rich as can be seen in a detailed view, shown in Fig. 2. Stable multiply charged clusters in charge states up to q=5 are observed. With the aid of a fitting procedure mass spectra for individual charge states can be obtained as shown in Fig. 3 for q=3. From these spectra we can determine the limit of stability, the so-called appearance size which depends on the cluster temperature.[7] For q=3 this limit is reached at n=10 to 11 for pure C_{60} clusters (open bars) as well as for those containing one C_{70} molecule (full bars). Evidently, the integration of C_{70} molecules into the cluster does not strongly change the cluster properties, as has been shown already before for clusters in lower charge states.[4] This similarity appears also in the increased intensity at n=13 and n=19, indicating that the geometrical structure is the same for pure and mixed fullerene clusters, independent of the charge state. The observed 'magic numbers' agree with those known from other van der Waals type clusters, like Ar_n^+, Kr_n^+ or Xe_n^+. In the present experiment the shell effects become more visible for higher charges, as the transferred energy increases with the degree of ionization and thus favours the evaporation of fullerene molecules.

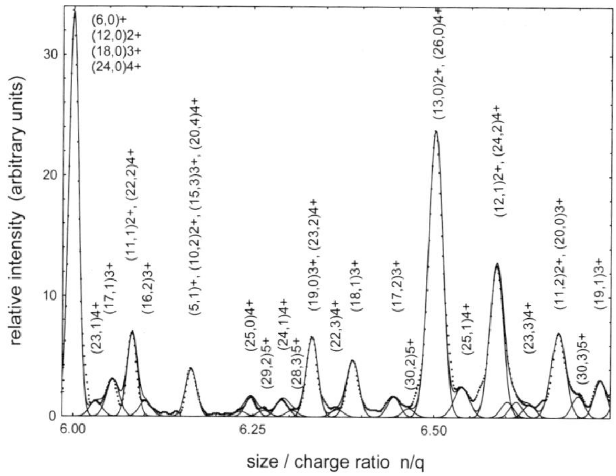

Fig. 2. Detailed view of the mass spectrum obtained at $T_{\text{oven}} = 565$ °C. The vertical notations (n,m)q+ identify contributions from clusters $((C_{60})_n(C_{70})_m)^{q+}$.

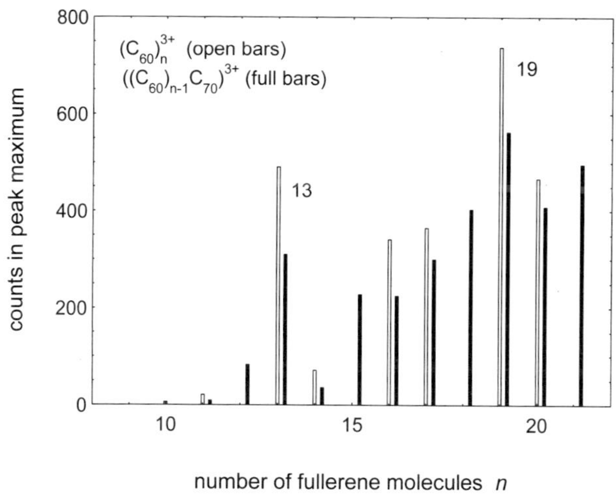

Fig. 3. Threshold region for triply charged pure and mixed fullerene clusters: $(C_{60})_n^{3+}$ (open bars: those which are superimposed to singly charged clusters are not shown) and $((C_{60})_{n-1}C_{70})^{3+}$ (full bars).

Table I. Appearance sizes of multiply charged clusters. a) present result; b) from Ref. 2; c) from Ref. 9; d) from Ref. 8.

q	$(C_{60})_n^{q+}$ a)	$(C_{60})_n^{q+}$ b)	$(C60)_{n-1}C_{70}^{q+}$ a)	Ar_n^{q+} c)	Na_n^{q+} d)
2	5	~25	5	91	25
3	10	-	11	226	49
4	21	-	22	-	72
5	(33)	-	(33)	-	92

The appearance sizes are compared in Table I with those measured for other van der Waals or metal clusters. The values observed for fullerenes are much lower due to the large size of the constituents and due to their large polarizabilities. Thus, 5 fullerenes are found to be necessary to stabilize 2 charges, whereas somewhat more than 30 molecules are required for 5 excess charges. Double ionization by highly charged ions and by laser irradiation leads to very different appearance sizes (ions: 5, laser: ~25). This reflects the different energy transfers in the two interactions and underlines the strong dependence of the appearance size on the internal cluster temperature.[8]

So far we have discussed the part of the spectrum which shows singly and multiply charged fullerene clusters. Cluster fragments and possible contributions from ion collisions with monomers and dimers, which are also present in the primary cluster beam, are shown in Fig. 4a. The inclusive spectrum (all multiplicities are included) obtained with clusters of fullerenes is very different compared to that obtained with C_{60} molecules (Fig. 4b). First, the charge state of intact fullerenes is more limited: in Fig. 4a the maximum fullerene charge seems to be ~3, where in the ion collision with the monomer (Fig. 4b) charge states up to $q = 10$ have been reported.[10] Second, the peak width of the intact C_{60} molecules is rather wide and an intense 'evaporation series' of C_{60-2l}^{+} ions is observed, which is totally absent in the monomer case (Fig. 4b). Finally, small size fragments C_k^+ are measured with k going up to 30 with a relatively narrow peak width. In the monomer case (Fig. 4b), the fragments sizes are much smaller (with the highest intensity at C^+) and the corresponding peak widths (for k = 9-11) are much larger. We will discuss this finding further below. However, when we select a class of events of the spectrum (a), where only one ion has been detected per event (single stop spectrum) we obtain a spectrum (see Fig. 4c), which becomes nearly identical with that of Fig. 4b. The peak widths are reduced, the evaporation peaks have disappeared as well as the small-size fragments for $k > 10$ approximately. Therefore, we attribute this spectrum to the interaction of the ions with monomers which are present in the primary cluster beam. In fact, we can vary its relative fraction in the spectrum (a) by varying the oven temperature and the flow of the He gas in the cluster source, thus varying the monomer fraction in the cluster beam.

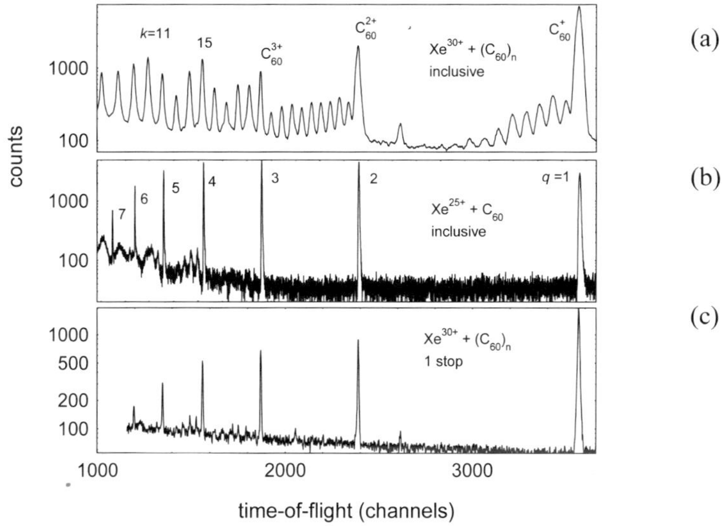

Fig. 4. Time-of flight mass spectra. (a) Inclusive spectrum obtained in Xe^{30+} - $(C60)_n$ collisions at 600 keV (oven temperature: 565 °C). (b) Inclusive spectrum for Xe^{25+} - C_{60} collisions. (c) Single stop events extracted from the spectrum (a).

In order to discuss further the question of charge mobility or localization, we will study the fragment correlation in the spectra with higher multiplicity. In Fig. 5 we investigate the correlation between two fragments as measured in the 2-stops spectrum. In the upper part we show the distribution of the fragments measured in correlation with a C_{60}^{+} fragment, in the lower part that one which is correlated with C_{60}^{3+}. It becomes clear, that the two fragments tend to have the same charge. Thus, C_{60}^{+} is preferentially detected with another fullerene in charge states one and two, C_{60}^{3+}, however, with a fragment C_{60}^{q+} with $q = 2,3$ and 4. It seems that the charge can be easily distributed among the fullerene molecules. From the intensity variation with the oven temperature, we attribute these 2-stops spectra mainly to reactions with fullerene dimers.

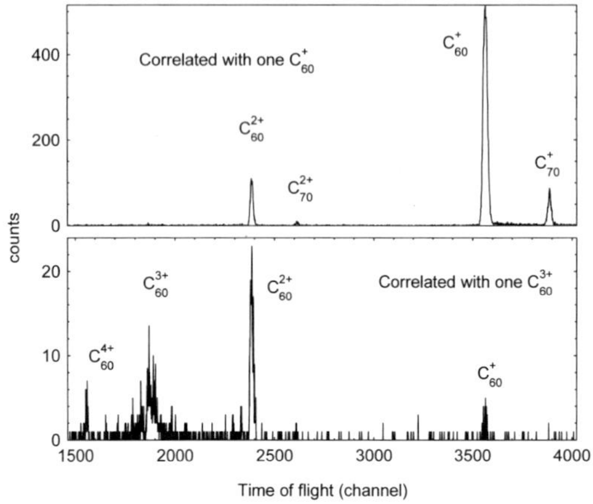

Fig. 5. Correlation between two charged fragments analysed in the 2 stops-spectrum (Xe^{30+} - $(C_{60})_n$; T_{oven} = 565 °C). Upper part: correlation with C_{60}^{+} ; lower part: correlation with C_{60}^{3+}.

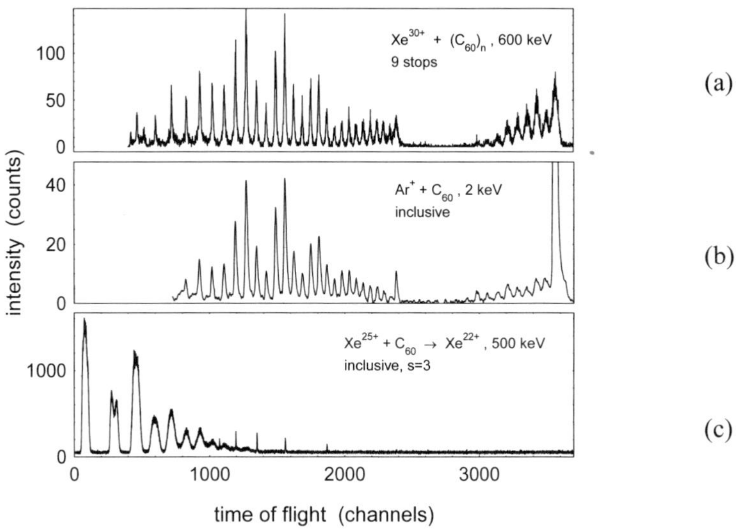

Fig. 6. (a) 9-stops spectrum obtained in Xe^{30+} - $(C_{60})_n$ collisions (T_{oven} = 565 °C); (b) Mass spectrum observed in Ar^{+} - C_{60} collisions (2 keV); (c) Mass spectrum obtained in Xe^{25+} - C_{60} collisions (500 keV), where three electrons have been stabilized at the projectile, corresponding approximately to a 9-time ionization of the fullerene.[11]

When increasing the multiplicity of the event (number of charged fragments), the observed spectrum changes dramatically. In the case of 9 stops (see Fig. 6a) we find singly charged fullerenes, which have most likely lost several C_2 units, or narrow peaks of singly charged small carbon clusters C_k^+, with k ranging up to 30. The correlation of 9 particles is measured preferentially within the two different groups of particles, i.e. either small fragments or nearly intact fullerenes are observed, most probably depending on the transferred energy.

The fragment distribution is very similar to that obtained in low-energy Ar^+ - C_{60} collisions (see Fig. 6b, Ref. 12). In that case, the inclusive spectrum corresponds more or less to a 1-stop spectrum as the fullerene is only singly (or, to a lower extent, doubly) ionized. In both spectra the peaks are very narrow and the distribution shows maxima at k=11 and 15. This spectrum is not induced by a high excess charge, but is due to thermally activated fragmentation. A charge induced fragmentation spectrum is shown in Fig. 6c. This has been obtained in collisions of Xe^{25+} ions with C_{60}, measured in coincidence with an outgoing Xe^{22+} projectile. It has been shown that this case corresponds to an average fullerene charge of ~9 see Ref. 11. Clearly the spectrum peaks at much smaller fragment sizes and the peak widths are strongly increased due to the Coulomb repulsion. Thus we conclude that in the fullerene cluster case both the charge and the excitation energy are well distributed over the constituents thus finally leading to the thermal fragmentation of fullerene molecules in low charge states.

How can the charge move in van der Waals clusters which are normally insulators? The distance between the centers of two neighboring molecules is about 10Å in the neutral system. It is expected to become smaller when one molecule is singly charged due to the charge-induced dipole interaction. When we treat two fullerene molecules as conducting spheres and develop the classical over-the barrier model to describe the processes of charge transfer[13] we obtain critical distances for the transfer of the first electron, which are 9.4, 9.8, 10.3, 11.0 and 11.7 Å for the charge states 1 to 5, respectively. Indeed, experimental studies with C_{60}^{4+} projectiles which collide with C_{60} have demonstrated the validity of this model, and have proven that a full electrical contact where both fullerenes share the charge equally is established at a center-center distance of 9.7 Å [14]. Thus, charged fullerenes in charge states ≥ 2 can communicate with their neighbors even at the equilibrium distance of the neutral fullerene cluster.

Even for the charge 1 we might expect charge mobility when taking into account a slight reduction of the equilibrium distance due to the charge.

4. Conclusions

In this work we have shown that clusters of fullerenes can sustain excess charges up to q=5 for a limited number of constituents (n=33). The presently observed appearance size for doubly charged clusters (n=5) is much lower than the one found previously in laser irradiation experiments as the transferred energy is rather low for distant ion collisions. The present observation of magic numbers for highly charged clusters (q=3,4,5) does not depend on the charge and the exact cluster composition and is consistent with those expected for van der Waals clusters. In spite of this, we observe correlation between intact C_{60} fragments of similar charges and fragmentation patterns which resemble those for thermally activated fragmentation of monomers of *low charge,* although a large number of electrons have been taken from the fullerene cluster. From these observations we conclude that the van der Waals clusters become electrically conducting once they are (multiply) charged. This view is supported by an over-the-barrier treatment yielding critical distances for electron transfer in the vicinity of the C_{60}-C_{60} distance in neutral van der Waals clusters of fullerenes. Finally, we note that the situation encountered here is similar to the one observed for Fullerite (van der Waals crystal with 10 Å C_{60}-C_{60} distance) which is electrically insulating in its pure form but becomes conducting when it is doped with electron accepting or donating atoms.

Acknowledgements

This work was initiated under the European network LEIF (HPRI-CT-1999-40012). It was supported by the Danish Research Foundation through the research center ACAP (Aarhus Center for Atomic Physics) and by the Swedish Research Council (521-2001-2226).

References

1. W. Tappe *et al*, *Phys. Rev. Lett.* **88**, 143401 (2002).
2. T.P. Martin, U. Näher, H. Schaber and U. Zimmermann, *Phys. Rev. Lett.* **70**, 3079 (1993).
3. W. Branz, N. Malinowski, H. Schaber and T.P. Martin, *Chem. Phys. Lett.* **328**, 245 (2000).
4. K. Hansen, R. Müller, H. Hohmann and E.E.B. Campbell, *Z. Phys.* D **40**, 361 (1997).
5. J.P.K. Doye, D.J. Wales, W. Brantz and F. Calvo, *Phys. Rev.* B **64**, 235409 (2001).
6. T. Bergen *et al*, *Rev. Sci. Instrum.* **70**, 3244 (1999).
7. F. Chandezon *et al*, *Phys. Rev. Lett.* **74**, 3784 (1995).
8. F. Chandezon *et al*, *Phys. Rev. Lett.* **87**, 153402 (2001).
9. O. Echt *et al*, *Phys. Rev.* A **38**, 3236 (1988).
10. A. Brenac *et al*, *Physica Scripta*, **T80**, 195 (1999).
11. S. Tomita *et al*, *Phys. Rev.* A **65**, 053201 (2002).
12. A. Reinköster *et al*, *J. Phys. B: At. Mol. Opt. Phys.* **35**, 4989 (2002).
13. H. Zettergren *et al*, *Phys. Rev.* A **66**, 032710 (2002).
14. H. Cederquist *et al*, *Phys. Rev.* A **63**, 0252014 (2001).

FULLERENE COLLISION AND IONIZATION DYNAMICS

E.E.B. Campbell

Dept. of Experimental Physics, Göteborg University and Chalmers, SE-41296 Göteborg, Sweden.

E-mail: eleanor.campbell@fy.chalmers.se

An overview is given of experiments used to probe the dynamics of excited fullerenes. The reaction channels observed in collisions between fullerene ions and fullerenes are discussed with emphasis on the molecular fusion reaction. In a second example, experiments in which the energy dissipation in C_{60} is studied using fs lasers are described and the production of excited Rydberg states is discussed.

1. Introduction

Fullerenes are fascinating clusters (or molecules) that have been available in macroscopic amounts for over 10 years. They fit nicely into the intermediate range between individual small molecules and bulk material. Some of their properties such as thermionic-like electron emission[1] or the emission of black-body radiation[2] are more akin to the behaviour of bulk material, however one can also carry out high resolution spectroscopy and study the molecular nature of the systems in great detail.[3] This range of behaviour combined with their relative simplicity (only one atomic element, high symmetry) and beauty makes them excellent model systems to probe the dynamics of highly excited complex molecules. An additional bonus, which makes them popular with atomic physicists, is the relative ease with which they can be handled and evaporated into vacuum to produce a relatively dense beam of neutral or ionized projectiles or targets for collision experiments. Many of the other papers in this volume provide illustrations of the range of experiments that can be carried out and the properties that can be probed in this way.

Low energy collision experiments, where the energy is predominantly transferred to the vibrational excitations of the fullerenes, have been extensively studied over the years.[4] These experiments have mainly been concerned with collisions between fullerenes and atoms or atomic ions and the main interest has been the investigation of reactions or the study of the fullerene fragmentation pattern. In a series of experiments dating back to 1992, we have studied the various reaction channels that can be observed in low energy (50 - 5000 eV in the center of mass frame of reference) collisions between fullerene ions and neutral fullerenes. Perhaps the most fascinating product of such collisions is the molecular fusion of the collision partners to produce a larger fullerene. The collision energy dependence of this reaction channel will be described in the following text and comparisons will be drawn between colliding fullerenes and colliding nuclei.

As stated above, the low energy collisions transfer energy predominantly to the vibrational degrees of freedom. A similar situation arises in experiments where the fullerenes are excited by nanosecond laser pulses. Although the energy is initially absorbed by the electrons, there is sufficient time between the absorption of consecutive photons for the energy to couple to the vibrational modes. In this way, very many photons can be accommodated leading to a very highly vibrationally excited molecule. The fragmentation patterns observed from nanosecond excitation are very similar to those found in collision experiments and can be explained as a purely statistical break-up of the vibrationally excited carbon cage.[5] The situation changes if the excitation occurs on a time scale that is short compared to the time needed for coupling between the electronic and vibrational degrees of freedom.[6] This can be studied using femtosecond laser pulses and is the topic of the second part of the paper.

2. Molecular Fusion in Fullerene-Fullerene Collisions

The fullerene-fullerene collision experiments were inspired by theoretical calculations showing fusion between two colliding sodium clusters carried out by R. Schmidt and co-workers.[7] The calculations showed very close similarities between cluster fusion and the fusion of two atomic nuclei. Unfortunately it is not experimentally feasible to carry out collisions between two beams of mass-resolved sodium clusters. It is, however, possible to attain enough intensity in fullerene beams to carry out such studies experimentally.

The experiments and theoretical interpretation have been described in a number of publications.[8-14] A positively charged fullerene ion beam is produced by evaporating fullerene powder from an oven at a temperature of about 500 °C and ionizing by electron impact at 200 eV. The ions are extracted into a second differentially pumped vacuum chamber by a pulsed electric field. The ions with the desired mass-to-charge ratio are mass-selected and directed into a scattering cell. The scattering cell is a cylindrical oven with a circular entrance of 2mm diameter and a horizontal exit slit of 2mm height allowing the detection of scattered ions at laboratory angles of up to 80°. Fullerene powder of high purity (commercially available, > 99.4% C_{60} or C_{70}) is evaporated inside the cell to form the target gas. The energy spread of the parent ion beam was measured to be 5 % (FWHM) of the laboratory collision energy, and the angular spread was measured to be 2 ± 0.5° (FWHM). The positively charged products of the collision reaction are detected by a time-of-flight reflectron mass spectrometer. The reflectron can be rotated around the scattering cell allowing the determination of the angular distribution of ions. This was not considered in the early experiments[8,10,11] and in these papers the cross sections for fusion were underestimated at high energies where substantial fragmentation of the fusion product occurs. Newer measurements determined the angular distributions of the fusion products and integrated over these to obtain corrected values of the absolute cross sections.[14] The new, corrected values are shown in Fig. 1 (triangles), together with the results of molecular dynamics simulations, reported in Ref. 14 (circles). Note that the error bars shown are relative errors in the measurement. There is a large error with respect to the absolute cross section due mainly to uncertainty in the fullerene vapour pressure.[4] The early classical MD simulations using empirical model potentials (open circles) are in very good agreement with the experiment as far as the threshold energy for fusion is concerned, but vastly overestimated the absolute magnitude of the cross section and the energy window for which fusion products can be observed. The quantum MD simulations using density functional theory with the local density approximation do, however, give excellent agreement with the experimental data.

The collision energy dependence of the cross section bears some resemblance to the energy dependence of nuclear fusion reactions. The comparison with a simple phenomenological "absorbing sphere" or "line-of-centres" model is shown in Fig. 2. The model assumes that the magnitude and energy dependence of the fusion cross section will be limited by the condition that the energy available along the line of centres of the two fullerenes exceeds this threshold. At an initial relative translational energy E and impact parameter

b the energy along the line of centres is $E(1-b^2/R^2)$, where R is the centre-to-centre distance.

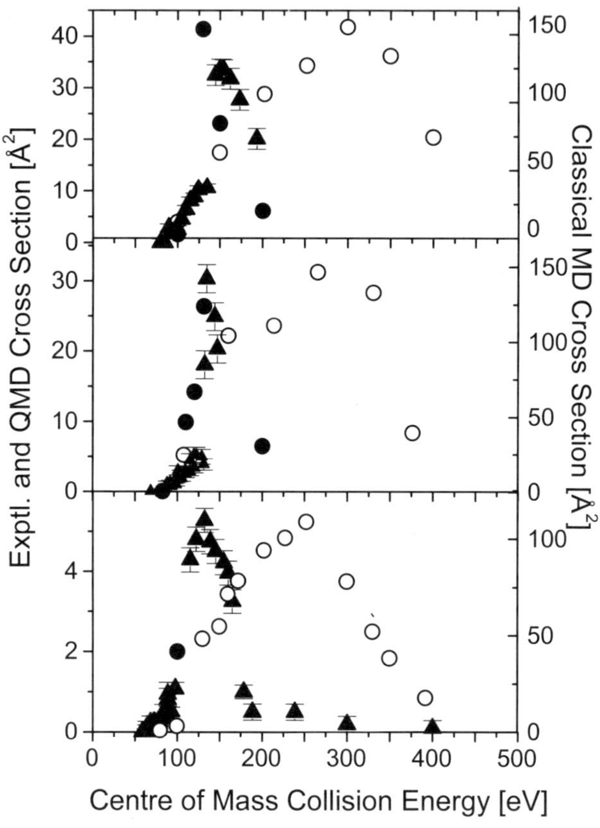

Fig. 1. Cross sections for fusion between two fullerenes (a) $C_{60}^+ + C_{60}$ (b) $C_{60}^+ + C_{70}$ and $C_{70}^+ + C_{60}$ (c) $C_{70}^+ + C_{7b0}$. Triangles: data corrected for fragment scattering. Filled circles: density functional MD calculations. Open circles (right hand axis): empirical potential MD calculations. Adapted from Ref. 14.

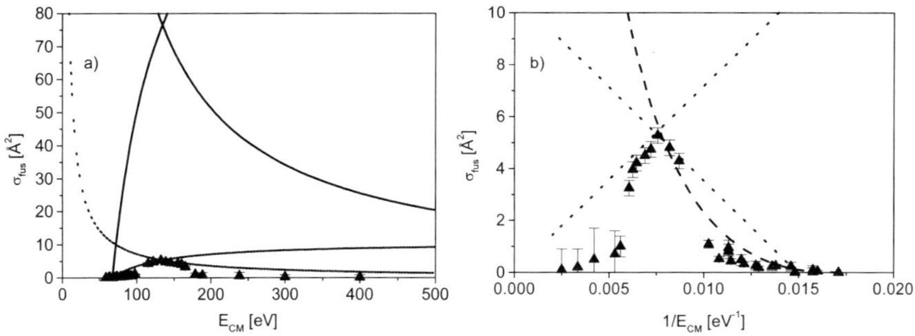

Fig. 2. Cross sections for the fusion reaction $C_{60}^+ + C_{60} \rightarrow C_{120}^+$. Triangles: experimental data. Full line: line-of-centres model and centrifugal cut-off. Dotted lines: full lines scaled by 0.07. Dashed line: steric model. (a) As function of centre of mass energy (b) As function of the reciprocal centre of mass energy. Adapted from Ref. 13.

Locating the barrier to fusion at $R = d \approx 7Å$, the maximal impact parameter that allows the crossing of the barrier is determined by $E(1 - b_m^2 / R^2) = E_{\text{bfus}}$ or

$$\sigma_{\text{fus}} = \begin{cases} 0 & , E \leq E_{\text{bfus}} \\ \pi d^2 \left(1 - \dfrac{E_{\text{bfus}}}{E}\right) & \end{cases} \qquad \text{line of centres model.} \tag{1}$$

The cross section rises rather sharply to a maximum and then falls steeply to higher energies. In nuclear physics this decrease in the probability of fusion is associated with the compound nucleus not being formed. The impact parameter becomes so high that the effective potential $V(R) + Eb^2 / R^2$ no longer supports a well. Taking the case of $C_{60}^+ + C_{60}$, the estimated stability of the peanut shaped isomer of C_{120}^+ is about that of the reactants.[15] So C_{120}^+ has a barrier for dissociation which is the fusion barrier whose height is about 80 eV. To fill in the well to the left of the barrier at R =7Å and a collision energy of 150eV requires impact parameters above 5 Å. The energy dependence of the cross section due to such a cut off is

$$\sigma_{\text{fus}} = \pi R^2 (E_{\text{bfus}} / E) \qquad \text{centrifugal cut off for } E > E_{\text{bfus}} \tag{2}$$

The measured fusion cross section differs from the simple model in several key respects. The rise at the threshold is not quite linear in $1/E$ and the slope is quite wrong (Fig 2). Also, the geometrical forefactor is nowhere near as large as πd^2. It is well known in conventional chemical kinetics that the size of the cross section is over estimated by the line of centres model. One tends to ascribe this to steric requirements of the reaction and the simplest correction is to lower the probability of reaction by a 'steric factor' $p, p<1$. The forefactor is then $\pi d^2 p$ and fitting values of p to the experimental data (corrected for scattering losses, Fig. 2, Ref. 14) yields values from 0.07 for fusion in $C_{60}^{+}+C_{60}$ collisions to 0.4 for $C_{60}^{+}+C_{70}$ or $C_{70}^{+}+C_{60}$ and 0.45 for $C_{70}^{+}+C_{70}$ collisions. A more refined approach makes the steric constraint depend also on the impact parameter, but to compute it requires knowledge of the barrier as a function of the approach geometry. In earlier work, the experimental data was fitted with a dependence which implies that the steric factor, p, rises linearly with collision energy.[4] This could be rationalized if different relative orientations of the fullerenes on impact have different fusion barriers. However, density functional computations do not support a strong dependence of the barrier on the detailed geometry of the two fullerenes at the moment of impact[16] although the calculated fusion probability is significantly lower than the simple model predictions and in good agreement with the experimental data (Fig. 1).

In order to at least partially explain the discrepancies with the simple model we have considered the possibility of competing reaction channels.[13] In the threshold regime the available phase space for the inelastic scattering channel is larger than that for the formation of a very hot compound fullerene, say C_{120}^{+}. Such a competing channel can be observed in molecular dynamics simulations. An interesting analogy to this (and a fun experiment to try at home) is to smash two eggs together. Under "normal smashing conditions" it is always the case that only one egg will smash. To estimate the branching ratio for fullerene-fullerene collisions we require the density of vibrational states of a polyatomic molecule. For our purpose a simple RRK estimate is sufficient because we only need ratios and not absolute numbers. For an n-atomic species with s vibrational modes, $s=3n-6$, and for a (geometric) mean vibrational frequency $\bar{\nu}$ ($2.7\text{x}10^{13}$ Hz = 900 cm^{-1} for C_{60}) the number of vibrational states below the energy E is

$$N(E)=\frac{1}{s!}\left(\frac{E}{h\bar{\nu}}\right)^{s} \qquad \text{harmonic count} \tag{3}$$

The Stirling approximation $s! \approx \sqrt{2\pi s}(s/e)^s$ allows us to focus attention on the variable of interest, namely $(E/h\bar{\nu})/(s/e) = e(E/sh\bar{\nu})$. When this variable exceeds unity, we are in the vibrational quasi-continuum where the density of states is exponentially large. The essential difference between the two channels is that C_{120} has (about) 6 times as many vibrational modes as does C_{60}. Once the energy is high enough to reach the vibrational quasi-continuum of C_{120}, it will have overwhelmingly more states than C_{60}. But it takes less energy to reach the quasi-continuum of C_{60}.[13] In the immediate post threshold region we thus expect that many collisions do not end up in fusion but in inelastic collisions. At higher energies the situation is reversed and, because of the steep rise of the density of states with energy, once formation of C_{120} is the dominant channel, it is the overwhelmingly dominant one. This very simple consideration leads to a much steeper rise of the cross section close to the threshold. The increasing phase space that becomes available for fusion at energies past the barrier meaning that the fusion cross section rises more rapidly than as $1 - E_{\text{bfus}}/E$, as seen in Fig. 1 and 2.

Fusion is operationally defined as formation of a compound species accompanied by an extensive dissipation of the internal energy. It is, however, possible for the hot compound species to fragment promptly. This process is very clearly seen in molecular dynamics simulations where many trajectories leading to C_{120} formation also show it breaking apart in one or two vibrations. For such short times energy is not fully dissipated and these events are not detected in the experiments as "'fusion". The competition with the prompt fragmentation (with incomplete energy equilibration) means that the fusion cross section will decrease strongly with energy, past the threshold for three-particle production. What we therefore require is an estimate for the cross section for direct (i.e. prompt) collision induced dissociation (CID). The CID cross section can be computed if all the excess energy is disposed of as kinetic energy of the three outgoing particles, The result is[13]

$$\sigma_{\text{CID}} = A\frac{(E - E_{\text{CID}})^{2.5}}{E} \tag{4}$$

where A is a geometrical factor and E_{CID} is the threshold for CID via fusion. The branching ratio for fusion is $(\sigma_{\text{com}} - \sigma_{\text{CID}})/\sigma_{\text{com}}$ where σ_{com} is the cross section for complex (=compound species) formation, given by the capture cross section of Eq. (1). The fusion cross section is

$$\sigma_{\text{fus}} = \sigma_{\text{2-body}}\left((\sigma_{\text{com}} - \sigma_{\text{CID}})/\sigma_{\text{com}}\right) \tag{5}$$

where $\sigma_{\text{2-body}}$ incorporates the centrifugal cut-off, Eq. (2), and accounts for the competition with scattering close to the threshold, as discussed above. The rapid decline of the fusion cross section beyond E_{CID}, as given by Eq. (5), is in good agreement with the experimental results.[13] This is shown in Fig. 3 for $C_{70}^+ + C_{70}$ collsions (the calculations do not take the range of initial internal energies into account which would give a more "gentle" behaviour at the wings).

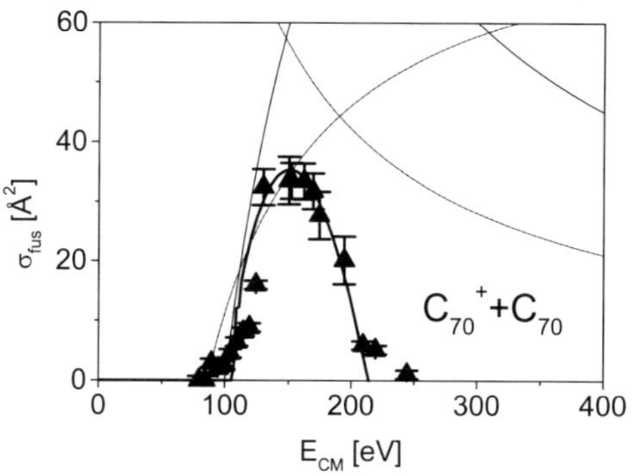

Fig. 3. Fusion cross section for $C_{70}^+ + C_{70}$ collisions. Triangles: experimental data. Thin full lines: line-of-centres model with centrifugal cut-off. Dotted lines: as full lines but scaled to fit the data by a factor of 0.45. Thick full line: fusion cross section considering scattering and collision induced dissociation as competing channels. Adapted from Ref. 13.

The model works equally well for $C_{60}^+ + C_{70}$ and $C_{70}^+ + C_{60}$ collisions but still overestimates the $C_{60}^+ + C_{60}$ cross section by a factor of 6 although the form of the cross sections is correctly reproduced. The reason for the exceptionally small cross section in $C_{60}^+ + C_{60}$ collisions is still not fully understood.

3. Fullerene Ionisation with Ultrashort Laser Pulses

The study of the photoionization of fullerenes has led to many interesting and occasionally unexpected results. One of the early observations made when exciting fullerenes with ns laser pulses was the delayed ionization of the neutral molecule on a microsecond time scale.[1] This can be interpreted in terms of statistical, thermionic emission of electrons from the vibrationally hot molecule and has been observed from many systems where the electron detachment energy is sufficiently far below the dissociation energy.[6] Recent experiments using femtosecond laser pulses have shown very different ionization behavior as the excitation time scale is varied from 25 fs to 5 ps.[17] For very short laser pulses (< 70 fs), the ionization is predominantly via direct multiphoton ionization showing above threshold ionization. For intermediate pulse duration the ionization is predominantly statistical but without strong vibrational heating of the molecule and, finally, for pulse duration beyond a few hundred femtoseconds the thermionic microsecond delayed ionization begins to appear.[18] This can be clearly seen in Fig. 4 where the mass spectra and photoelectron spectra are shown for three different laser pulse durations but with the same fluence (J/cm^2). The intermediate range has been explained in terms of ionization from the thermalized hot electron bath before the energy has been coupled to vibrational degrees of freedom.[19] Although this is a thermal emission (as clearly seen from the photoelectron spectra) it happens promptly on the time scale of the mass spectrometer (sub-ns). Similar behaviour has been observed for sodium cluster ions.[20] The statistical model used to explain the photoelectron spectra in the intermediate ionization regime can also reproduce the fluence dependence of the ion intensities and provide information on the density of states.[19] The model uses a time constant for coupling of the electronic to vibrational degrees of freedom of 240 fs. This was found by using the model to fit Penning ionization results[21,22] where the internal excitation energy of the fullerene is well defined. Recent pump-probe measurements confirm a time constant on the order of 200-300 fs.[23] An interesting result from the statistical model is that a log-log plot of the calculated ion yields as a function of fluence gives gradients that correspond to exactly the number of photons needed for direct multiphoton ionization.[19] One should therefore be careful when interpreting high log-log gradients in terms of direct multiphoton processes.

The electron emission is however not entirely statistical in nature. Superimposed on the thermal emission for electron kinetic energies in the range 0.5-1.55 eV is a rich structure of peaks.[24] This is shown in Fig. 5 for excitation with 800 nm pulses and a pulse duration of 1.5 ps. The structure is due to the excitation and single-photon ionization of excited Rydberg states within the duration of the laser pulse. Although the structure is best resolved for ps laser pulses it can still be observed for pulse durations of less than 100 fs. The intensity of the laser is too low in these experiments (ca. 10^{12} W/cm^2) for

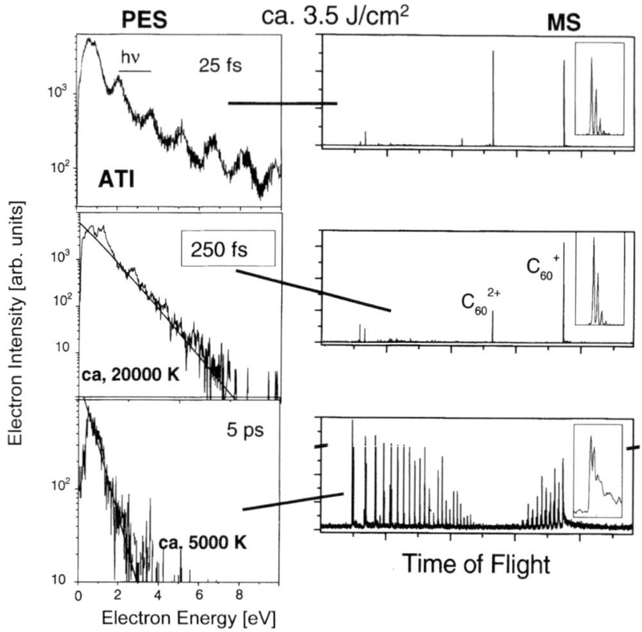

Fig. 4. Photoelectron spectra (left) and positive ion mass spectra (right) from the interaction of ultrashort pulses of different duration but the same fluence with gas phase C_{60}. Data from Ref. 17.

accessing the wide range of energies covering over 1eV by ponderomotive effects or field induced resonances. Instead, we interpret the formation of the Rydberg series in terms of the breakdown of the Born-Oppenheimer approximation. The states that we observe (n = 4-15) have principle quantum numbers that correspond to the range where the recurrence time is comparable to the vibrational frequency. In this range the Born-Oppenheimer coupling term $\langle \Psi_{av} | \hat{T}_N | \Psi_{bv'} \rangle$ becomes important and allows mixing between Rydberg vibronic states. This implies that the initial vibrational energy in the fullerene beam (almost 5 eV at a temperature of 500 °C) is necessary for exciting the states. To test this we have also recently carried out experiments with the fullerene molecules in the ground vibrational state. This is achieved by cooling the molecules in a gas-aggregation-type source by means of many collisions with helium gas at liquid nitrogen temperature.[25] The experiments did indeed show that the structure in the photoelectron spectra due to the Rydberg states disappeared when the molecular beam was initially in the vibrational ground state.[26] Other excitation scenarios may be possible and this is the subject of ongoing investigations.

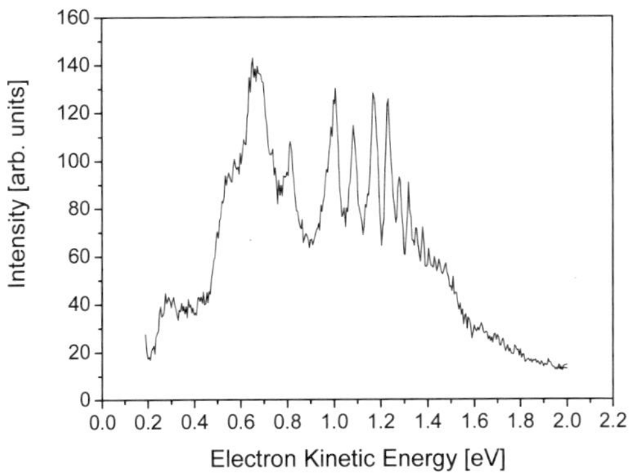

Fig. 5. Photoelectron spectrum from C_{60} ionised with 1.5 ps pulses at a wavelength of 800 nm. Data from Ref. 24.

Acknowledgements

Many people have contributed to the work presented here. I would particularly like to thank the PhD students involved in the fullerene-fullerene collision work, Rudolf Ehlich, Frank Rohmund and Alexei Glotov. The fs laser work has been carried out at the Max-Born-Institute in Berlin together with I.V. Hertel and the PhD students Kai Hoffmann and Mark Boyle. I would also like to thank Klavs Hansen and Martin Hedén for their contributions to the laser work. My two colleagues R. Schmidt (TU Dresden) and R.D. Levine (Hebrew Univ., Jerusalem) have provided essential theoretical support. Finally, financial support from the EU, DFG, Vetenskapsrådet and the Göran Gustafsson Foundation is gratefully acknowledged.

References

1. E. E. B. Campbell, G. Ulmer and I. V. Hertel, *Phys. Rev. Lett.* **67**, 1986 (1991).
2. R. Mitzner and E. E. B. Campbell, *J. Chem. Phys.* **103 No. 7** (1995).
3. R. E. Haufler, Y. Chai, L. P. F. Chibante, M. R. Fraelich, R. B. Weisman, R. F. Curl and R. E. Smallea, *J. Chem. Phys.* **95**, 2197 (1991).
4. E. E. B. Campbell and F. Rohmund, *Rep. on Progress in Physics* **63**, 1061 (2000).
5. E. E. B. Campbell, T. Raz and R. D. Levine, *Chem. Phys. Lett.* **253**, 261 (1996).
6. E. E. B. Campbell and R. D. Levine, *Annu. Rev. Phys. Chem.* **51**, 65 (2000).
7. R. Schmidt, G. Seifert and H. O. Lutz, *Phys. Lett.* A **158**, 231 (1991).
8. E. E. B. Campbell, V. Schyja, R. Ehlich and I. V. Hertel, *Phys. Rev. Lett.* **70 (3)**, 263 (1993).
9. E. E. B. Campbell, F. Rohmund and A. V. Glotov, *Il Nuovo Cimente* (1997).
10. F. Rohmund, A. V. Glotov, K. Hansen and E. E. B. Campbell, *J. Phys. B: At. Mol. Opt. Phys.* **29**, 5143 (1996).
11. F. Rohmund, E. E. B. Campbell, O. Knospe, G. Seifert and R. Schmidt, Phys. Rev. Lett. **76**, 3289 (1996).
12. A. V. Glotov, F. Rohmund and E. E. B. Campbell, *in Proc. of Int. Symp. on similarities and differences between atomic nuclei and microclusters: Unified developments for cluster sciences*, edited by Y. Abe and S.-M. Lee, 1997).
13. E. E. B. Campbell, A. Glotov, A. Lassesson and R. D. Levine, *C.R. Physique* **3**, 341 (2002).
14. A. Glotov, O. Knospe, R. Schmidt and E. E. B. Campbell, *Eur. J. Phys.* D **16**, 33 (2001).
15. D. L. Strout, R. L. Murry, C. Xu, W. C. Eckhoff, G. K. Odom and G. E. Scuseria, *Chem. Phys. Lett.* **214 (6)**, 576 (1993).
16. A. Glotov, *PhD Thesis*, Göteborg University (2000).
17. E. E. B. Campbell, K. Hansen, K. Hoffmann, G. Korn, M. Tchaplyguine, M. Wittmann and I. V. Hertel, *Phys. Rev. Lett.* **84**, 2128 (2000).
18. E. E. B. Campbell, K. Hoffmann and I. V. Hertel, *E. Phys. J.* D , in press (2001).
19. K. Hansen, K. Hoffmann and E. E. B. Campbell, *J. Chem. Phys.* **119**, 2513 (2003).
20. R. Schlipper, R. Kusche, B. von Issendorf and H. Haberland, *Appl. Phys.* A **72**, 255 (2001).
21. J. M. Weber, K. Hansen, M. W. Ruf and H. Hotop, *Chem. Phys.* **239**, 271 (1998).
22. B. Brunetti, P. Candori, R. Ferramosche, S. Falcinelli, F. Vecchiocattivi, A. Sassara and M. Chergui, *Chem. Phys. Lett.* **294**, 584 (1998).
23. M. Boyle et. al., unpublished results (2003).
24. M. Boyle, K. Hoffmann, C. P. Schulz, I. V. Hertel, R. D. Levine and E. E. B. Campbell, *Phys. Rev. Lett.* **87**, 273401 (2001).
25. K. Hansen, R. Mueller, P. Brockhaus, E. E. B. Campbell and I. V. Hertel, *Z. Phys.* D **42**, 153 (1997).
26. M. Boyle, C. P. Schulz, I. V. Hertel, M. Heden and E. E. B. Campbell, , in preparation (2003).

MULTIPLE IONIZATION AND FRAGMENTATION OF C_{60} IN COLLISIONS WITH FAST IONS

N. M. Kabachnik

Fakultät für Physik, Universität Bielefeld, D-33615 Bielefeld, Germany
and
Fritz-Haber-Institut der Max-Planck-Gesellshaft, D-14195 Berlin, Germany
and
Institute of Nuclear Physics, Moscow State University, 119992 Moscow, Russia
E-mail: nkabach@mail.ru

A. Reinköster

Fakultät für Physik, Universität Bielefeld, D-33615 Bielefeld, Germany
and
Fritz-Haber-Institut der Max-Planck-Gesellshaft, D-14195 Berlin, Germany

U. Werner and H. O. Lutz

Fakultät für Physik, Universität Bielefeld, D-33615 Bielefeld, Germany

A short overview of recent experimental and theoretical results on multiple ionization and fragmentation of C_{60} in collisions with singly and multiply charged ions is presented. Experimentally, the produced electrons and positively charged ions have been detected and mass analyzed using a time- and position-sensitive multi-particle detector. The projectile ions ranged from H^+ to Ar^{z+} (z = 1–3) with velocities of about 1 a.u.. Cross sections of multiple ionization, evaporation rates of C_2 fragments, asymmetric fission, and multi-fragmentation patterns are discussed.

1. Introduction

One of the perspective methods for investigation of the properties of fullerenes is a study of fullerene-ion collisions.[1] Due to a variety of projectiles and the possibility of varying their energies and charges one can choose different ranges of interaction from very soft glancing collisions, which do not strongly perturb the target-fullerene, up to the violent collisions which completely destroy it. On the other hand, the dynamics of ion-fullerene interaction is still not well understood. Not much is known about the mechanism of energy transfer from the projectile to the fullerene and even less is known about energy distribution and redistribution among various degrees of freedom in the fullerene. Therefore ion-fullerene collisions are of fundamental importance. Since experiments with fullerenes are comparatively simple they may be used as a model study for the interaction of ions with clusters and other nano-objects.

During the last decade the study of fullerene-ion collisions was one of the major directions of investigation in our group in Bielefeld university. The experimental efforts were directed to the study of multiple ionization and fragmentation of fullerenes in collisions with light ions from hydrogen and helium to neon and argon. We were interested in the collision velocity range below and about 1 a.u., that is in the region where both mechanisms of energy transfer, electronic and nuclear, are supposedly equally important. Analyzing the energy dependence of various channels of fullerene fragmentation we were able to determine the dominant mechanism. Our experimental results stimulated the theoretical investigations which shed some light on the collision dynamics.

2. Experiment

Details of the experimental technique have been published elsewhere.[2–4] Briefly, a collimated beam of H^+, D^+, He^+, Ne^+, and Ar^{z+} ions from a 350 keV accelerator interacts with the C_{60} vapor target which is provided by an oven heated to 550 °C. Electrons and slow ions generated in the collision are separated by a weak homogeneous electric field (330 V/cm) perpendicular to the primary beam. Electrons are detected in a channeltron at one side of the interaction region. Positive ions are accelerated towards a position- and time-sensitive multi-particle detector[2] at the other side. After passing a field-free time-of-flight region, the ions are post-accelerated by a voltage of 5 kV to increase the detector efficiency. The first collision-induced electron registered in the channeltron serves as a start pulse for the coincidence electronics, and time of flight and position on the multi-particle detector are recorded for each positive fragment. As a consequence of the applied cross-wire principle, the detector is capable of resolving particles which arrive "at the same time" on different wires. This "zero dead-time" feature is particularly useful for the study of multi-fragmentation processes, where several correlated fragments with equal masses occur. With this technique we observed up to seven charged fragments in coincidence.

3. Results and Discussion

In a typical TOF spectrum of fragments produced in ion-fullerene collisions three different groups of the peaks can be distinguished (see e.g. Refs. 5 and 6): (i) strong and narrow peaks that correspond to multiply ionized "parent" fullerenes C_{60}^{q+} with $q = 1$–5; each of them is followed by a sequence of (ii) smaller peaks corresponding to even-mass fullerene-type "daughter" ions C_{60-2m}^{q+} ($m = 1$–7) which are mainly formed by successive emission ("evaporation") of neutral dimers C_2; and finally (iii) the numerous broadened peaks corresponding to mostly singly charged small mass fragments C_n^+ ($n = 1$–19). The latter group corresponds to multi-fragmentation processes which are associated with a complete breakdown of the fullerene. It was found that the relative strength of each group depends dramatically on the projectile ion[5,6] and strongly varies with the collision velocity.[7,8] The multi-fragmentation part of the spectrum is very weak when protons are used as projectiles. He ions produce a broad spectrum of small singly charged fragments although parent and daughter fullerene ions are still present in the spectrum. For heavier Ar ions the multi-fragmentation becomes a dominant feature with the main contribution from the lowest mass fragment ions. Some features of the above discussed groups are considered below.

3.1. *Multiple ionization by proton impact*

In proton-C_{60} collisions the very small contribution of multi-fragmentation is negligible and thus the sum of the yields of C_{60}^{q+} and the fullerene-type ions with the same charge produced by a successive C_2 evaporation provide information on the multiple ionization cross section in the collision. Our results obtained in the collision energy range[9] 50–300 keV combined with the published results at higher energies[10] are shown in Fig. 1a. The multiple ionization cross section decreases with increasing energy and increasing degree of ionization. In the analysis of the multiple ionization cross section at smaller energies[11] it was found that its velocity dependence resembles that for the electronic stopping power. On the other hand, such behavior is typical for the multiple ionization of atoms and small molecules where the statistical energy deposition model (SED)[12,13] describes well the experimental data.[14,15] Within the SED model the process of multiple ionization

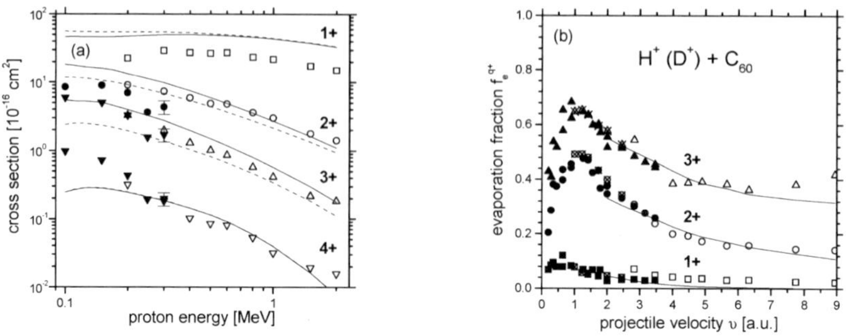

Fig. 1. C_{60} ionization cross section (a) and evaporation fraction (b) after H^+ impact. Shown are experimental (symbols) and theoretical (lines) results. Hollow symbols are results from Tsuchida *et al.*, crossed symbols indicate results with D^+ collisions. The theoretical results are shown with (—) and without (- - -) delayed processes.

is considered as proceeding in two stages. First, the projectile transfers energy to electronic degrees of freedom of the target system. In the second stage, the "heated" target emits a certain number of electrons with a probability proportional to the corresponding volume in the phase space.

We used this model to calculate the multiple ionization cross sections in proton-C_{60} collisions. In order to calculate the deposited energy we applied the so-called local electronic density approximation, well-known in the stopping-power theory. It is based on the ideas of Lindhard and Scharff[16] who suggested considering the electrons of the target as a free electron gas with a density equal to the actual electronic density in the target. An application of this model to proton-C_{60} collisions gives very realistic results for the stopping power of fullerenes[9] (see also Ref. 17). With this energy deposition we calculated the multiple ionization cross sections within the SED model. The results are shown in Fig. 1a as dashed curves. The solid curves show the same results, but corrected for the delayed electron emission (see below). In general the model results reproduce well the experiment. Some discrepancy exists for the singly charged fullerenes (see discussion in Ref. 9).

3.2. *Evaporation of C_2 fragments*

The analysis of the experimental results shows that the production of the "daughter" fullerene-type fragments, which are supposedly produced by a consecutive emission of neutral C_2 fragments from the parent ions, strongly depends on the projectile energy. Typical results for proton(deuteron)-fullerene collisions[9] are shown in Fig. 1b where the evaporation fraction determined as

$$f_e^{q+} = \frac{\sum_{m\geq 1} I(C_{60-2m}^{q+})}{I(C_{60}^{q+}) + \sum_{m\geq 1} I(C_{60-2m}^{q+})} \tag{1}$$

is shown as a function of projectile velocity. Here $I(C_n^{q+})$ is the measured intensity of the C_n^{q+} clusters. The striking feature of these results is a strong difference between the values of the evaporation fraction for different charges of the fullerene. We have suggested explaining this phenomenon by taking into account the delayed electron emission. Since the time of observation in a typical experiment is about 1 μs and the produced fragment-fullerene still has large inner energy (temperature) there is a possibility that it emits an electron additionally to the primary electrons emitted in the collision. This process, which is similar to thermo-emission from solids, can change the charge distribution of fragments. We have suggested a model for the time evolution of fragment-fullerenes which accounts for the C_2 evaporation and the delayed electron emission. Using realistic parameters characterizing both processes and solving the system of rate equations we were able to reproduce the experimental data (see Fig. 1b).

3.3. *Ion-induced asymmetric fission*

The multi-hit capability and position resolution of the detector allow a detailed study of C_{60} fission: $X^{z+} + C_{60} \rightarrow C_{60-2m}^{q+} + C_n^{+} + (C_x)$ with two charged fragments detected in coincidence. This strongly asymmetric fission was studied in experiments with hydrogen and helium ion projectiles.[18] The yield of fragment-fullerenes produced in fission processes increases with charge state q, which is intuitively clear since the fission barrier decreases with increasing charge state q. No fission was observed for doubly charged C_{60}^{2+}. The triply charged fragment-fullerene are formed at least an order of magnitude more often by C_2 evaporation than by fission (see Fig. 2a), whereas for C_{60}^{4+} fission is even slightly more probable than the loss of C_2 (see Fig. 2b). Consequently, the height of the fission barrier for C_{60}^{4+} ions should be similar to the activation energy needed for C_2 evaporation (about

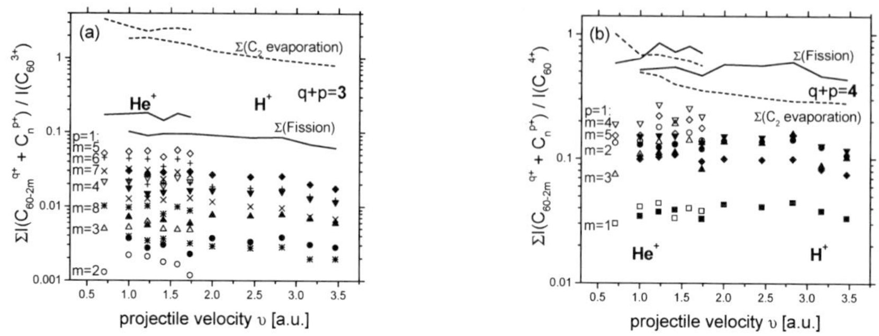

Fig. 2. The intensity of fission processes for parent charge states of 3 (a) and 4 (b) as function of the H^+/He^+ velocity. The solid lines correspond to the sum of the intensities of all fission processes, the dashed lines to the evaporation intensity with the same C_{60} parent charge state.

10 eV). We have found no evidence for successive fission (triple coincidences) as well as for emission of doubly charged small fragments.

3.4. *Size of fragments in multi-fragmentation*

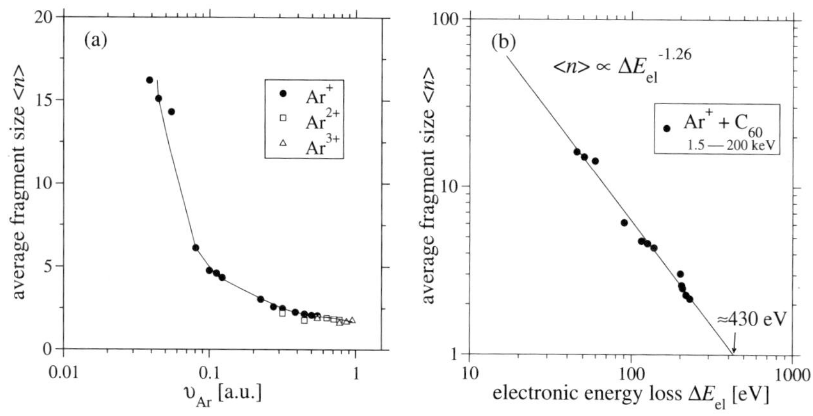

Fig. 3. The average fragment size after C_{60} multi-fragmentation as function of the projectile velocity (a) and the electronic energy loss (b).

Multi-fragmentation, i.e. a complete breakdown of the fullerene cage, is a dominant process in collisions with heavier ions. We have studied this process in Ar^{z+} + C_{60} collisions.[19] The measured fragment mass-spectrum shows a very strong dependence on the velocity of the projectile. In general, larger C_n^+ fragments are more often produced in collisions with slow ions. On the contrary, in collisions with fast Ar ions nearly exclusively very small C_n^+ fragments ($n < 5$) are observed. An important observation concerns the size distribution of the small singly charged fragments. For collisions with fast argon ions, the C_n^+ yield decreases with increasing cluster size n according to a power law $n^{-\lambda}$ where λ shows values of ≈ 2 (see Ref. 5). A power low distribution has stimulated a discussion of possible critical behavior in the dynamical phase of cluster fragmentation.[20, 21]

In order to discuss the dependence of a fragment size on the projectile velocity more quantitatively it is convenient to define the average fragment size $< n >$ in multi-fragmentation:

$$< n >= \sum_n n \cdot I(C_n^+) \Big/ \sum_n I(C_n^+) \,. \qquad (2)$$

The measured velocity dependence of $< n >$ in collisions with Ar ions is shown in Fig. 3a. The average size of C_n^+ fragments decreases with increasing projectile velocity. This effect is very pronounced at low energies, whereas at higher energies (100 – 300 keV$\cdot z$) the average fragment size becomes nearly constant and reaches $< n > \approx 2$, which is still larger than the asymptotic value $< n >=1$. A similar saturation effect was observed for the C_{60} multi-fragmentation by $He^{+,2+}$ impact.[7, 8] This effect is difficult to explain if one supposes that the multi-fragmentation is determined by the electronic energy loss in the collision. According to conventional calculations in the velocity region studied here, a higher electronic energy loss is expected for faster ions. If all the electronic energy is transferred to the vibrational degrees of freedom it should therefore result in additional multi-fragmentation for faster ions, which is not observed in our study. In contrast, recent more elaborate non-adiabatic quantum molecular dynamic (NA-QMD) calculations[22] predict a saturation effect for the electronic excitation at projectile velocities of about $v = 0.2 - 0.45$ a.u. and almost vanishing nuclear excitation in this velocity region. The observed saturation of the average fragment size $< n >$ is explained fairly well by the NA-QMD calculations. In Fig. 3b we show the average fragment size $< n >$ as a function of the electronic energy loss $\Delta E_{\rm el}$, taken from Ref.[22] for penetrating collisions. It turns out that this relation follows a power law and that $< n >$ is approximately proportional to the inverse value of the electronic energy loss. Interestingly, the extrapolation to the fragment size $<n>$=1, which means the total destruction of the C_{60} molecule, yields an energy loss value which is close to the total binding energy of the C_{60} molecule.

4. Conclusions

The multi-coincidence time-of-flight method which was developed in our laboratory is a powerful tool for studying the multiple ionization and fragmentation of fullerenes in collisions with ions. It allows separation and study of various individual processes such as asymmetric fission, C_2 fragment evaporation, multi-fragmentation etc. We have studied the dependence of the characteristics of these processes on the projectile type and velocity. We have found that in many cases the dynamics of the process is determined by the energy transfer from the projectile to the fullerene. Further investigations into the mechanism of energy transfer and energy dissipation in the fullerene are desirable.

Acknowledgments

We are grateful to B. Siegmann for the help in the measurements and to B. Huber and R. Schmidt for numerous illuminating discussions. This work was supported by the Deutsche Forschungsgemeinschaft(DFG) and has been carried out within the framework of the European network LEIF (HPRI-CT-1999-40012). NMK and AR gratefully acknowledge the hospitality and the financial support of the Fritz-Haber Institute of the Max-Planck Society.

References

1. E.E.B. Campbell and F. Rohmund, *Rep. Prog. Phys.* **63**, 1061 (2000).
2. J. Becker, K. Beckord, U. Werner, and H.O. Lutz, *Nucl. Instrum. Methods* A **337**, 409 (1994).
3. U. Werner, J. Becker, K. Beckord, and H.O. Lutz, *Nucl. Instrum. Methods* B **124**, 298 (1997).
4. U. Werner, K. Beckord, J. Becker, and H.O. Lutz, *Phys. Rev. Lett.* **74**, 1962 (1995).
5. U. Werner, V.N. Kondratyev, and H.O. Lutz, *Nuovo Cimento* **110**, 1215 (1997).
6. A. Reinköster, U. Werner, and H.O. Lutz, *Europhys. Lett.* **43**, 653 (1998).
7. T. Schlathölter, O. Hadjar, R. Hoekstra, and R. Morgenstern, *Phys. Rev. Lett.* **82**, 73 (1999).
8. T. Schlathölter, O. Hadjar, J. Manske, R. Hoekstra, and R. Morgenstern, *Int. J. Mass Spectrom.* **192**, 245 (1999).

9. A. Reinköster, U. Werner, N.M. Kabachnik, and H.O. Lutz, *Phys. Rev.* A **64**, 023201 (2001).
10. H. Tsuchida, A. Itoh, Y. Nakai, K. Miyabe, and N. Imanishi, *J. Phys.* B **31**, 5383 (1998).
11. J. Opitz, H. Lebius, S. Tomita, B.A. Huber, P. Moretto-Capelle, D. Bordenave-Montesquieu, A. Bordenave-Montesquieu, A. Reinköster, U. Werner, H.O. Lutz, A. Niehaus, H.T. Schmidt, and H. Cederquist, *Phys. Rev.* A **62**, 022705 (2000).
12. A. Russek and J. Meli, *Physica* **46**, 222 (1970).
13. C.L. Cocke, *Phys. Rev.* A **20**, 749 (1979).
14. N.M. Kabachnik, V.N. Kondratyev, Z. Roller-Lutz, and H.O. Lutz, *Phys. Rev.* A **56**, 2848 (1997).
15. N.M. Kabachnik, V.N. Kondratyev, Z. Roller-Lutz, and H.O. Lutz, *Phys. Rev.* A **57**, 990 (1998).
16. J. Lindhard and M. Scharff, *Mat. Fys. Medd. Dan. Vid. Selsk.* **27**, no. 15 (1953).
17. P. Moretto-Capelle, D. Bordenave-Montesquieu, A. Rentenier and A. Bordenave-Montesquieu, *J. Phys.* B **34**, L611 (2001).
18. A. Reinköster, B. Siegmann, U. Werner, and H.O. Lutz, accepted by *Radiation Physics and Chemistry.*
19. A. Reinköster, B. Siegmann, U. Werner, B.A. Huber, and H.O. Lutz, *J. Phys.* B **35**, 4989 (2002).
20. V.N. Kondratyev and H.O. Lutz, *Z. Phys.* D **40**, 210 (1997).
21. V.N. Kondratyev, H.O. Lutz, and S. Ayik, *J. Chem. Phys.* **106**, 7766 (1997).
22. T. Kunert and R. Schmidt, *Phys. Rev. Lett.* **86**, 5258 (2001).

ELECTRON AND ION IMPACT ON FULLERENE IONS

D. Hathiramani, H. Bräuning, R. Trassl and E. Salzborn
Institut für Kernphysik, Universität Giessen, D-35392 Giessen, Germany
E-mail: Erhard.Salzborn@strz.uni-giessen.de

P. Scheier
Institut für Ionenphysik, Universität Innsbruck, A-6020 Innsbruck, Austria

A.A. Narits and L.P. Presnyakov
P.N. Lebedev Physical Institute, 117924 Moscow, Russia

Employing the crossed-beams technique, we have studied the interaction of fullerene ions both with electrons and He^{2+}-ions. Electron-impact ionization cross sections for C_{60}^{q+} (q=1,2,3) have been measured at electron energies up to 1000 eV. Unusual features in shape and charge state dependence have been found, which are not observed for atomic ions. The evaporative loss of neutral C_2 fragments in collisions with electrons indicates the presence of two different mechanisms. In a first-ever ion-ion crossed-beams experiment involving fullerene ions, absolute cross sections for charge transfer in the collision systems C_{60}^{+}+He^{2+} and C_{60}^{+}+C_{60}^{2+} have been obtained.

1. Introduction

Fullerenes in general and fullerene ions in particular offer new opportunities to study the interaction of extended structures. The low ionization energy, even for multiply charged fullerene ions, as compared to atomic ions should lead to effects normally expected only in ion-solid surface interactions. However, most collision experiments performed have studied either the interaction of a charged particle with a neutral fullerene or of a fullerene ion with a neutral particle.

In this paper we report on our studies of the interaction of positively

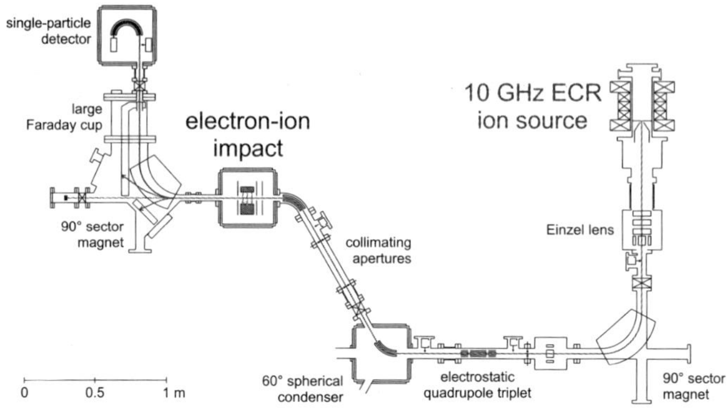

Fig. 1. Schematic setup of the electron–ion collision experiment

charged fullerene ions with electrons. We show cross sections for electron-impact ionization and propose a new model to explain the observed behaviour for electron-impact induced C_2-fragmentation. We also present our very first data of collisions between singly charged fullerene ions and He^{2+}-ions as well as C_{60}^{2+} ions.

2. Electron–Ion Collisions

Using the electron-ion crossed-beams setup[1] (see Fig. 1) we have measured absolute cross sections for electron-impact ionization of fullerene ions. A commercially available mixture of fullerenes, mainly containing C_{60} and C_{70}, was evaporated in an electrically heated oven. The neutral vapor was introduced into the plasma of a 10 GHz Electron Cyclotron Resonance (ECR) ion source,[2] where the positively and negatively charged fullerene ions were produced. After mass and charge analysis the ion beam was collimated to 2×2 mm^2 and crossed with an intense electron beam[3] with electron currents up to 450 mA. The energy of the electrons can be varied between 10 and 1000 eV. After the electron–ion interaction, the product ions were separated from the incident ion beam by a 90°-magnet and detected in a single-particle detector. The current of the parent ion beam was measured simultaneously in a Faraday cup. Employing the dynamic crossed-beams technique introduced by Müller *et al.*,[4] where the electron beam is moved through the ion beam with simultaneous registration of the primary and the product ion intensities, absolute cross sections were obtained. The total experimental uncertainties are typically ±10% at the

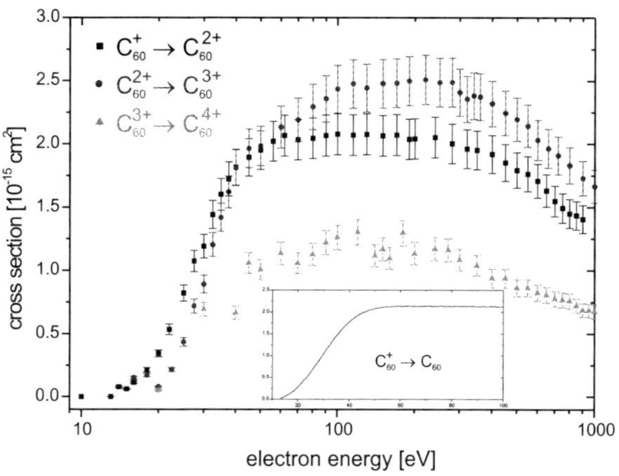

Fig. 2. Single–ionization cross sections for C_{60}^{q+} ($q = 1, 2, 3$). The error bars indicate the total experimental uncertainties.

maximum of the cross sections with systematic errors of about 8% and statistical errors at the 95% confidence level.

Single–ionization cross section measurements for fullerene ions C_{60}^{q+} have been performed for charge states $q = 1, 2, 3$. The results are shown in Fig. 2. These cross sections show a behaviour that has not been observed with any atomic or light molecular ion. After the typical increase above the ionization threshold, the cross sections remain constant over a large energy range. Furthermore, the cross sections do not show any structure which could be attributed to indirect ionization processes (see inset in Fig. 2, which was measured using a high-resolution energy-scan method[5]). It should be noted that the cross section for single-ionization of C_{60}^{2+} exceeds that of C_{60}^{+} at energies above about 60 eV. The reason for this unexpected behaviour is not yet understood.

When energetic electrons collide with a C_{60}^{q+} molecule, fragmentation may occur in addition to ionization. This mainly happens through the evaporation of neutral C_2 dimers. The measured cross section for the process $C_{60}^{+}+e^{-}\rightarrow C_{58}^{+} + \ldots$ is shown in Fig. 3. Cross sections for the same kind of reaction but with doubly and triply charged ions show the same shape differing mainly in the absolute value. The energy dependence suggests the presence of two different mechanisms. The high-energy part of the cross section for the reaction $C_{60}^{q+}\rightarrow C_{58}^{q+}$ can be fitted by the Lotz formula[6] which is able to describe direct ionization only. However, since the charge state of

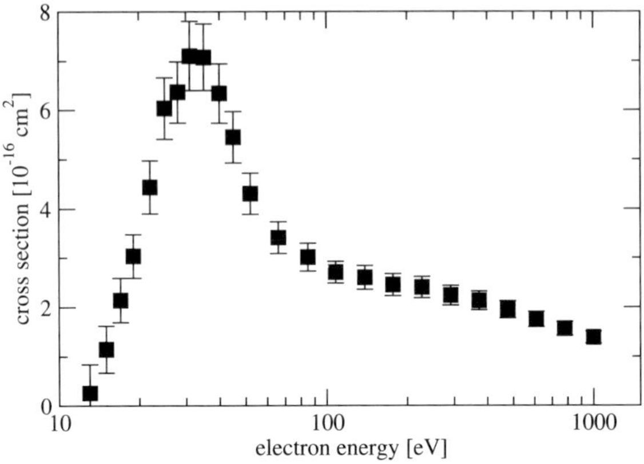

Fig. 3. Absolute cross section for C_2 fragmentation from C_{60}^+ by electron impact: $C_{60}^+ + e^- \rightarrow C_{58}^+ + \ldots$ The error bars indicate the total experimental uncertainties.

the reaction product is the same as that of the parent ion, the behaviour can be explained as an "unsuccessful ionization". This means that the outgoing electron has an energy which is low enough to be recaptured by the fullerene. The shape of the cross section at low energies is very similar to the giant plasmon resonance observed by Hertel.[7] It therefore suggests that at low electron impact energies, the fragmentation process induced by electron impact is predominantly caused by a plasmon excitation.

In a collision between an energetic electron and a negative fullerene-ion, ionization also plays an important role besides the detachement. Figure 4 shows the measured cross sections for double, triple and four-fold ionization of C_{60}^-, C_{70}^- and C_{84}^-. The dependence of the cross section on the size of the fullerene increases with the degree of ionization. For double ionization the maximum cross section for C_{84}^- is only about 20% larger than for C_{60}^-. For four-fold ionization however, the ratio is already about a factor of two. The increase of the cross section with the number of carbon atoms suggests a dependence on the geometrical size of the fullerene cage.[8]

3. Ion-Ion Collisions

In this paper we present first data on the non-resonant charge transfer for the collision system $C_{60}^+ + He^{2+} \rightarrow C_{60}^{2+} + He^+$ and on charge transfer and fragmentation in the resonant collision system $C_{60}^+ + C_{60}^{2+} \rightarrow C_{60}^{2+} + C_{60}^+$.

The Giessen ion-ion crossed-beams setup[9,10] as shown in Fig. 5 provides the ideal tool to enable studies of a wide range of ion-ion collision systems.

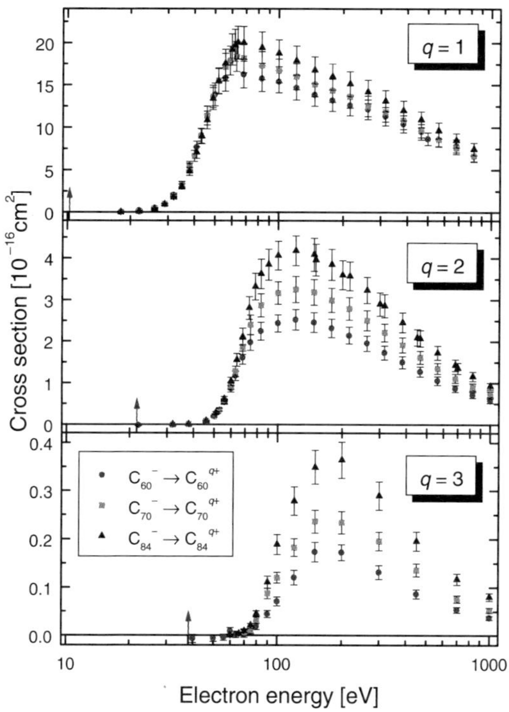

Fig. 4. Absolute cross sections for electron impact ionization of negatively charged fullerene ions. The arrows represent the expected threshold for electron impact ionization.

Two Electron Cyclotron Resonance (ECR) ion sources provide a C_{60}^+-ion beam with 3 keV energy and a $^3He^{2+}$ or C_{60}^{2+}-ion beam, respectively, with energies in the range of 80 keV up to 180 keV. The ion beams are crossed at an angle of 17.5° in an ultra-high vacuum collision chamber immediately after passing two electrostatic beam cleaners. The reaction products are separated from the parent beams by electrostatic analyzers. While the parent beams are collected in Faraday cups for normalization, the product ions are detected in a Channeltron detector for the low energy beam line and a position sensitive Micro Channel Plates (MCP) detector for the high energy beam line, respectively.

To minimize the background contribution from collisions with the residual gas in the interaction region, a pressure of better than 10^{-10} mbar is maintained. In spite of this low pressure, the background contribution on the detectors is still a factor 10^3 to 10^4 (depending on the ion) higher than

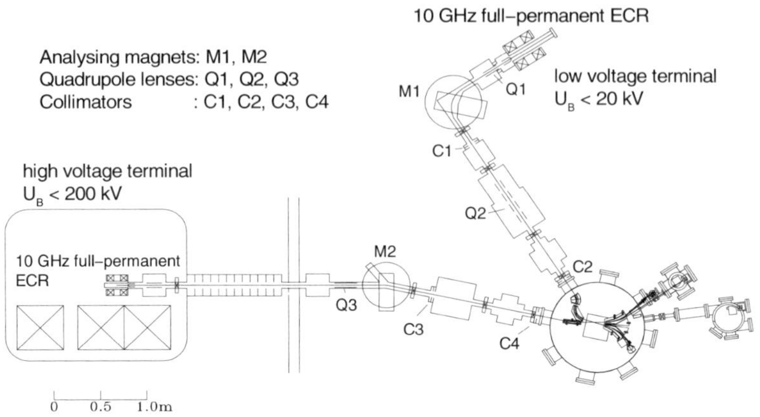

Fig. 5. Schematic setup of the ion–ion collision experiment.

the reaction rate of 0.4 true events per second for typical fullerene beam currents of about 1 to 2 nA in the resonant reaction $C_{60}^{+}+C_{60}^{2+} \rightarrow C_{60}^{2+}+C_{60}^{+}$. While the beam current for the $^{3}He^{2+}$ beam is typically more than an order of magnitude larger, the non-resonant character of the respective reaction also results in similar reaction rates. However, the background contribution can be strongly suppressed using the time coincidence technique, which provides a powerful tool for signal recovery and gives a clear signal of the electron transfer reaction.

The use of a position sensitive detector for the C_{60}^{+} reaction products of the reaction $C_{60}^{+} + C_{60}^{2+} \rightarrow C_{60}^{2+} + C_{60}^{+}$ enables the measurement of the deflection angle of these ions in the electrostatic field of the analyzers, which mainly contains information about possible fragmentation and the scattering angle. The latter, however, is very small.

For the nonresonant reaction $C_{60}^{+}+He^{2+} \rightarrow C_{60}^{2+}+He^{+}$ the cross section decreases with increasing collision energy from $(2.52 \pm 0.15) \times 10^{-15}$ cm^2 at 38.5 keV center-of-mass energy to $(3.82 \pm 0.25) \times 10^{-16}$ cm^2 at 196.2 keV. The error given is the statistical error of the 67% confidence interval only. The systematical error is mostly determined by the uncertainties in the detection efficiencies and amount to 25%. In contrast, for the resonant reaction the cross section does not change within the error bars over the energy range from 25 to 70 keV. Surprisingly, the average cross section in this region of $(9.2 \pm 1.0) \times 10^{-15}$ cm^2 virtually coincides with the data by Rohmund *et al.*[11] for the reaction $C_{60}^{+} + C_{60} \rightarrow C_{60} + C_{60}^{+}$ at much lower collision energies between 0.5 and 1.9 keV. It also compares favourably with the measured cross section of $(6.3 \pm 1.8) \times 10^{-15}$ cm^2 for double electron transfer in C_{70}^{3+} + C_{60} collisions at 45.2 keV center-of-mass energy, which is dominated by the cross section for single-electron transfer in the C_{70}^{2+} + C_{60}^{+} intermediate state.[12] A detailed comparison of the experimental data with theoretical calculations based on over- and under-barrier transitions of hole particles through the potential barrier of two conducting spheres is given in Ref. 13.

Acknowledgments

This work was supported by BMBF/GSI and DFG. L.P. and A.N. gratefully acknowledge support from the Russian Foundation of Basic Research (Grant 02-02-16274).

References

1. K. Tinschert, A. Müller, G. Hofmann, K. Huber, R. Becker, D.C. Gregory, and E. Salzborn. *J. Phys. B: At. Mol. Opt. Phys.* **22**, 532 (1989).
2. M. Liehr, M. Schlapp, R. Trassl, G. Hofmann, M. Stenke, R. Völpel, and E. Salzborn. *Nucl. Instr. Meth.* B **79**, 697 (1993).
3. R. Becker, A. Müller, C. Achenbach, K. Tinschert, and E. Salzborn. *Nucl. Instr. Meth.* B **9**, 385 (1985).
4. A. Müller, K. Tinschert, C. Achenbach, and E. Salzborn. *Nucl. Instr. Meth.* B **10/11**, 204 (1985).
5. A. Müller, K. Tinschert, G. Hofmann, E. Salzborn, and G.H. Dunn. *Phys. Rev. Lett.* **61**, 70 (1988).
6. W. Lotz. *Zeitschrift für Physik* **206**, 205 (1967).
7. I.V. Hertel, H. Steger, J. de Vries, B. Weisser, B. Kamke C. Menzel, and W. Kamkeet. *Phys. Rev. Lett.* **68**, 784 (1992).
8. D. Hathiramani, K. Aichele, W. Arnold, K. Huber, E. Salzborn, and P. Scheier. *Phys. Rev. Lett.* **85**, 3604 (2000).
9. S. Meuser, F. Melchert, S. Krüdener, A. Pfeiffer, K.v. Diemar, and E. Salzborn. *Rev. Sci. Instrum.* **67**, 2752 (1996).
10. F. Melchert and E. Salzborn. In Y. Itikawa, editor, *The Physics of Electronic and Atomic Collisions*, pages 478–494. American Insitute of Physics (2000).
11. F. Rohmund and E.E. Campbell. *J. Phys. B: At. Mol. Opt. Phys.* **30**, 5293 (1997).
12. H. Shen, P. Hvelplund, D. Mathur, A. Barany, H. Cederquist, N. Selberg, and D.C. Lorents. *Phys. Rev.* A **52**, 3847 (1995).
13. H. Bräuning, R. Trassl, A. Diehl, A. Theiß, E. Salzborn, A.A. Narits, and L.P. Presnyakov. 'Resonant electron transfer in collisions between two fullerene ions'. Submitted to *Phys. Rev. Lett.*

Clusters on a Surface

COLLISIONS OF ELECTRONS AND PHOTONS WITH SUPPORTED ATOMS, SUPPORTED CLUSTERS AND SOLIDS: CHANGES IN ELECTRONIC PROPERTIES

V. M. Mikoushkin, S.Yu. Nikonov, V.V. Shnitov and Yu.S. Gordeev

A.F.Ioffe Physical-Technical Institute Russian Academy of Science, 194021 St.Petersburg, Polytechnicheskaya str.26, Russia
E-mail: V.Mikoushkin@mail.ioffe.ru

Photoionization, electron scattering and excitation of silver and copper have been studied by a set of electron spectroscopy methods. Regularities of changes of electronic properties of matter in the conversion "supported atom – supported cluster of increasing size – solids" have been revealed. Evidence has been obtained that the shell electron structure of flat supported clusters is formed according to the same magic numbers as that of the spherical free clusters. The "cluster – solid" transition has been studied for fullerenes C_{60} when these carbon clusters condense into crystal and when condensed fullerenes decay under electron or synchrotron radiation and are transformed into amorphous carbon. Photo- and electron- induced fission and transformation of fullerenes into amorphous carbon is accompanied by a decrease in electron binding energies and radical increase in the relaxation energy as well as in the case of metal clusters.

1. Introduction

A lot of attention has been paid in the last decade to clusters. Most of these studies have been devoted to free clusters in the gas phase.[1,2] The technique used enables investigation of clusters with definite numbers of atoms, but by a restricted number of tools because of low density of target. We will discuss here another group of experiments with clusters supported by the surfaces. Systems of supported clusters are characterized by spread in sizes, but they give an opportunity to use a wide range of electron spectroscopy methods developed in surface science. Interesting results related to properties of supported clusters were obtained by these methods before our first publications.[3-6] For instance, shifts of Auger and photoelectron lines in clusters with respect to those in solids were revealed in Ref. 6. Several mechanisms of the shifts were discussed in literature. Among them were changes in electronic structure ($\Delta\varepsilon$) and relaxation energy (ΔR), charging of a cluster due to ionization in collision process, and the cluster-substrate interaction. However, no reliable model of size-related shifts was suggested before our research. The problem was the lack of information about detailed dependences of the studied parameters on cluster sizes and the lack of complementary methods used. In addition, there were no studies of the electronic properties of supported clusters with a known number of atoms. The goal of this research was to study the transformation of the electronic properties of matter in the conversion "supported atom - supported cluster of increasing size – solids" by using photoionization, electron scattering, and electron excitation processes. Transition "cluster – solids" have been analyzed for fullerenes C_{60} when these carbon clusters condense into van der Waals crystal and when condensed fullerenes decay under electron or synchrotron radiation and are transformed into amorphous carbon.

2. Methods

In this research, electron spectroscopy methods, a combination of X-ray photoemission spectroscopy (XPS) and X-ray Auger electron spectroscopy (AES) were used first.[3-6] This combination allows the gathering of direct information about the relaxation energy (ΔR) of cluster electrons in electron emission process. Electron impact AES and Electron Energy Loss Spectroscopy (EELS) were also available. X-ray Al K_α line ($h\nu$ = 1486.6 eV) and primary electrons with energies E = 1.5 -3.0 keV were used. Ag- and Cu- clusters were formed *in situ* on two types of substrates: pyrolytic graphite and Si(100, 111) single crystals. The graphite surface was cleaned by cleavage in dry air with subsequent annealing in high vacuum. The surface of the Si-crystals was cleaned by heating in high vacuum or by ion etching with subsequent annealing. The average cluster size was increased by sequential uniform deposition of metal over the entire area of the substrate. The deposited doses n_s of metals were measured by monitoring the intensities of the adsorbate and substrate lines. Three spectrometers were used: LHS-11 (Leybold-AG) photoelectron spectrometer, spectrometer with rotable energy analyzer[11] and multichannel in angle electron spectrometer.[12] Russian–German Synchrotron Radiation beamline and facilities were available both for irradiation of supported fullerenes and for their characterization by means of high-resolution photoelectron spectroscopy.

3. Transformation of Electron Structure of Silver in Conversion "Supported Atom - Supported Cluster of Increasing Size – Solid"

Shifts of Ag $3d_{3/2,5/2}$ photoelectron lines (ΔE_{k}) and Ag $M_{4,5}VV$ Auger lines (ΔE_A) have been observed when the average size of silver clusters increases in the course of sequential uniform deposition of silver atoms on the surfaces of pyrolytic graphite and silicon.[3,5] In both cases, these lines monotonously approach the values of the bulk silver metal with increasing deposited dose (n_s). Figure 1 shows the core level binding energy $E_B(n_s) = h\nu - E_k(n_s) - e\varphi$ (a) and Auger electron kinetic energy $E_A(n_s)$ (b), which were obtained in one of the photoemission experiments, versus the cluster size. The Auger energy dependence proved to be very similar to those obtained in the electron impact experiments. This evidences the reproducibility of the technique used for the cluster system fabrication. The first important conclusion on the nature of the line shifts follows from Fig.1: the variation in cluster charging in an emission event due to an increase in cluster radius (capacity) cannot be an essential reason for the observed shifts. Indeed, Fig.1 shows a rapid decrease in binding energy E_B and invariability of Auger electron energy E_A at the early stage of cluster formation. This behavior excludes charging, because the charge after Auger decay is twice as large as that after potoemission, and this would lead to inverse dependences.

The important problem of controlling the average number (N) of atoms in supported clusters has been solved in the research.[7] Deposited doze n_s of metal atoms has been connected with the average number N of atoms by the relation $N = n_s/n_o$, where n_o is the concentration of clusters. Concentration n_o equals the concentration of the surface defects, around which clusters are formed. It can be estimated as the end of the "shoulder" in the left part of the binding energy curve (Fig. 1a, $n_s < n_o = 5*10^{13}$ atoms/cm^2, ~1/20 of Ag monolayer), which corresponds to the stage of fixing metal atoms on the surface defects, when only separate atoms are on the surface. Scale N is presented at the top of Fig. 1.

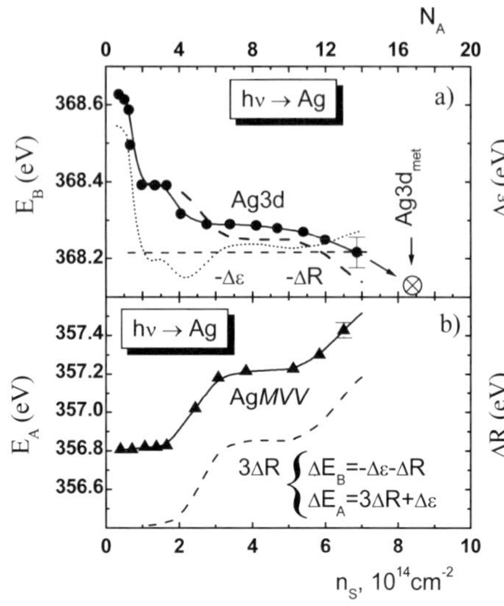

Fig. 1. (a) Ag 3d core-electron binding energy (points) and Ag 3d core-electron level energy (Δε, dashed line) versus cluster-size. (b) Ag MVV Auger electron kinetic energy (points) and relaxation energy (ΔR, dashed line) versus cluster size.

The similar character of the size-related line shifts for two different substrates (graphite and Si) evidences weak influence of these surfaces on the cluster electronic properties.[8] Variations in the weak cluster-surface interaction cannot be a reason for the discussed energy shifts. The remaining reasons for the line shifts must be associated only with the changes in the energy structure (Δε) and relaxation energy (ΔR). To distinguish between these two reasons, the relationships for binding and Auger electron energy shifts, which are known as Auger-parameter formalism, were used: $\Delta E_B = - \Delta\varepsilon - \Delta R$ and $\Delta E_A = \Delta\varepsilon + 3\Delta R$. Thus, size dependences have been obtained directly for the energy level ε (Fig. 1a, dashed curve) and for the relaxation energy R (Fig. 1b, dashed curve). The conclusion has been made that the dominant role in the observed binding energy shift belongs to a sharp increase in the core level energy ε at the early stage of cluster formation. The comparison of size dependences of the binding and Auger electron energies with that of the relaxation energy R shows that these curves are almost completely described by the same dependences of the relaxation energy at the next stages of cluster growth. The increasing role of the relaxation energy in the cluster size increase is physically caused by enlargement of the density of states near the Fermi level and of the degree of collectivization of valence electrons, which leads to an increase in the extra-atomic relaxation energy.

Two possible mechanisms for the transformation of the electron energy structure of a supported cluster due size increase can be suggested: (1) promotion of quasimolecular orbitals in the process of merging atoms into cluster and cluster growth, and (2) charge exchange between the cluster and supporting surface. The first mechanism is characterized by different energy shifts of different levels, and the second mechanism by equal shifts according to the model of metal sphere. Measurements of the energy shifts of Ag 3s, 3p, 3d and 4d levels in conversion “supported atom-supported cluster” showed different energy shifts for different levels. It means that the radical increase in the energy levels in merging separate atoms into dimers is caused by the promotion of quasimolecular orbitals. For Ag3d level, they are 5gσ and 6hσ orbitals.

It was shown experimentally that the electron structure of a supported cluster consists of shells.[9,10] This structure manifests itself as stepwise peculiarities in the size dependences of the relaxation energy (or Auger energy) measured in a wide range of cluster sizes. The positions of the observed steps coincide with the "magic" numbers for spherical free clusters: 2, 8, 20, and 46. Some of the steps can be seen in Fig. 1. The conclusion has been made that the shell electron structure of supported flat clusters is formed according to the same regularities as that of the spherical free ones. This result and the effect of nonmonotonous variation in spin-orbital splitting of core levels[10] also confirms the reliability of the suggested method of estimation of cluster size. The sharpness of steps is the evidence of low dispersion of cluster sizes in our experiment.

4. Size-dependent Local Plasmons in Ag-clusters

Collective oscillations of valence electrons (plasmons) are a very important characteristic of the condensed matter. Excitation of local plasmons in a single Ag-cluster has been observed in the research.[5,7] To clarify the nature of the local plasmon excitation, transformation of plasmon spectra with variation in cluster size has been investigated in two series of experiments, when clusters are irradiated by electrons and by X-rays. Figure 2 shows size dependences of the probability (per atom) of excitation of the local Ag-cluster plasmon by electrons (a) and by 3d-photoelectrons (b). Black and open circles correspond to the silicon and graphite surfaces. By comparison, the relaxation energy curve (ΔR) is also plotted in panel (a) of Fig. 2. Figure 2 illustrates a very important result: a local plasmon is excited in clusters consisting, on average, of only a few atoms (N = 2 ÷ 4). This means that plasma oscillations can be excited in systems consisting of only a few electrons.

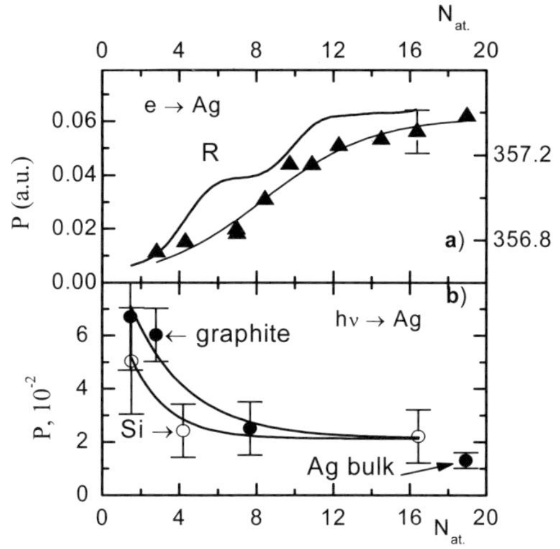

Fig. 2. The probability (per atom) of excitation of local Ag-cluster plasmons by electrons (a) and by 3d-photoelectrons (b) versus cluster size. Black and open points correspond to the silicon and graphite surfaces. Relaxation energy curve (ΔR) is plotted in panel (a).

As Fig. 2 shows, the probability of the local plasmon excitation is hardly dependent on the cluster size in the case of particles containing more than 20 atoms, both for reflected external electrons and for photoelectrons. For smaller clusters, the dependences are directly opposite: the probability of excitation of a local plasmon by an external electron decreases with cluster size decrease, while that for a photoelectron increases sharply. The first dependence is predictable and correlates well with the size dependence of the relaxation energy: the lower relaxation energy is caused by a lower degree of electron collectivization in the cluster, which is also responsible for the lower probability of excitation of collective oscillations. The anomalous increase in the probability of excitation of a plasmon by a photoelectron has been qualitatively explained by the polarization of the outer 5s shell of the silver atom.[7]

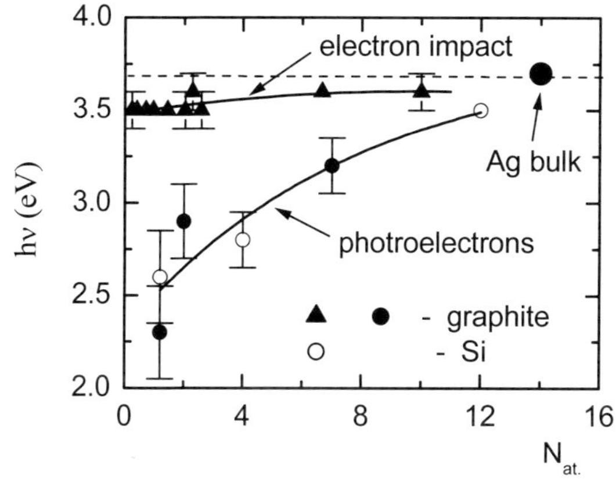

Fig. 3. Average energy of the local Ag-plasmon excited by electrons (a) and by 3d-photoelectrons (b) versus cluster size. Black and open circles correspond to the silicon and graphite surfaces.

The existence of low-energy modes in local collective oscillations has been revealed experimentally as a "red" shift of the average plasmon energy. Figure 3 demonstrates a decrease in the average energy of the local Ag-plasmon excited both by electrons (a) and by 3d-photoelectrons (b) with decreasing cluster size. Black and open circles correspond to the silicon and graphite surfaces respectively. Notice that the data obtained for two different substrates are very close to each other. This fact confirms the conclusion that the clusters are quasi-isolated and that the plasmons observed characterize the electron system of a cluster only, without contribution of the substrate's electrons. In the photoemission experiment, the "red" shift observed in tiny clusters reaches about one third of the plasmon energy of large clusters. The "red" shift in the case of photoelectron excitation is considerably larger than that in the case of external electron excitation. This fact leads to the conclusion that surface plasmon modes with low momenta are excited more efficiently by photoelectrons and in small clusters.

5. Transformation of Electronic Properties of Carbon in Conversion "Cluster C_{60} – Solid"

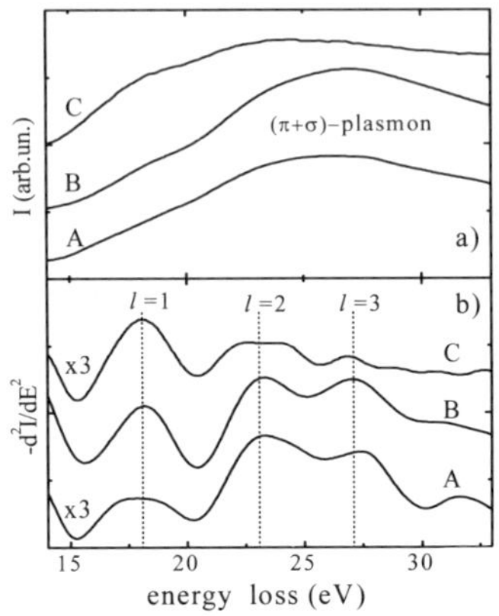

Fig. 4. Ordinary (a) and doubly differentiated (b) EEL spectra of fullerite C_{60} obtained in three different experiments (A, B and C) with different samples and primary electron energies.

Conversion "cluster C_{60} - solid" has been considered when fullerenes C_{60} condense into Van-der-Waals crystal, and when condensed fullerenes decay under electron or synchrotron radiation (SR) into amorphous carbon. Electron scattering and plasma oscillation properties of free fullerene has been studied in the electron- C_{60} collision experiment.[11] In particular, diffraction of an electron on a free fullerene has been revealed. The diffraction manifests itself in the oscillations of the differential cross sections of elastically and inelastically scattered electrons. Plasma oscillations in free fullerenes have also been observed. The $(\pi+\sigma)$-plasmon spectra have been measured and decomposed into the elementary components corresponding to the dipole (l=1), quadrupole (l=2), and octupole (l=3) modes. The experimental energies of the plasmon modes proved to be 18.1 ± 0.3 eV, 24.3 ± 0.3 eV, 29.0 ± 0.3 eV for l = 1, 2, 3, respectively. These modes are described by the well known formula for metal cluster in the Mie theory.

Clusters of C_{60} condensed into solid (fullerite) have been studied in an electron reflection experiment by measuring electron energy loss spectra (EELS). Figure 4 shows ordinary (a) and doubly differentiated (b) EEL spectra of fullerite C_{60} in the region of $(\pi+\sigma)$-plasmon, which were obtained in three different experiments (A, B and C) with different samples and primary electron energies. Differentiation makes the spectra peculiarities more evident. Instead of one broad (π+σ)-plasmon peak, these spectra exhibit at least three peaks at the energies of 18.0 ± 0.2 eV, 23.1 ± 0.3 eV, and 26.9 ± 0.3 eV. The energy positions of these peaks proved to be very close to those of the dipole, quadrupole, and octupole modes of a separate C_{60} molecule. The conclusion has been made that the plasma oscillations in the solid C_{60} condensate exist in separate molecules without essential contribution of neighboring ones. Thus, no radical changes in the plasmon structure have been observed in the conversion "separate cluster C_{60} - solid".

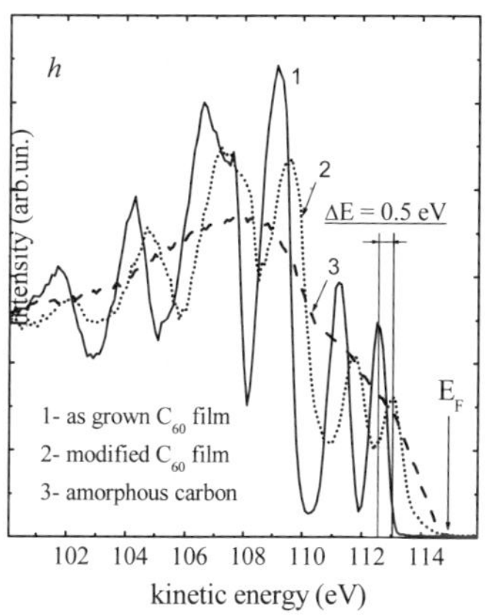

Fig. 5. Photoelectron spectra of valence electrons of pristine fullerenes (1), of modified fullerenes (2) and of amorphous carbon (3).

Radical changes in electronic properties of condensed fullerenes have been observed under electron- and synchrotron radiation (SR). The first stage of the radiation-induced modification of fullerenes is mainly associated with the creation of chemical bonds between carbon atoms of neighboring molecules. Finally, after a long-term electron irradiation, fullerenes are fragmented and converted into a phase of amorphous carbon, which can be created by heavy ion bombardment of graphite or fullerenes.[12-14] Similar processes have been observed for the first time for SR. Figure 5 illustrates the modification of photoelectron spectra of valence electrons of pristine fullerenes under a nonmonochromatic SR provided with "zero diffraction order" mode. The spectrum of pristine fullerenes (1) shows prominent molecular peaks (HOMO) and very low continuous background. Comparing this spectrum with that of irradiated fullerenes (2), one can see that the intensities of molecular peaks diminish under SR irradiation, and continuous background increases towards the spectrum of amorphous carbon (3). In addition, "blue" shift of the molecular peaks is observed. The revealed transformation of the spectra of valence electrons evidences the transformation of electron and atomic structure of C_{60} clusters towards that of amorphous carbon. Photo- and electron-induced fission of fullerenes is accompanied by a decrease in electron binding energies and a radical increase in the relaxation energy as in the case of metal clusters discussed above.

Acknowledgements

The research was partly supported by INTAS grant no 2136, and NWO grant no 047.009.012.

References

1. For review, see W.A.de Heer, *Rev.Mod.Phys.*, **65,** 611 (1993).
2. *Physics and Chemistry of Finite Systems: From Clusters to Crystal*, ed. by P.Jena, S. N. Khanna, B.K. Rao, Kluwer, Boston, London (1992).
3. Yu.S. Gordeev, M.V. Gomoyunova, A.K. Grigor'ev, V.M. Mikoushkin, I.I. Pronin, S.E. Sysoev, V.V. Shnitov, and N.S. Farajev, *Phys.Solid State* **36**, 1298 (1994).
4. Yu.S. Gordeev, M.V. Gomoyunova, V.M. Mikoushkin, I.I. Pronin, and S.E. Sysoev, *Tech.Phys.Lett.* **20,** 142 (1994).
5. Yu.S. Gordeev, M.V. Gomoyunova, V.M. Mikoushkin, et al., *Phys.Solid State* **36,** 973 (1994).
6. M.V. Gomoyunova, Yu.S. Gordeev, V.M. Mikoushkin, I.I. Pronin, S.E. Sysoev, V.V. Shnitov, *Phys. Low-Dim. Struct.* **4/5,** 11 (1996).
7. V.M. Mikoushkin, S.E. Sysoev, *Phys.Solid State* **38**, 305 (1996).
8. Yu.S. Gordeev, M.V. Gomoyunova, V.M. Mikoushkin, I.I. Pronin, S.E. Sysoev, *Int. Symp. "Nanostructures"*, St.Petersburg, Russia, p.300 (1996).
9. V.M. Mikoushkin, *Contributed Papers of the 6th EPS Conference on Atomic and Molecular Physics,* Inv. 41, Siena, Italy (1998).
10. V.M. Mikoushkin, *Technical Physsics* **44**, 1077 (1999).
11. L.G. Gerchikov, P.V. Efimov, V.M. Mikoushkin, A.V. Solov'yov, *Rhys.Rev.Lett.* **81,** 2707 (1998).
12. S.Yu. Gordeev, V.M. Mikoushkin, V.V. Shnitov, *Phys. Sol. State* **42**, 381 (2000).
13. Yu.S. Gordeev, V.M. Mikoushkin and V.V. Shnitov, *Molecul. Mater.* **13**, 1 (2000).
14. V.V. Shnitov, V.M. Mikoushkin, S.Yu. Gordeev, *Phys. Sol. State* **44**, 428 (2002).

DEPOSITION AND STM OBSERVATION OF SIZE-SELECTED PLATINUM CLUSTERS ON SILICON(111)-7X7 SURFACE

H. Yasumatsu

Cluster Research Laboratory, Toyota Technological Institute: In East Tokyo Laboratory, Genesis Research Institute, Inc., Futamata, Ichikawa, Chiba 272-0001, Japan
E-mail: yasumatsu@clusterlab.jp

T. Hayakawa, S. Koizumi

East Tokyo Laboratory, Genesis Research Institute, Inc., Futamata, Ichikawa, Chiba 272-0001, Japan

T. Kondow

Cluster Research Laboratory, Toyota Technological Institute: In East Tokyo Laboratory, Genesis Research Institute, Inc., Futamata, Ichikawa, Chiba 272-0001, Japan

We investigated individual size-selected platinum clusters deposited on a silicon(111)-7x7 surface by means of a scanning-tunneling microscope (STM). The sample was prepared by collision of size-selected platinum cluster cations on the surface at a collision energy of 0.5-3 eV per platinum atom. It was found that the clusters were deposited firmly on the surface without dissociation nor aggregation. The real diameters of the clusters were estimated by considering the radius of the STM tip used for the measurement and the tunneling distance between the cluster and the tip.

1. Introduction

A cluster deposited on a solid surface has novel physical and chemical properties, which characteristically change with its size, geometric and electronic structures, and are modified by their interaction with the surface.[1] In order to elucidate the specific properties, it is necessary to observe individual clusters with known sizes. Many studies on STM observation of clusters on solid surfaces have been reported,[2] but in most cases, the clusters are not size-selected. Obviously the size-selection is the most important in such studies.

A size-selected cluster can be deposited on a solid surface by collision of a size-selected cluster ion onto a solid surface, if the size does not change in the course of the cluster collision. Harbich and his coworkers[3] and Schneider and his coworkers[4] have studied deposition of size-selected metal clusters on rare-gas layers prepared over a metal surface and a bare metal surface maintained at 25-120 K by the collision of size-selected metal cluster ions at a collision energy 0.1-30 eV per cluster. They have shown that the clusters are deposited stably on the surfaces at 25-120 K, but when the surface temperature is increased, the clusters move on the surfaces into aggregates. We have demonstrated that size-selected platinum and silver clusters are deposited on a graphite surface maintained at 300 K with forming a firm bonding to the surface by choosing an appropriate collision energy and a solid surface.[5] Kanayama and his coworkers have also shown that hydrogenated silicon clusters are deposited firmly on a silicon surface.[6]

In this paper, we reported STM observation of a size-selected platinum cluster deposited on a silicon(111)-7x7 surface. We found that the deposited cluster is stuck on the surface firmly and gives stable and reproducible results of the STM observation.

2. Experimental

Figure 1 shows a schematic view of the apparatus used in the present experiments, which consists of a cluster-ion source, ion-guides with ion-focusing lenses, a quadrupole mass-filter, a cluster-deposition equipment, an STM, devices for treatment of a sample and an STM tip, and a load-lock system.[5] Cluster ions are produced by a magnetron sputtering source (Kurt J. Lester TRS-2CV MM) mounted in an aggregation chamber filled with helium gas at a pressure of ~30 Pa,[7] and are admitted into the first octopole ion guide, in which the cluster ions are 'cooled' by collision with helium gas (~1 Pa). The cluster ions are transported further through another two sets of octopole ion-guides and passed through the quadrupole mass-filter (ABB Extrel MEXM-9000) for the size selection. Cluster ions after the size-selection are transported further through the fourth octopole ion-guide, and allowed to collide onto a silicon(111)-7x7 surface at a collision energy, E_{col}, of 0.5-3 eV per platinum atom under the ambient pressure of $6x10^{-8}$ Pa. The dose of the cluster deposition is monitored by measuring a cluster-ion current arriving at the surface by using a pico-ammeter (Keithley 487). The surface deposited with a given dose is transferred to an STM head (Unisoku) located at the bottom of the STM chamber maintained at a pressure less than $5x10^{-9}$ Pa. The STM head is isolated vibrationally by suspending the STM head with three coil springs made of beryllium-cupper alloy. The STM chamber itself is placed on a pneumatic table, whose characteristic frequency is as low as 2 Hz. In addition, the vibration of the STM head is damped by three ribbons made of Cu-Mn-Fe-Mg alloy (M2052) placed in parallel to the suspension coil spring. The STM head is designed to be cooled down to 6.8 K, although the measurements in the present work were performed at the sample temperatures of 160 and 300 K. The silicon surface was prepared by repeating cycles of flushing (1500 K) and annealing (1150-900 K) under an ambient pressure less than $1x10^{-7}$ Pa. The surface thus prepared was found to give atomically-resolved STM images and LEED patterns (Vacuum Generators RVL-900/03) of the 7x7 structure.

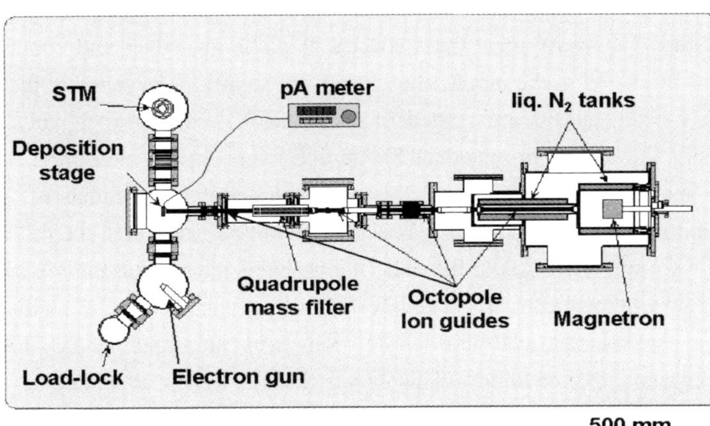

Fig. 1. Schematic view of the apparatus used for cluster deposition and STM observation.

3. Results

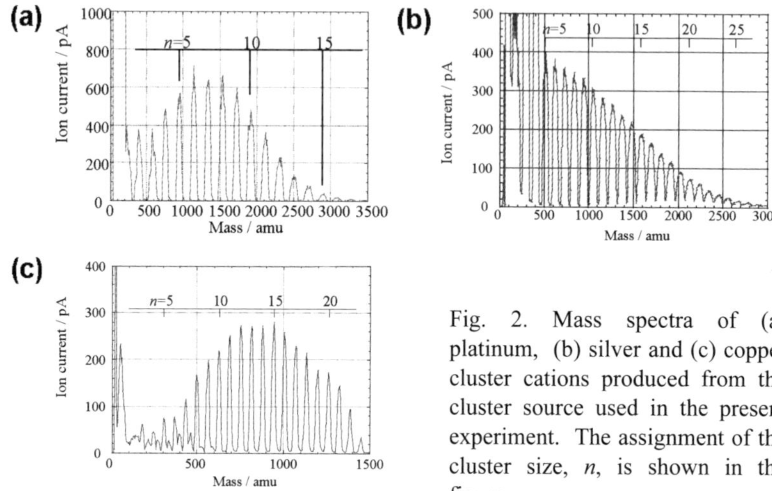

Fig. 2. Mass spectra of (a) platinum, (b) silver and (c) copper cluster cations produced from the cluster source used in the present experiment. The assignment of the cluster size, n, is shown in the figure.

Figure 2 shows typical mass spectra of platinum, silver and copper cluster cations produced in the magnetron cluster-ion source. The size and the currents of the cluster cations are ranged in 1-50 and 0.1-1 nA, respectively, with an intensity fluctuation of less than 5% in one hour. In the measurements of the mass spectra shown in Fig. 2, the mass resolution was traded off with the transmittance of the mass filter. With a higher mass resolution, the full widths at half maximum (FWHM) of 0.5 and 10 amu were achieved at the masses of 200 and 2000 amu, respectively. The FWHM of the translational-energy distribution of the size-selected platinum cluster ions was measured to be 5 eV by a retarding-potential method. This broad distribution is attributed to off-axial motions of the cluster ions by the rf field in the quadrupole mass filter. A profile of the cluster-ion beam is found to be expressed by a two-dimensional Gaussian function having an FWHM of 5 mm.

Figure 3 shows STM images of $(Pt)_5$, $(Pt)_{10}$ and $(Pt)_{35}$ deposited on the silicon surface at E_{col}=3, 0.75 and 0.5 eV, respectively. These images were obtained with a tungsten tip at a constant tunneling-current (I) of 500 pA and V_s=-2.00 V (topographic image) and at the sample temperatures of 160 K for $(Pt)_{35}$ and 300 K for $(Pt)_5$ and $(Pt)_{10}$. The STM images of the silicon surface are observed with an atomic resolution, but the clusters in sight do not exhibit any atomic structure. The STM images of the surface deposited with the clusters show that they are not distributed in particular locations such as dislocation lines. The STM images of the clusters in sight did not change as a result of scanning the tip many times over the clusters for at least a week.

Figure 4 shows a correlation between the apparent diameter (D) and height (H) obtained from the STM image of $(Pt)_5$ deposited on the silicon surface at E_{col}=3 eV (see also panel (a) of Fig. 3). They have an almost monotonic correlation. These profiles did not change with V_s.

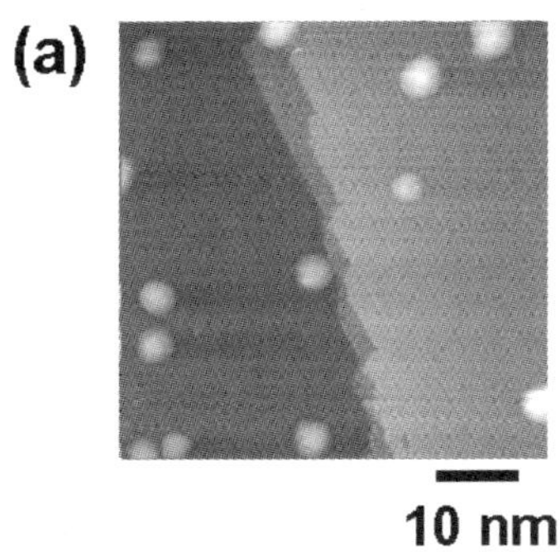

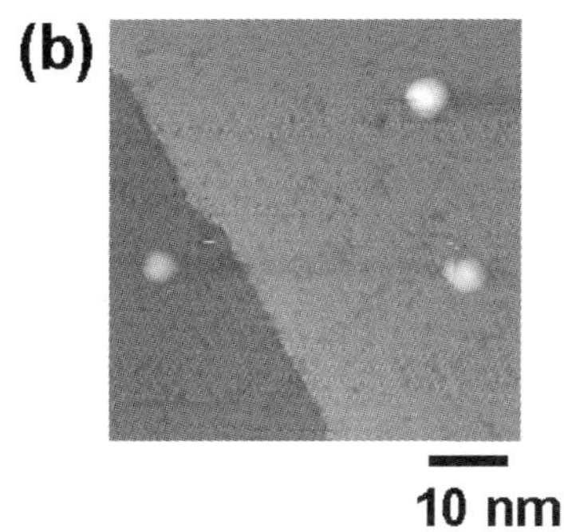

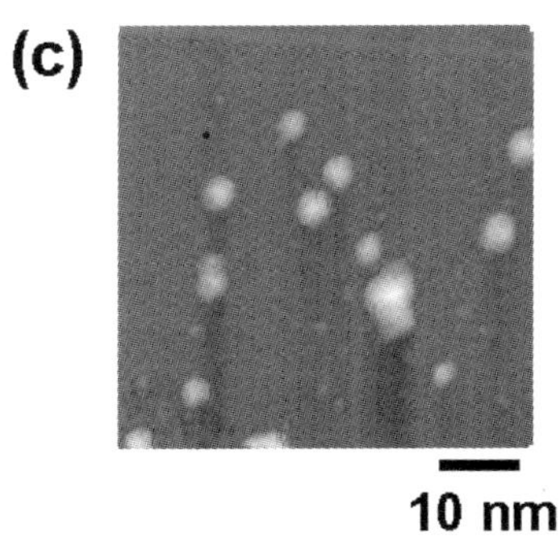

Fig. 3. STM images of (a) $(Pt)_5$, (b) $(Pt)_{10}$ and (c) $(Pt)_{35}$ deposited on a silicon(111)-7x7 surface at the collision energy of 3, 0.75 and 0.5 eV per platinum atom. The images were obtained at a pressure less than $5x10^{-9}$ Pa by using a tungsten tip at a constant current mode (topographic images), where the sample temperatures were 160 K for $(Pt)_{35}$ and 300 K for $(Pt)_5$ and $(Pt)_{10}$. The scales are shown in the figure.

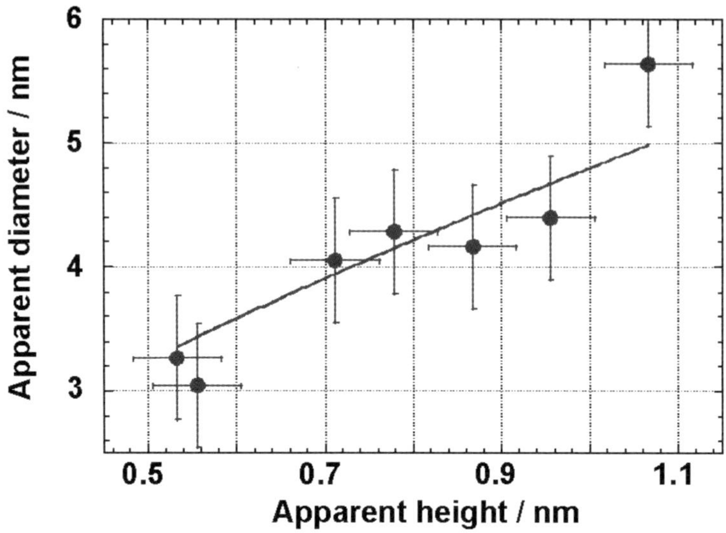

Fig. 4. Correlation between the apparent diameter and the height of $(Pt)_5$ deposited on the silicon(111) surface at a collision energy of 3 eV per platinum atom.

4. Discussion

4.1. *Cluster deposition by cluster impact*

Platinum clusters observed by the STM look firmly stuck on the silicon surface, because their STM images did not change by repeated scans of the tip over the clusters. Furthermore, the clusters are neither dissociated nor aggregated on the surface during and after deposition, because the observed number density of the clusters on the surface is almost equal to that calculated from the current of the cluster ions arriving at the surface, the area and the time for the deposition. It is likely that the platinum clusters are bonded chemically with surface atoms by the cluster impact, because the effective temperature of the clusters rises as high as ~10000 K by the impact.[8,9]

4.2. *Shape of cluster observed by STM*

As shown in Fig. 3, the STM images exhibit an atomically resolved profile corresponding to the clean silicon(111)-7x7 surface and structureless profiles attributable to the clusters. The structureless profiles for the clusters are explained in such a manner that the clusters look larger than the silicon surface by the STM tip, because a larger volume of the tip is involved in observing an object protruding more highly from the surface, and hence the effective diameter of the tip becomes larger when the STM images of the clusters are measured. Therefore, the apparent diameters (D~5 nm) of $(Pt)_5$ are observed to be larger than the actual diameter (~0.5 nm) as shown in Fig. 4. In addition to the larger effective diameter of the tip, a finite tunneling distance between the cluster and the tip also gives the larger apparent diameters observed as described below. The influences of the finite tip diameter and the finite tunnelling distance are estimated by assuming that (1) both of the cluster and the tip are hemispherical with radii of r and R, respectively, while the surface is flat and (2) the nearest-neighbor distances between the tip and the cluster and between the tip and the surface are maintained at d and h, respectively, during every STM scan.[10] For $d>h$, D and H are calculated as

$$D=2\sqrt{H\{H+2(h+R)\}} \tag{1}$$

and

$$H=r+d-h. \tag{2}$$

Equations (1) and (2) show that both the apparent diameter and the apparent height are observed to be larger than real ones. As shown by the curves in Fig. 4, the experimental results are reproduced well by the calculation according to Eq. (1) with the best-fit parameter of $h+R$=2.2 nm when a single $(Pt)_5$ is observed. If h is substituted by a typical value (1 nm), the present tip has effective diameter of 1.2 nm. By employing an ideal value of R (=0) together with h=1 nm, the apparent diameter of a platinum cluster having H=0.5 nm is calculated to be 2.2 nm, which is the experimental limit for the determination of the cluster diameter by the present STM.

As shown in Fig. 4, the correlation between the apparent diameter and the apparent height is deviated slightly from the theoretical one (solid line in Fig. 4). This arises probably because the clusters are deviated differently from a hemispherical shape due to a broad distribution in the collision energies (FWHM=5 eV).

Acknowledgements

We are grateful to Professor H. Haberland and his co-workers (Universität Freiburg) for providing us with useful information on the cluster-ion source. We are indebted to Dr. W. S. Yun (Korea Research Institute of Standard and Science) for his contribution in the development of the STM machine. Thanks are also due to Dr. Hashizume and Dr. Fujimori (Advanced Research Laboratory, Hitachi Ltd.) for their illuminating advice on the improvement of the STM machine. We would like to acknowledge Chuo Spring Co., Ltd. for providing us the special coil springs employed in the STM machine. Dr. Kawahara (Seishin Co.) and Professor Mio (the University of Tokyo) are also acknowledged for providing us detailed information on the special damper to suppress vibration of the STM head. This research was supported by the Special Cluster Research Project of Genesis Research Institute, Inc.

References

1. In K.-H. Meiwes-Broer (ed.) *Metal Clusters at Surfaces*, Ed. by K.-H. Meiwes-Broer, Springer-Verlag, Heidelberg, pp. 158-173 (2000).
2. A. Piednir, E Perrot, S. Granjeand, A. Humbert, C. Chapon and C. R. Henry, *Surf. Sci.* **391**, 19 (1997).
3. R. Schaub, H. Jödicke, F. Brunet, R. Monot, J. Buttet and W. Harbich, *Phys. Rev. Lett.*, **86**, 3590 (2001).
4. S. Messerli, S. Schintke, K. Morgenstern, A. Sanchez, U. Heiz and W.-D. Schneider, *Surf. Sci.*, **465**, 331 (2000).
5. H. Yasumatsu, T. Hayawaka, S. Koizumi and T. Kondow, *Trans. Mat. Res. Jpn.*, **27**, 209 (2002).
6. L. Bolotov, N. Uchida and T. Kanayama, *Eur. Phys. J.* D **16**, 271 (2001).
7. H. Haberland, M. Mall, M. Moseler, Y. Qiang, T. Reiners and Y. Thurner, *J. Vac. Sci. Technol.* A, **12**, 2925 (1994).
8. C. L. Cleveland and U. Landman, *Science*, **257**, 355 (1992).
9. H. Yasumatsu, S. Koizumi, A. Terasaki and T. Kondow, *J. Phys. Chem.* A, **102**, 9581 (1998).
10. D. Klyachko and D. M. Chen, *Surf. Sci.*, **446**, 98 (2000).

SILICON CLUSTER LATTICE SYSTEM (CLS) FORMED ON AN AMORPHOUS CARBON SURFACE BY SUPERSONIC CLUSTER BEAM IRRADIATION

M. Muto, M. Oki, and Y. Iwata

Cluster Advanced Nanoprocesses CRT, National Institute of Advanced Industrial Science and Technology, Tsukuba Central 2, Tsukuba 305-8568, Japan.
E-mail: m-muto@aist.go.jp

H. Yamauchi, H. Matsuhata and S. Okayama

Nanoelectronics Research Institute, National Institute of Advanced Industrial Science and Technology, Tsukuba Central 2, Tsukuba 305-8568, Japan.

Y. Ikuhara and T. Iwamoto

Institute of Engineering Innovation, School of Engineering, The University of Tokyo, 2-11-16 Yayoi, Bunkyo, Tokyo 113-8656, Japan.

T. Sawada

Department of Advanced Material Sciences, Graduate School of Frontier Sciences, The University of Tokyo, 7-3-1 Hongo, Bunkyo, Tokyo 113-8656, Japan.

We demonstrate the silicon cluster lattice system (CLS) assembling on an amorphous carbon film by using a well-defined supersonic neutral silicon cluster beam generated in a newly exploited spatiotemporal confined cluster source (SCCS). The SCCS has successfully achieved a characteristically narrow size distribution of a silicon cluster beam with $\Delta N/N$<5%. Neutral Si_N clusters collided with the substrate target at an energy of 1.1eV/Si atom, which is several times smaller than the cohesive energy at 4.0-4.2eV of silicon clusters. Images of the silicon clusters of 2-3nm in diameter taken by the high angle annular dark field (HAADF) method of a scanning transmission electron microscope (STEM) show a good contrast with a carbon background of the substrate. A unit cluster monolayer (CML) covers fully the substrate surface with silicon clusters of 2.3nm in diameter in the density of $1.0 x 10^{13} cm^{-2}$. The CLS formed on the substrate surface was developed spontaneously as the deposition density increases. The silicon clusters made pairs randomly oriented at 0.20CML. In progress of the deposition density around 0.67CML, silicon clusters had a tendency to form partially hexagonal closed packed structures. In further progress up to 1.0CML, the silicon clusters lined up spontaneously to form a tetragonal structure with a lattice constant of 4.0nm. The ultra high resolution transmission electron microscope (UHRTEM) imaged clearly a crystallographic atomic lattice structure of the individual silicon clusters. It was proved that the third power of the inter-cluster distance R_C and the product of the HAADF diffraction intensities $I_i{*}I_j$ of all isolated cluster pairs had a linear relation. The correlation shows a possible conclusion that the induced dipole potential between silicon clusters coming close to each other dominantly makes them ordered spontaneously to form a silicon CLS.

1. Introduction

In the gas-phase cluster assembly process by using cluster beam deposition (CBD), clusters play a role as elementary nanoblocks to build up ordered nanostructures. Such "nanoarchitecture" in blocks of clusters uniquely enables one to construct three-dimensionally ordered cluster lattice systems (3DCLS) efficiently. The new conceptual nanostructured materials exhibit possibilities to particularly extend electronic, optical, and mechanical functions.

Molecular beam epitaxy (MBE) has currently attracted one's attention as the most practicable method for nanostructure synthesis on a well-defined crystal surface.[1] Atoms deposited onto the crystal surface aggregate to form clusters in alignment with the atomic rows of the crystal lattice. Such clusters growing on the substrate surface, called "hut clusters", are extensively arranged in plane order to form nanopatterns. In the CBD process, on the contrary, clusters themselves form a stable crystallographic structure before landing on the substrate surface. The deposited clusters migrate onto the substrate surface at a significantly high speed, comparable with the atomic diffusion process.[2] Juxtaposing with neighbour clusters, the deposited clusters are extensively arranged in both plane and three-dimensional order to construct a new conceptual 3DCLS.

In either case of landing in CBD or growing in MBE, adsorbed clusters on a crystal surface induce periodic charge density modulations resulting in a variety of superstructures of the crystal lattice.[3] The strain field arising in the growth of stacked multilayers of heteroepitaxial islands forms a tensile deformed area in the substrate surface lattice, which possibly acts as a preferential nucleation sites resulting in self-patterning of a two-dimensional array of ordered clusters.[4] On the other hand, it can be considered that a dipole is induced in the individual adsorbed clusters that undergo charge transfer from the substrate surface. The induced dipole causes the long range interaction between clusters in a potential of P_iP_j/r_{ij}^3, where P_i is the induced dipole moment and r_{ij} is the distance between the interacting clusters. When the magnitude of the induced dipole moment, which is proportional to the number of cluster constituent atoms (cluster size), becomes uniform, equivalent periodic potentials work on the clusters so that long range ordering of clusters is formed on the surface. The interaction between adsorbed clusters possibly plays an important role as an alternative mechanism in spontaneous ordering to construct the 3DCLS.

In this paper, we demonstrate the silicon CLS assembling by using a well-defined supersonic neutral silicon cluster beam. We investigate the interaction between the adsorbed silicon clusters on an amorphous carbon film surface, which is inert for most molecules and guarantees that the clusters are not aligned with the atomic rows of the substrate surface. Then the amorphous carbon film is sutable for examination of the cluster-cluster interaction without any large influence of the substrate surface.

2. Well-defined Supersonic Silicon Cluster Beam

Well-defined stable clusters are very important in controlling the order of the blocks of clusters. Spatiotemporal confined cluster source (SCCS) has been newly exploited to achieve cluster growth under the well-defined thermo-dynamical conditions controlled by a laser-induced shock wave. We describe a principle of the SCCS and brief results of the cluster beam qualities. The details appear elsewhere.[5]

The cluster growth cell of the SCCS has an ellipsoidal shape with a couple of focal points separated 20mm apart. On the center axis of the ellipsoidal cell, a small hole of 0.7mm in diameter is opened for introduction of the laser beam and extraction of the cluster beam. A silicon target was set so that the ablation point on the target surface is located at the focal point. We carefully poured a helium gas continuously into the cluster growth cell with no turbulent flow. The typical helium gas pressure in the cell was P_{He}=130Pa. A pulsed Nd:YAG laser (wave length 532nm, pulse duration 10ns, energy 50-300mJ/pulse) was introduced along the center axis, and irradiates the target surface at a beam spot of 0.8mm in diameter. A laser-ablated dense vapor ejects from the target surface symmetrically with the center axis. The vapor ejection induces a shock wave in the symmetric helium gas flow after forming a Knudsen layer. The shock front propagates symmetrically to the wall of the ellipsoidal cell, and then locally locally on the alternate focal point after reflection on the wall. The vapor front travelling initially at a slower velocity of 8.0×10^3m/s than the shock front is stopped at the focal point by colliding with the reflected shock front. At the contact boundary of both gas phases, a mixed dense gas layer of vapor and helium was formed in a narrow space of 0.4mm in thickness, where clusters can grow. A sufficient number of atomic collision times in the locally-confined mixed gas layer possibly complete uniform thermo-dynamical conditions for well-defined cluster growth. Generated silicon clusters in the SCCS were extracted into vacuum following the helium gas flow through a skimmer set at 20mm away from the cluster source. The ion components were suppressed to pass through the skimmer in a higher electrostatic potential. The size distribution of the generated neutral clusters was analyzed by a two-step-acceleration type time of flight mass spectrometer (TOFMS).

A typical size distribution of the silicon clusters generated in the new principal SCCS is shown in Fig.1. In the region of cluster size from N=9 to 34, stable Si_{23} clusters including $Si_{23}H_3$ and $Si_{23}H_6$ clusters show characteristically higher abundance of 46% compared with $Si_{19}H_x$, $Si_{21}H_x$ clusters in abundance of 14% and 12%, respectively. The SCCS has successfully achieved such a characteristically narrow size distribution of a Si_N cluster beam as with $\Delta N/N<5\%$, where ΔN is the full width at the half maximum in the size distribution.

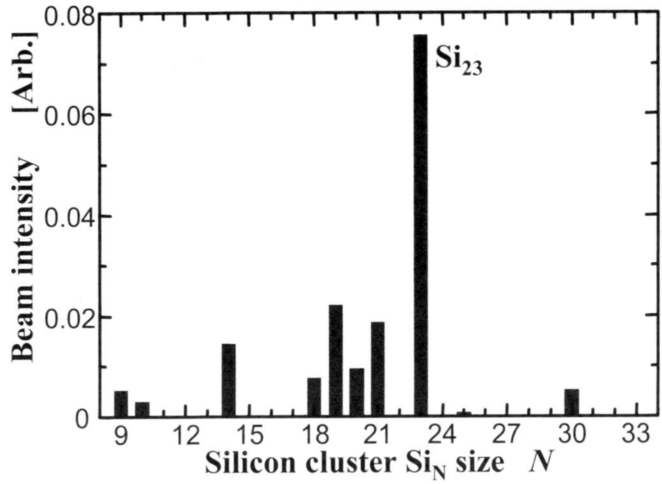

Fig. 1. A typical size distribution of silicon clusters generated in the SCCS.

3. Silicon Cluster Lattice System (CLS)

3.1. Cluster beam irradiation

The well-defined silicon cluster beam generated in the new SCCS was deposited in an ultra-high vacuum onto an amorphous carbon thin film of a triafol micro grid system that was set 85mm away from the cluster source. Neutral Si_N clusters flying with a supersonic helium gas flow at a velocity of 2.8km/s collided with the substrate target at an energy of 1.1eV/Si atom, which is several times smaller than the cohesive energy at 4.0-4.2eV of silicon clusters.[6] Accordingly, the Si_N clusters landed on the substrate surface without dissociation. A great portion of the kinetic energy might be transferred to the substrate lattice or might be dissipated for the translation energy of cluster migration on the substrate surface.

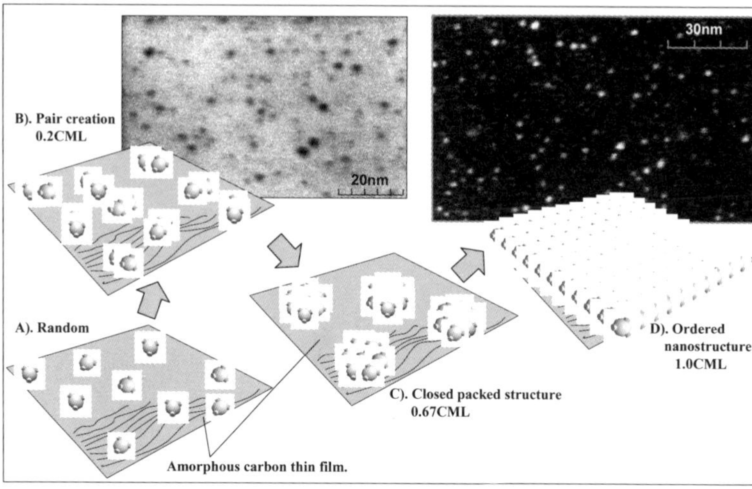

Fig. 2. The silicon CLS on an amorphous carbon thin film developed spontaneously as the deposition density increased, random orientation A), pair creation at 0.2CML B), partially hexagonal closed packed structure at 0.67CML C), ordered CLS at 1.0CML D).

3.2. Nanostructures of silicon clusters on an amorphous carbon surface

After irradiation of the silicon cluster beam, the target substrate was once taken out into an atmospheric circumstance for diagnosis of the CLS formed on the carbon thin films. Images of the silicon clusters of 2-3nm in diameter taken by the high angle annular dark field (HAADF) method of a scanning transmission electron microscope (STEM, HD2000 Hitachi) show a good contrast with a carbon background of the substrate [Fig. 2 D]. The silicon component of the imaging particles was identified by the energy dispersive X-ray spectrometry (EDS). Ultra high resolution transmission electron microscope (UHRTEM, resolution=1.0Å) imaged clearly a crystallographic structure of the individual silicon clusters forming atomic lattices. A unit cluster mono layer (CML) covers fully a substrate surface with silicon clusters of 2.3nm in diameter in the density of $1.0x10^{13}cm^{-2}$. The CLS of the isolated silicon clusters formed on the substrate surface were developed spontaneously as the deposition density increased. On the lower deposition density stage at 0.20CML, the silicon clusters made randomly oriented pairs [Fig. 2 B]. In the process of the deposition density around 0.67CML, the silicon clusters had a tendency to form partially hexagonal closed packed structures [Fig. 2 C]. In further progress up to 1.0CML, the silicon clusters lined up spontaneously to form a tetragonal structure with a lattice constant of 4.0nm [Fig. 2 D].

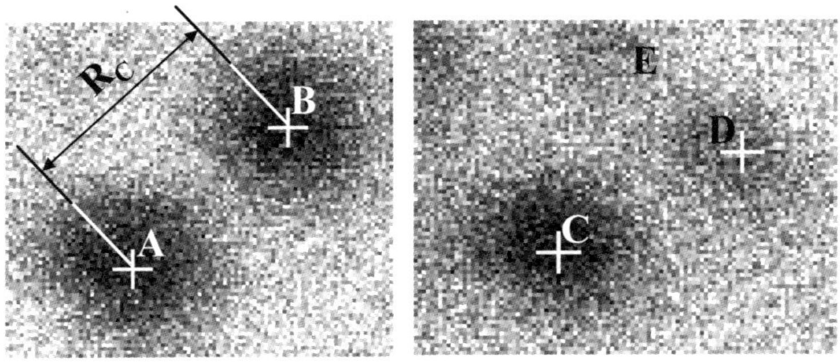

Fig. 3. Isolated silicon cluster pair of A and B at a distance R_C (left), and non-isolated paring system of C, D, and E (right).

4. Discussion

Turning back again to the stage forming cluster pairs in the coverage of 0.20CML, we investigate the interaction between clusters. When clusters form an isolated pair, the interaction between the clusters can be solved accurately as a two-body problem. Accordingly, isolated cluster pairs should be defined here by a typical pair of clusters A and B in Fig. 3. A and B are mutually the nearest neighbor of the other, in contrast with D forming pairs with both of C and E. In the coverage stage of 0.20CML, more than 78% of the total clusters came close to each other to form an isolated pair with an inter-cluster distance R_C below $R_C/[R_m]=0.5$, where $[R_m]$ is the standard inter-cluster distance in supposing regular interval deposition.

Dipole interactions can be considered as a most probable interaction between clusters forming a cluster pair. The dipole moment P_i induced on each cluster is proportional to the cluster size N_i, and the interaction potential between the induced dipole moments has a form of P_iP_j/r_{ij}^3. In the HAADF images, the diffraction intensity I_i from each cluster is also proportional to N_i. Then it can reveal the mechanism of the ordering process of silicon clusters to examine the relation of the third power of the inter-cluster distance R_C with the product of the diffraction intensities I_i*I_j of all isolated cluster pairs. The relations of R_C and

$(I_i*I_j)^{1/3}$ summarized in Fig. 4 prove the linearity of the third power of R_C with the product of their cluster size N_1 and N_2. The correlation shows a possible conclusion that the dipole potential between the close silicon clusters in 1.0CML dominantly makes them ordered spontaneously. The interaction potential between clusters may possibly enable the construction of 3DCLS in blocks of clusters.

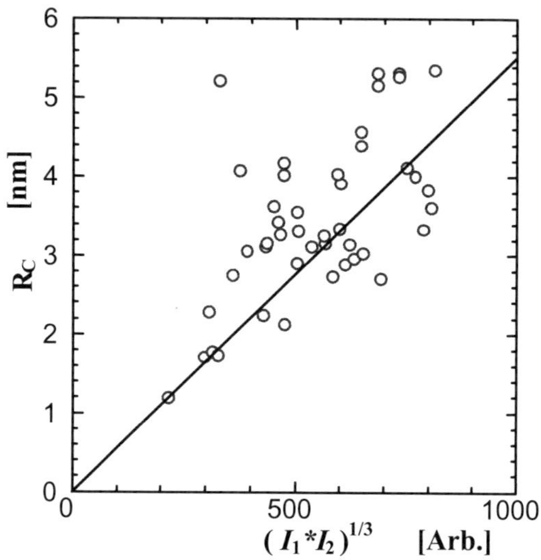

Fig. 4. Relation of the inter-cluster distance R_C with the product of the diffraction intensities I_i*I_j of the HAADF images of all isolated cluster pairs.

5. Conclusions

We have observed spontaneous ordering of silicon clusters in the cluster beam deposition (CBD) processes on an amorphous carbon film surface. A unit CML covers fully the substrate surface with silicon clusters of 2.3nm in diameter in the density of $1.0x10^{13}cm^{-2}$. The cluster lattice system (CLS) formed on the substrate surface was developed spontaneously as the deposition density increased. The silicon clusters made pairs randomly oriented at 0.20CML. In the progress of the deposition density around 0.67CML, silicon clusters had a tendency to form partially hexagonal closed packed structures. In further progress up to 1.0CML, the silicon clusters lined up spontaneously to form a tetragonal structure with a lattice constant of 4.0nm. The UHRTEM imaged clearly a crystallographic atomic lattice structure of the individual silicon clusters. Investigation of the isolated cluster pairs could reach a possible conclusion that the induced dipole potential between the close silicon clusters dominantly makes them ordered spontaneously. The interaction potential between clusters may possibly enable the construction of 3DCLS in blocks of silicon clusters.

Acknowledgements

The authors thank Dr. H. Oyanagi for useful discussions and Dr. S. Kanemaru for offer convenient occasions for STEM observation. These studies have been carried out using the grants of the government and private matching funds of METI.

References

1. Y.-W. Mo, D. E. Savage, B. S. Swartzentruber, and M. G. Lagally, *Phys. Rev. Letters* **65**, 1020 (1990).
2. P. Jensen, *Rev. Mod. Phys.* **71**, 1695 (1999).
3. J. Xhie, K. Sattler, U. Muller, N. Venkateswaran, and G. Raina, *Phys. Rev.* B **43**, 8917 (1991).
4. G. Capellini, M.De Seta, C. Spinella, F. Evangelisti, *Appl. Phys. Letters* **82**, 1772 (2003).
5. Y. Iwata, M. Kishida, M. Muto, S. Yu, T. Sawada, A. Fukuda, T. Takiya, A. Komura, K. Nakajima, *Chem. Phys. Letters* **358**, 36 (2002).
6. D. K. Yu, Q. Zhang, and S. T. Lee, *Phys. Rev.* B **65**, 245417-1 (2002).

LIST OF PARTICIPANTS

Prof. Adoui, Lamri
CIRIL - Caen
CIRIL Rue Claude Bloch BP5133 cedex 5
F-14070 CAEN
France
E-mail: *adoui@ganil.fr*

Dr. Alcami, Manuel
Departamento de Química Facultad de Ciencias C-IX Universidad Autónoma de Madrid
28049 Madrid
Spain
E-mail: *manuel.alcami@uam.es*

Dr. Andersson, Mats
Department of Experimental Physics
Chalmers University of Technology
SE-41296 Goteborg
Sweden
E-mail: *mats.andersson@fy.chalmers.se*

Prof. Dr. Awaya, Yohko
Musashino Art University
1-736 Ogawa, Kodaira
187-8505 Tokyo
Japan
E-mail: *awaya@musabi.ac.jp*

Professor Broyer, Michel
Universite Lyon1
Laboratoire de Spectrométrie Ionique et Moléculaire, LASIM Bâtiment A. Kastler Université Lyon1 43, Boulevard du 11 Novembre 1918
69622 Cedex VILLEURBANNE Cedex
FRANCE
E-mail: *broyer@lasim.univ-lyon1.fr*

Dr. Basalaev, Alexei
Ioffe Physical-Technical Institute
Politechnicheskaya 26
194021 St.Petersburg
Russia
E-mail: *a.basalaev@mail.ioffe.ru*

Prof.Dr. Becker, Uwe
Fritz-Haber-Institut
Faradayweg 4-6
14 195 Berlin
Germany
E-mail: *becker_u@fhi-berlin.mpg.de*

Dr. Beroff-Wohrer, Karine-Caroline
CNRS-Université Paris 6
Groupe de Physique des Solides Université Paris 6 2, place Jussieu
75251 Paris Cedex 05
France
E-mail: *wohrer@gps.jussieu.fr*

Prof. Bréchignac, Catherine
C.N.R.S. Laboratoire Aimé Cotton
Laboratoire Aimé Cotton, Bat 505, Campus d'Orsay
91405 ORSAY Cedex
FRANCE
E-mail: *catherine.brechignac@lac.u-psud.fr*

Dr. Cocke, Charles
Kansas State University
Physics Department Cardwell Hall Kansas State University
66506 Manhattan,KS
USA
E-mail: *cocke@phys.ksu.edu*

Prof. Campbell, Eleanor
Goeteborg University
Dept of Experimental Physics School of Physics and Engineering Physics
Goeteborg University and Chalmers
41296 Goeteborg
Sweden
E-mail: *eleanor.campbell@fy.chalmers.se*

Prof. Castleman, A. Welford
Penn State University
Department of Chemistry and Physics 152 Davey Lab
16802 University Park
USA
E-mail: *awc@psu.edu*

Dr. Chabot, Marin
Institut de Physique Nucléaire
IPNO
91406 Cedex Orsay
France
E-mail: *chabot@ipno.in2p3.fr*

Prof. Cherepkov, Nikolai A.
State University of Aerospace Instrumentation
Bolshaya Morskaya str. 67
190000 St.Petersburg
Russia
E-mail: *nik_cherepkov@yahoo.com*

Prof. Fabrikant, Ilya
University of Nebraska
Department of Physics and Astronomy
University of Nebraska 116 Brace
68588 Lincoln, NE
USA
E-mail: *ifabrikant1@unl.edu*

Prof. Connerade, Jean-Patrick
Imperial College London
Quantum Optics and Laser Science Group
Physics Department Imperial College
Prince Consord Road
SW7 2BW London
UK
E-mail: *atomic_clusters@compuserve.com*

Mr. Di Paola, Cono
Dept. Chemistry, University of Rome "La Sapienza"
Piazzale Aldo Moro 5
00185 Roma
Italy
E-mail: *dipaola@caspur.it*

Mr. Diaz-Tendero, Sergio
Departamento de Quimica
Departamento de Quimica Facultad de Ciencias C-IX Universidad Autonoma de Madrid
28049 Madrid
Spain
E-mail: *sergio.diaztendero@uam.es*

Mr. Doronin, Yuriy
Verkin Institute for Low Temperature Physics & Engineering NASU
47, Lenin ave.
61103 Kharkov
Ukraine
E-mail: *HC27@HC.SCANA.COM.UA*

Dr. Gochitashvili, Malkhaz
Tbilisi State University, Physics Department,
Chavchavadze av.3
380028 Tbilisi
Georgia
E-mail: *mgochitashvili@hotmail.com*

Dr. Fielicke, Andre
FOM Institute Rijnhuizen
Edisonbaan 14
3439 MN Nieuwegein
The Netherlands
E-mail: *fielicke@rijnh.nl*

Dr. Gerchikov, Leonid
St. Petersburg Technical University
St.Petersburg Technical University,
Polytechnicheskaya 29,
195251 St.Petersburg
Russia
E-mail: *leonid@priv.hop.stu.neva.ru*

Dr. Giglio, Eric
CIRIL/GANIL
Rue Claude Bloch - BP 5133
14070 Caen
France
E-mail: *giglio@ganil.fr*

Prof. Gianturco, Franco A.
Universita' di Roma "La Sapienza"
Dipartimento di Chimica
Piazzale Aldo Moro 5
00185 Rome
ITALY
E-mail: *a.gianturco@caspur.it*

Prof. Huber, Bernd
CEA - CIRIL
Rue Claude Bloch
F-14070 Caen
France
E-mail: *huber@ganil.fr*

Prof. Dr. Greiner, Walter
Institute für Theoretische Physik der
Universität Frankfurt am Main
Robert-Mayer Str. 8-10
D-60054 Frankfurt am Main
Germany
E-mail: *greiner@th.physik.uni-frankfurt.de*

Dr. Gspann, Jürgen
Universität Karlsruhe
Institut für Mikrostrukturtechnik Postfach
3640 Universität Karlsruhe
76021 Karlsruhe
Germany
E-mail: *juergen.gspann@imt.fzk.de*

Prof. Guet, Claude
CEA
CEA Ile de France Département de
Physique Théorique et Appliquée
91680 Bruyéres-le-Chatel
France
E-mail: *claude.guet@cea.fr*

Prof. Haberland, Helmut
Department of Physics, University of
Freiburg, Germany
VF, Stefan-Meierstrasse 19
79104 Freiburg
Germany
E-mail: *haberland@physik.uni-freiburg.de*

Prof. Dr. Kabachnik, Nikolay
Institute of Nuclear Physics, Moscow State
University
Vorobjevy Gory
119992 Moscow
Russia
Email: *nkabach@mail.ru*

Dr. Ichihashi, Masahiko
Toyota Technological Institute
Cluster Research Laboratory, Toyota Technological Institute: in East Tokyo Laboratory, Genesis Research Institute, Inc., 717-86 Futamata
272-0001 Ichikawa, Chiba
Japan
E-mail: *ichihashi@clusterlab.jp*

Dr. Ipatov, Andrei
St.Petersburg State Technical University
Experimental Physics Department
Polytechnicheskaja 29
195251 St.Petersburg
Russia
E-mail: *andrei_ipatov@mail.ru*

Prof. Ivanov, Vadim
Saint-Petersburg State Politechnical University
Experimental Physics Depatment
Polithechnicheskaya str. 29,
195251 St.Petersburg
Russia
E-mail: *ivanov@tuexph.stu.neva.ru*

Dr. Jellinek, Julius
Argonne National Laboratory
Chemistry Division Argonne National laboratory 9700 S. Cass Ave.
60439 Argonne, Illinois
USA
E-mail: *jellinek@anl.gov*

Prof. Karpeshin, Fedor
St. Petersburg State University,
Institute of Physics
Ulianovskaya 1, Petrodvorets
198504 St Petersburg
Russia
E-mail: *karpesh@nuclpc1.phys.spbu.ru*

Dr. Kashenock, Galina
Saint-Petersburg State Politechnical University
Experimental Physics Depatment
Polithechnicheskaya str. 29,
195251 St.Petersburg
Russia

Prof. Kaya, Koji
Director-General, Institute for Molecular Science
Naka-38, Aza Saigou, myoudaiji-Machi
444-8585 Okazaki
Japan
E-mail: *kaya@ims.ac.jp*

Dr. Kebaili, Nouari
CNRS - Laboratoire Aimé Cotton
Laboratoire Aimé Cotton, Bat 505,
Campus d`Orsay
91405 Orsay
FRANCE
E-mail: *nouari.kebaili@lac.u-psud.fr*

Mr. Kidun, Oleg
Max-Planck Institut
Weinberg 2
06120 Halle
Germany
E-mail: *okidun@mpi-halle.de*

Dr. Korol, Andrey
A.F. Ioffe Physical-Technical Institute
Politechnicheskaja 26
194021 St Petersburg
Russia
E-mail: *korol@rpro.ioffe.rssi.ru*

Prof. Kondow, Tamotsu
Toyota Technological Institute
717-86 Futamata
272-0001 Ichikawa, Chiba
Japan
E-mail: *kondow@clusterlab.jp*

Prof. Kresin, Vitaly
University of Southern California
Department of Physics and Astronomy
University of Southern California
90089-0484 Los Angeles, CA
USA
E-mail: *kresin@usc.edu*

Ms. Lando, Aurelie
CNRS
Laboratoire Aimé Cotton, Bat 505,
Campus d'Orsay
91405 Orsay
France
E-mail: *aurelie.lando@lac.u-psud.fr*

Prof. Dr. Lutz, Hans O.
Universität Bielefeld
Universitätsstr. 25
D-33615 Bielefeld
GERMANY
E-Mail: *Lutz@physik.uni-bielefeld.de*

Dr. Lahreche-Gomri, Rahima
Département de Physique Université de
Annaba BP 12 23000 Annaba
Algeria
E-mail: *grah_a@yahoo.fr*

Dr. Lapicki, Gregory
East Carolina University
Department of Physics East Carolina
University
27858 Greenville
USA
E-mail: *lapicki@physicist.net*

Dr. Levin, Jon
University of Tennessee
Department of Physics and Astronomy 401
Nielsen Physics Building
37996 Knoxville, TN
USA
E-mail: *jlevin@utk.edu*

Prof. Lievens, Peter
K.U.Leuven
Laboratorium voor Vaste-Stoffysica en
Magnetisme Departement Natuurkunde en
Sterrenkunde Celestijnenlaan 200D
B-3001 Leuven
Belgium
E-mail: *Peter.Lievens@fys.kuleuven.ac.be*

PhD. Lomsadze, Ramaz
Tbilisi State University
Tbilisi State University, Department of
Physics, Chavchavadze Avenue 3
380028 Tbilisi
Georgia
E-mail: *Lomsadze86@hotmail.com*

Mr. Matveentsev, Anton
A.F. Ioffe Physical-Technical Institute
Politechnicheskaya 26,
194021 St. Petersburg
Russia
E-mail: *matveentsev@mailru.com*

Dr. Lyalin, Andrey
Institute of Physics, St Petersburg State University
Ulianovskaya str. 1
198504 St Petersburg, Petrodvorez
Russia
E-mail: *lyalin@th.physik.uni-frankfurt.de*

Prof. Märk, Tilmann
University Innsbruck
Technikerstr.25
A-6020 Innsbruck
Austria
E-mail: *tilmann.maerk@uibk.ac.at*

Prof. Manninen, Matti
University of Jyvaskyla
Department of Physics University of Jyväskylä PO Box 35
FIN-40014 Jyväskylä
Finland
E-mail: *matti.manninen@phys.jyu.fi*

Prof. Manson, Steven
Georgia State University
Department of Physics and Astronomy
Georgia State University
30303 Atlanta, Georgia
USA
E-mail: *smanson@gsu.edu*

Prof. Dr. Meiwes-Broer, Karl-Heinz
University of Rostock
Universitaetsplatz 3
18051 Rostock
Germany
E-mail: *meiwes@physik.uni-rostock.de*

Mr. Majima, Takuya
Kyoto University
Dept. of Nuclear Engineering, Kyoto University, Yoshidahonmachi, Sakyo-ku
601-8501 Kyoto
Japan
E-mail: *majima@nucleng.kyoto-u.ac.jp*

Dr. Martin, Fernando
Departamento de Química
Departamento de Química Facultad de Ciencias C-IX Universidad Autonóma de Madrid
28049 Madrid
Spain
E-mail: *fernando.martin@uam.es*

Mr. Martinet, Guillaume
IPN Orsay
13 Les Avelines
91940 Les Ulis
France
E-mail: *martinet@ipno.in2p3.fr*

Prof. Masson, Albert
C.N.R.S.
L.A.C.
91400 ORSAY
FRANCE
E-mail: *albert.masson@lac.u-psud.fr*

Mrs. Muto, Makiko
National Institute of Advanced Industrial Science and Technology
Cluster Advanced Nanoprocesses CRT, National Institute of Advanced Industrial Science and Technology, Tsukuba Central 2, Umezono 1-1-1
305-8568 Tsukuba
Japan
E-mail: *m-muto@aist.go.jp*

Prof. McConkey, William
University of Windsor
Physics Department, University of Windsor, Sunset Avenue
N9B 3P4 Windsor, Ontario
Canada
E-mail: *mcconk@uwindsor.ca*

Prof. Mikoushkin, Valeri
A.F. Ioffe Physical-Technical Institute
Politechnicheskaya 26,
194021 St. Petersburg
Russia
E-mail: *V.Mikoushkin@pop.ioffe.rssi.ru*

Prof. Mishoustin, Igor
The Kurchatov Institute, Russian Research Center
Moscow 123182,
Russia
E-mail: *mish@mbslab.kiae.ru*

Dr. Msezane, Alfred
Clark Atlanta University
223 James P. Brawley Drive, SW CTSPS, Box 92
30314 Atlanta
USA
E-mail: *amsezane@ctsps.cau.edu*

Dr. Obolensky, Oleg
A.F. Ioffe Institute
Politechnicheskaja 26
194021 Saint Petersburg
Russia
E-mail: *oleg@mail.ioffe.ru*

Prof. Nakajima, Atsushi
Keio University, Department of Chemistry
3-14-1 Hiyoshi, Kohoku-ku,
223-8522 Yokohama
Japan
E-mail: *nakajima@sepia.chem.keio.ac.jp*

Dr. Nakamura, Masato
Nihon University
7-24-1, Narashino-dai
274-8501 Funabashi
Japan
E-mail: *mooming@phys.ge.cst.nihon-u.ac.jp*

Mr. Narits, Alexander
P.N.Lebedev Physical Institute
Leninskiy prospect, 53
119991 Moscow
Russia
E-mail: *presn@sci.lebedev.ru*

Dr. Nesterenko, Valentin
Bogoliubov Laboratory of Theoretical Physics
Joint Institute for Nuclear Research,
Dubna, Moscow region, 141980
Russia
E-mail: *nester@thsun1.jinr.ru*

Mr. Prasalovich, Syargey
Göteborg University / Chalmers
Fysikgränd 3, Atomic Physics Group, School of Physics and Engineering Physics, Göteborg University & Chalmers University of Technology
41296 Göteborg
Sweden
E-mail: *f3apras@fy.chalmers.se*

Prof. Ostrovsky, Valentin
Institute of Physics, St Petersburg State University
Ulianovskaya str. 1
198504 St Petersburg, Petrodvorez
Russia
E-mail: *valentin.ostrovsky@pobox.spbu.ru*

Dr. Parneix, Pascal
Laboratoire de Photophysique Moléculaire, CNRS
Laboratoire de Photophysique Moleculaire
Bat. 210 Université Paris-Sud
91405 ORSAY
FRANCE
E-mail: *pascal.parneix@ppm.u-psud.fr*

Dr. Politis, Marie-Franzoise
G.P.S Université paris 6
GPS 2 place Jussieu
75251 Paris cedex 05
France
E-mail: *politis@gps.jussieu.fr*

Dr. Polozkov, Roman
Saint-Petersburg State Politechnical University
Experimental Physics Depatment
Polithechnicheskaya str. 29
195251 Saint-Petersburg
Russia
E-mail: *polozkov@tueph.stu.neva.ru*

Prof. Schmidt, Rüdiger
TU Dresden
TU Dresden Institute of Theoretical Physics
01062 Dresden
Germany
E-mail: *schmidt@physik.tu-dresden.de*

Prof. Rost, Jan Michael
MPG
Max Planck Institute for the Physics of Complex Systems
01187 Dresden
Germany
E-mail: *rost@mpipks-dresden.mpg.de*

Prof. Salzborn, Erhard
University of Giessen,Germany
Institut für Kernphysik Leihgesterner Weg 217
D-35392 Giessen
Germany
E-mail: *salzborn@strz.uni-giessen.de*

Dr. Samarin, Sergey
University of Western Australia,
35, Stirling Hwy.
6009 Crawley WA
Australia
E-mail: *samar@physics.uwa.edu.au*

Prof. Sato, Yukinori
IMRAM, Tohoku University
Institute of Multidisciplinary Research for Advanced Materials (IMRAM), Tohoku University 2-1-1 Katahira, Aoba-ku
980-8577 Sendai
Japan
E-mail: *satoyuki@tagen.tohoku.ac.jp*

Prof. Shaginyan, Vasily
Petersburg Nuclear Physics Institute
188300 Gatchina
Russia
E-mail: *vrshag@thd.pnpi.spb.ru*

Prof. Schweikhard, Lutz
University of Greifswald
Institute of Physics University of Greifswald
D-17487 Greifswald
Germany
E-mail: *Lutz.Schweikhard@physik.uni-greifswald.de*

Mr. Scifoni, Emanuele
University of Genova
Department of Chemistry
Via Dodecanneso 31
16146 Genova
Italy
E-mail: *scifoni@caspur.it*

Mr. Sebastianelli, Francesco
University of Rome "La Sapienza"
Department of Chemistry
Piazzale Aldo Moro 5
00185 Roma
Italy
E-mail: *f.sebastianelli@caspur.it*

Dr. Solovjev, Igor
A.F. Ioffe Physical-Technical Institute
Politechnicheskaya 26,
194021 St. Petersburg
Russia
E-mail: *igor@rpro.ioffe.rssi.ru*

Dr. van der Burgt, Peter
National University of Ireland, Maynooth
Department of Experimental Physics
National University of Ireland
Maynooth, County Kildare
Ireland
E-mail: *Peter.vanderBurgt@may.ie*

Prof.Dr. Solovyov, Andrey
A.F. Ioffe Physical-Technical Institute
Politechnicheskaya 26,
194021 St. Petersburg
Russia
E-mail: *solovyov@rpro.ioffe.rssi.ru*

Mr. Solovyov, Ilia
J.W. Goethe University
J.W. Goethe University Institute for Theoretical Physics
Robert-Mayer Str. 8-10
60054 Frankfurt am Main
Germany
E-mail: *ilia@th.physik.uni-frankfurt.de*

Prof.Dr. Stöcker, Hörst
Institiut für Theoretische Physik,
J.W. Goethe Universität
Robert-Mayer Straße 8-10
60054 Frankfurt am Main
Germany
E-mail: *stoecker@th.physik.uni-frankfurt.de*

Dr. Tomita, Shigeo
Univ. of Aarhus
Department of Physics and Astronomy
University of Aarhus
DK-8000 Aarhus
Denmark
E-mail: *tomita@phys.au.dk*

Assistant Prof. Yasumatsu, Hisato
Toyota Technological Insitute
Cluster Research Laboratory, Toyota Technological Institute: In East Tokyo Laboratory, Genesis Research Institute, Inc. 717-86 Futamata, Ichikawa, Chiba 272-0001, Japan
272-0001 Ichikawa
Japan
E-mail: *yasumatsu@clusterlab.jp*

Dr. von Issendorff, Bernd
Universitaet Freiburg
Stefan Meier Strasse 21
79104 Freiburg
Germany
E-mail: *bernd.von.issendorff@physik.uni-freiburg.de*

Prof. Williams, James
University of Western Australia
35 Stirling Highway Crawley
6009 Perth
Australia
E-mail: *jfw@physics.uwa.edu.au*

Prof. Wöste, Ludger
Freie Universitaet Berlin, FB Physik
Arnimallee 14
14195 Berlin
Germany
E-mail: *woeste@physik.fu-berlin.de*

Mr. Zettergren, Henning
Stockholm University
Physics Department, Stockholm University, AlbaNova
S-10691 Stockholm
Sweden
E-mail: *henning@physto.se*

Engineer Danylchenko, Oleksandr
Institute for Low Temperature Physics and Engineering
47 Lenin Ave.
61103 Kharkov
Ukraine
E-mail: danylchenko@ilt.kharkov.ua

Dr. Yamaguchi, Yasutaka
Department of Mechanophysics Engineering, Osaka University
2-1 Yamadaoka
565-0871 Suita, Osaka
Japan
E-mail: *yamaguchi@mech.eng.osaka-u.ac.jp*